Springer Series in Solid-State Sciences

Volume 190

Series editors

Bernhard Keimer, Stuttgart, Germany
Roberto Merlin, Ann Arbor, MI, USA
Hans-Joachim Queisser, Stuttgart, Germany
Klaus von Klitzing, Stuttgart, Germany

The Springer Series in Solid-State Sciences consists of fundamental scientific books prepared by leading researchers in the field. They strive to communicate, in a systematic and comprehensive way, the basic principles as well as new developments in theoretical and experimental solid-state physics.

More information about this series at http://www.springer.com/series/682

Dario Bercioux · Jérôme Cayssol
Maia G. Vergniory · M. Reyes Calvo
Editors

Topological Matter

Lectures from the Topological Matter
School 2017

Editors
Dario Bercioux
Theoretical Mesoscopic Physics
Donostia International Physics Center
Donostia/San Sebastián, Gipuzkoa, Spain

Jérôme Cayssol
Laboratoire Ondes et Matière d'aquitaine
Talence, France

Maia G. Vergniory
Donostia International Physics Center
Donostia/San Sebastián, Gipuzkoa, Spain

M. Reyes Calvo
CIC nanoGUNE
Donostia/San Sebastián, Gipuzkoa, Spain

ISSN 0171-1873 ISSN 2197-4179 (electronic)
Springer Series in Solid-State Sciences
ISBN 978-3-030-09477-5 ISBN 978-3-319-76388-0 (eBook)
https://doi.org/10.1007/978-3-319-76388-0

© Springer Nature Switzerland AG 2018
Softcover re-print of the Hardcover 1st edition 2018
This work is subject to copyright. All rights are reserved by the Publisher, whether the whole or part of the material is concerned, specifically the rights of translation, reprinting, reuse of illustrations, recitation, broadcasting, reproduction on microfilms or in any other physical way, and transmission or information storage and retrieval, electronic adaptation, computer software, or by similar or dissimilar methodology now known or hereafter developed.
The use of general descriptive names, registered names, trademarks, service marks, etc. in this publication does not imply, even in the absence of a specific statement, that such names are exempt from the relevant protective laws and regulations and therefore free for general use.
The publisher, the authors and the editors are safe to assume that the advice and information in this book are believed to be true and accurate at the date of publication. Neither the publisher nor the authors or the editors give a warranty, express or implied, with respect to the material contained herein or for any errors or omissions that may have been made. The publisher remains neutral with regard to jurisdictional claims in published maps and institutional affiliations.

This Springer imprint is published by the registered company Springer Nature Switzerland AG
The registered company address is: Gewerbestrasse 11, 6330 Cham, Switzerland

Preface

In the past few years, the Donostia International Physics Center (DIPC) has launched an extensive research effort to the investigation of topological states of matter (TSM). The education of graduate students and young postdoctoral fellows is an essential aspect of this effort. In this spirit, we have organised on a yearly basis a summer school which gathers worldwide experts on the subject—the Topological Matter School series. Our goal is to provide students with a pedagogical but comprehensive and up-to-date presentation of this quickly growing field. The lectures from the past 2017 edition of the school are now collected in this book, which aims to serve as an educational introduction to those newly approaching the study of topology in condensed matter systems. The volume includes chapters both on the fundamental theoretical aspects and on some of the latest experimental breakthroughs in the field.

Topology is a field of mathematics that has fed almost all the domains of physics, ranging from high-energy to condensed matter systems, from the pioneering work of Dirac on magnetic monopoles to modern gauge theories, classification of topological defects in ordered phases, Berezinskii–Kosterlitz–Thouless transition, spin chains, quantum Hall effects, and in the last decade topological insulators and semimetals. The topological effects discussed in this book are related to the particular winding properties of electronic Berry phase of Bloch electronic states in specific insulating and conducting materials. Therefore, it can be seen as a refinement of standard band structure theory where the primary interest was focused only on the energy level dispersion and gaps while ignoring the subtle properties of the quantum mechanical phases.

In condensed matter, topological effects are mainly related to the non-trivial winding of the phase of bulk Bloch states around the whole Brillouin zone. TSM manifests themselves in several new features: (i) the emergence of gapless conducting states confined at the external boundary of the system or between two systems with different windings, like the chiral edge states of quantum Hall (QH) insulators and Chern insulators, or the helical edge states or surface states of time-reversal invariant topological insulators (TIs); (ii) the existence of quantised response functions that are topologically protected from local perturbations, like the polarisation in 1D insulators or the transverse conductivity in QH insulators and 2D Chern insulators; (iii) quantum anomalies in Weyl semimetals (WSs); (iv) Majorana quasi-particle in hybrid systems combining topological insulators with superconductors, or even nanowires with strong spin–orbit, Zeeman coupling and proximity-induced superconductivity. Interestingly all these effects can be tested experimentally, and most of them have already been (or on the way to be) confirmed in real materials.

The implications of TSM go beyond condensed matter physics and have given rise to a highly multidisciplinary field of research that includes chemistry, photonics, atomic, polymer physics, etc. Besides the fundamental interest that the field has peaked in the scientific community, the realisation of topological materials also has important technological consequences. A crucial advantage common to all topological materials is the robustness of specific features (like the existence of metallic edge states) against local perturbations or details of sample preparation. The properties of topological materials are expected to lead to technological applications in electronics, spintronics and optoelectronics. For example, the spin-momentum locking of edge or surface states may serve to generate spin-polarised currents or the large quantised nonlinear optical effects can promote technologies like solar cells or photodetectors beyond their current limits. Another relevant aspect of TSM for electronics is related to quantum confinement (QC): to the date, QC has been typically obtained by complicated engineering processes. The advent of topological matter has changed the rules of the game; confinement spontaneously occurs at the surface of a topological insulator and can be thus optimised by the right choice of material, even if the available energy window for the quasi-particles is limited compared to the nanofabrication platforms.

The book is composed of ten contributions. The first three chapters are devoted to the general characterisation of the topology of condensed matter systems. Chapters 4 and 5 are devoted, respectively, to theoretical and experimental aspects of electronic transport in low-dimensional hybrid systems made of TIs and superconductors. Chapters 6 and 7 are devoted to the physics of 3D Weyl semimetal. Chapters 8, 9 and 10 describe various aspect of growth and characterisation of topological materials.

Chapter 1 "Band Theory Without Any Hamiltonians or 'The way Band Theory Should be Taught'" introduces the theory of Topological Quantum Chemistry. This new formalism predicts the presence or absence of topological phases by studying the behaviour of orbitals lying in some special positions of the crystal lattice in real space. Throughout the chapter, the main concepts of the theory will be analysed following a well-known example: graphene.

In Chap. 2 "Topological Crystalline Insulators", Titus Neupert and Frank Schindler introduce the concept of Wilson loop, which is Berry phase diagnostic for systems with band degeneracy. The authors show the connection of this topological quantity to the eigenvalues of the position operator, thus to the generalisation of the problem of polarisation in solids. The concept of Wilson loop is employed for investigating topological crystalline insulators and a new class of topological systems named higher-order topological insulators.

Dominik Gresch and Alexey Soluyanov in their contribution entitled "Calculating Topological Invariants with Z2Pack" (Chap. 3) present a general introduction to the concept of Chern number in non-interacting band systems. They describe an efficient procedure for extracting the Chern number in connection with the Berry phase and the Wilson loop. This efficient method is at the core of the Python code Z2Pack: through the use of several examples, the authors explain how to use this package for the evaluation of topological properties.

In the contribution "Transport in Topological Insulator Nanowires" by Jens H. Bardarson and Roni Ilan (Chap. 4), the authors make an in-depth analysis of the quantum transport properties of quasi-one-dimensional topological insulator quantum wire. They present the effect of magnetic field and disorder on the transport properties of these wires. The second part is devoted to proximity-induced superconductivity in topological insulator nanowires, with an emphasis on the emergence and possible detection of Majorana fermions in these hybrid junctions.

In his contribution to this book (Chap. 5), Erwann Bocquillon reviews the consequences of induced superconductivity in the surface states of a topological insulator, long predicted as a path for the generation of topological superconductivity and Majorana states. By using microwave excitation and detection techniques, Bocquillon and collaborators have detected the elusive signatures of Majorana bound states in Josephson junctions using HgTe as a weak link. In this chapter, the theoretical and technical aspects of their experiments are first introduced to provide the reader with the necessary background to understand the following detailed review or their results and prospective work in the field.

In Chap. 6, Adolfo G. Grushin presents how field theoretical tools borrowed from high-energy physics can be used to study low-energy/effective models of topological matter. First, a generic model for Weyl semimetals is interpreted as a Lorentz breaking theory for fermions in the continuum. Then, three different possibilities to promote such a field theory to live on a lattice are discussed touching on the importance of the Nielsen–Ninomiya (or fermion doubling) theorem. Finally, Adolfo G. Grushin emphasises that Weyl semi-metallic phases of matter and related systems be described by ambiguous field theories (theories predicting observable quantities that are finite but depend on the regularisation procedure), which highlight interesting aspects of their responses to external fields and make contact with quantum anomalies.

In Chap. 7, Alberto Cortijo explains how quantum anomalies, and in particular the chiral anomaly, arise in the recently discovered Weyl semimetals. In such semimetals, Weyl nodes appear in pairs with opposite chiralities (left-handed and right-handed Weyl fermions). For a given chirality, the density of Weyl fermions is not conserved in the presence of collinear electric and magnetic fields: this is the chiral anomaly. After reviewing the role of symmetries in field theory (both classical and quantum) briefly and defining quantum anomalies, Alberto Cortijo uses the semiclassical Boltzmann theory to derive the formula for the rate of change of left (respectively, right)-handed fermionic densities. The crucial ingredient is the introduction of Berry phase terms in the semiclassical equations of motion since each Weyl node behaves as a magnetic monopole in k-space, namely a source/sink of Berry curvature flux. The chapter is self-contained and also addresses the crucial role of internode relaxation processes and the positive magnetoconductivity of Weyl semimetals arising from the quantum anomaly.

In Chap. 8 "Topological Materials in Heusler Compounds", Felser and Sun will present the Heuslers compounds and all the different topological materials they can realise. The interplay of symmetry, spin–orbit coupling and magnetic structure allows for the realisation of a wide variety of topological phases through Berry curvature design, from Weyl semimetals to nodal lines or the recently discovered antiskyrmions.

In Chap. 9, Schoop and Topp introduce some basic concepts of solid-state chemistry and how they can help identify new topological materials, providing a short overview of common crystal growth methods and the most significant characterisation techniques available to identify topological properties. This chapter aims to provide a guide for implementing simple chemical principles in the search for new topological materials, as well as giving a basic introduction to the steps necessary to experimentally verify the electronic structure of a material.

In Chap. 10, Haim Beidenkopf presents an intuitive analogy between the real space topological screw dislocation in solids and the momentum space Weyl node structure of topological semimetals. Bulk-boundary correspondence results in unique surface features in the form of step edges at the surface of bulk with screw dislocations and surface Fermi arcs in the surface for Weyl semimetals. In both cases, Beidenkopf and his team apply scanning tunnelling microscopy to the study of these phenomena. The real space case of dislocations can be characterised by just topographical images of step edges in the surface. The detection of momentum surface Fermi arcs requires more advanced techniques such as quasi-particle interference (QPI), which allow extracting valuable information on the properties of Weyl fermions in materials such as TaAs.

The editors thank the authors of each contribution for making this volume possible and successful. We are also grateful to the staff of the DIPC and of Cursos de Verano of the University of Basque Country for the support during the running of the different editions of the TMS school: from basic to advanced.

Gipuzkoa, Spain	Dario Bercioux
Gipuzkoa, Spain	M. Reyes Calvo
Talence, France	Jérôme Cayssol
Gipuzkoa, Spain	Maia G. Vergniory

Contents

1	**Band Theory Without Any Hamiltonians or "The Way Band Theory Should Be Taught"**	1
	I. Robredo, B. A. Bernevig and Juan L. Mañes	
	1.1 Introduction	1
	1.2 Hexagonal Lattice	2
	1.2.1 Orbits for the Different q Points	6
	1.2.2 Adding Orbitals	7
	1.3 Adding p Orbitals at 2b Positions	8
	1.3.1 Spinless p Orbitals	8
	1.3.2 Spinful p Orbitals	10
	1.4 Inducing a Band Representation	12
	1.5 Little Groups at **k** Points in the First BZ	13
	1.6 Example of Band Representation	15
	1.6.1 Spinful Graphene	15
	1.6.2 Spinless Graphene	18
	1.7 Subducing the Band Representation	19
	1.7.1 Γ Point	19
	1.7.2 K Point	20
	1.7.3 M Point	20
	1.7.4 High-Symmetry Lines	20
	1.8 Conclusion	22
	References	29
2	**Topological Crystalline Insulators**	31
	Titus Neupert and Frank Schindler	
	2.1 Wilson Loops and the Bulk-Boundary Correspondence	31
	2.1.1 Introduction and Motivation	31
	2.1.2 Definitions	32
	2.1.3 Wilson Loop and Position Operator	33
	2.1.4 Bulk-Boundary Correspondence	41

2.2	Topological Crystalline Insulators		44
	2.2.1	2D Topological Crystalline Insulator	45
	2.2.2	Mirror Chern Number	47
	2.2.3	C_2T-Invariant Topological Crystalline Insulator	49
2.3	Higher-Order Topological Insulators		50
	2.3.1	2D Model with Corner Modes	50
	2.3.2	3D Model with Hinge Modes	55
	2.3.3	Interacting Symmetry-Protected Topological Phases with Corner Modes	57
References			60

3 Calculating Topological Invariants with Z2Pack 63
Dominik Gresch and Alexey Soluyanov

3.1	The Chern Number		63
	3.1.1	Topology in Non-interacting Materials	63
	3.1.2	Defining the Chern Number	67
3.2	The Z2Pack Code		72
	3.2.1	Introduction to the Code	72
	3.2.2	The Haldane Model	77
	3.2.3	Identifying Weyl Semimetals	81
	3.2.4	Convergence Options	82
3.3	Time-Reversal Symmetry: $\mathbb{Z}_2$ Classification		85
	3.3.1	Individual Chern Numbers	85
	3.3.2	Tight-Binding Example	88
References			91

4 Transport in Topological Insulator Nanowires 93
Jens H. Bardarson and Roni Ilan

4.1	Overview and General Considerations		93
4.2	Topological Insulator Nanowires: Normal State Properties		95
	4.2.1	Band Structure of a Clean Wire	96
	4.2.2	Aharonov–Bohm Effect and Magnetoconductance Oscillations	97
	4.2.3	Perfectly Transmitted Mode	100
	4.2.4	Wires in a Perpendicular Field: Chiral Transport	101
4.3	Topological Insulator Nanowires and Superconductivity		102
	4.3.1	Topological Superconducting Phases in One Dimension	103
	4.3.2	Boundaries and Interferences: Zero Modes	104
	4.3.3	Transport Signatures of Topological Superconductivity	106
4.4	Technical Details: Transfer Matrix Technique		108
4.5	Experimental Status and Outlook		110
References			111

5 Microwave Studies of the Fractional Josephson Effect in HgTe-Based Josephson Junctions ... 115
E. Bocquillon, J. Wiedenmann, R. S. Deacon, T. M. Klapwijk, H. Buhmann and L. W. Molenkamp
- 5.1 Gapless Andreev Bound States in Topological Josephson Junctions ... 116
  - 5.1.1 p Wave Superconductivity in 2D and 3D Topological Insulators ... 116
  - 5.1.2 Gapless Andreev Bound States in 2D and 3D Topological Insulators ... 117
  - 5.1.3 Fractional Josephson Effect ... 120
- 5.2 HgTe-Based Josephson Junctions and Experimental Techniques ... 122
  - 5.2.1 Fabrication of HgTe-Based Josephson Junctions ... 123
  - 5.2.2 Basic Properties of HgTe-Based Josephson Junctions ... 125
  - 5.2.3 Experimental Setups ... 127
- 5.3 Experimental Observation of the Fractional Josephson Effect ... 129
  - 5.3.1 Observation of Josephson Emission at $f_J/2$ ... 129
  - 5.3.2 Observation of Even Sequences of Shapiro Steps ... 132
- 5.4 Analysis: Assessing the Topological Origin of the Fractional Josephson Effect ... 135
  - 5.4.1 Modeling of a Topological Josephson Junction with 2π- and 4π-periodic Modes ... 136
  - 5.4.2 Time-Reversal and Parity Symmetry Breaking, and Landau–Zener Transitions ... 139
- 5.5 Summary, Conclusions, and Outlook ... 141
- References ... 143

6 Common and Not-So-Common High-Energy Theory Methods for Condensed Matter Physics ... 149
Adolfo G. Grushin
- 6.1 Introduction: What This Chapter Is and What It Is Not ... 149
- 6.2 Lorentz Breaking Field Theories ... 150
  - 6.2.1 One Useful Field Theory: Lorentz Breaking QED ... 150
  - 6.2.2 Generalizations of Lorentz Breaking Field Theories ... 156
- 6.3 Field Theories on The Lattice ... 158
  - 6.3.1 "Simple" Lattice Fermions ... 158
  - 6.3.2 Wilson Fermions ... 159
  - 6.3.3 Ginsparg–Wilson Fermions ... 160
- 6.4 Quantum Field Theories Can be Finite But Undetermined ... 161
  - 6.4.1 A 1+1 D Example: The Schwinger Model ... 162

		6.4.2 A 3+1 D Example: Lorentz Breaking QED	164

    6.4.2   A 3+1 D Example: Lorentz Breaking QED 164
    6.4.3   Connections to the Chiral Anomaly 168
  6.5 Beyond Weyl Fermions 171
  6.6 Conclusions 172
  References 173

7 Anomalies and Kinetic Theory 177
  Alberto Cortijo
  7.1 Introduction 177
  7.2 Chiral Anomaly in Weyl Semimetals 178
  7.3 Chiral Kinetic Theory 180
    7.3.1   Boltzmann Equation 181
    7.3.2   Semiclassical Equations of Motion 182
    7.3.3   The Chiral Anomaly 185
  7.4 Conclusions 195
  References 197

8 Topological Materials in Heusler Compounds 199
  Yan Sun and Claudia Felser
  8.1 Topological Insulators in Heusler Compounds 199
  8.2 Weyl Semimetal in Half-Heusler GdPtBi
       with External Field 200
  8.3 Tuneable Anomalous Hall Effect in Half-Metallic Topological
       Semimetal with Weyl Points and Nodal Lines 202
  8.4 AHE in Non-collinear AFM with Weyl Points 203
  8.5 Strong Anomalous Hall and Anomalous Nernst Effect in
       Compensated Ferrimagnets 206
  8.6 Antiskyrmions 208
  References 209

**9 Topological Materials and Solid-State Chemistry—Finding
  and Characterizing New Topological Materials** 211
  L. M. Schoop and A. Topp
  9.1 The Role of Solid-State Chemistry in the Search for
       Topological Materials 211
  9.2 Simple Rules from Solid-State Chemistry 212
    9.2.1   Counting Electrons in Solids 212
    9.2.2   Size of the Elements 215
    9.2.3   Bonding Type 215
    9.2.4   A Database for Inorganic Crystalline Compounds 216
    9.2.5   Linking Structures to Properties 216
  9.3 Topological Materials 216
    9.3.1   3D Analogs of Graphene—3D Dirac Semimetals 217
    9.3.2   Weyl Semimetals 220

		9.3.3 Nodal Line Semimetals	220
	9.4	Nonsymmorphic Symmetries	220
		9.4.1 The Problem with the Half-Filled Band	221
	9.5	The Cycle of Material Development	223
		9.5.1 Synthesis Methods	224
		9.5.2 Measuring the Electronic Structure of Materials—ARPES	226
		9.5.3 Example—The Nonsymmorphic Square-Net Compound ZrSiS	230
		9.5.4 Beyond ZrSiS	238
	9.6	Conclusion	241
	References		241
10	**Momentum and Real-Space Study of Topological Semimetals and Topological Defects**		**245**
	Haim Beidenkopf		
	10.1	Introduction	245
	10.2	Topological Screw Dislocations	246
	10.3	Topological Weyl Semimetals and Their Analogy to Screw Dislocations	248
	10.4	The Topological Weyl Semimetal TaAs	250
	10.5	Topological Bulk Origin of the Fermi-Arc States in TaAs	254
	10.6	Summary	255
	References		256
Index			**257**

Contributors

Jens H. Bardarson Department of Physics, KTH Royal Institute of Technology, Stockholm, Sweden

Haim Beidenkopf Condensed Matter Physics Department, Weizmann Institute of Science, Rehovot, Israel

B. A. Bernevig Department of Physics, Princeton University, Princeton, NJ, USA

E. Bocquillon Laboratoire Pierre Aigrain, École Normale Supérieure, Paris, France

H. Buhmann Physikalisches Institut (EP3), Institute for Topological Insulators, University of Würzburg, Würzburg, Germany

Alberto Cortijo Materials Science Factory, Instituto de Ciencia de Materiales de Madrid, CSIC, Madrid, Spain

R. S. Deacon Advanced Device Laboratory, Center for Emergent Matter Science, RIKEN, Wako-shi, Saitama, Japan

Claudia Felser Max Planck Institute for Chemical Physics of Solids, Dresden, Germany

Dominik Gresch ETH Zurich, Institut für Theoretische Physik, Zürich, Switzerland

Adolfo G. Grushin Institut Néel, CNRS and Université Grenoble Alpes, Grenoble, France; Department of Physics, University of California Berkeley, Berkeley, CA, USA

Roni Ilan Raymond and Beverly Sackler School of Physics and Astronomy, Tel-Aviv University, Tel-Aviv, Israel

T. M. Klapwijk Faculty of Applied Sciences, Kavli Institute of Nanoscience, Delft University of Technology, Delft, The Netherlands

Juan L. Mañes Condensed Matter Physics Department, University of the Basque Country UPV/EHU, Bilbao, Spain

L. W. Molenkamp Physikalisches Institut (EP3), Institute for Topological Insulators, University of Würzburg, Würzburg, Germany

Titus Neupert Department of Physics, University of Zurich, Zurich, Switzerland

I. Robredo Donostia International Physics Center, Donostia-San Sebastián, Spain; Condensed Matter Physics Department, University of the Basque Country UPV/EHU, Bilbao, Spain

Frank Schindler Department of Physics, University of Zurich, Zurich, Switzerland

L. M. Schoop Department of Chemistry, Princeton University, Princeton, NJ, USA

Alexey Soluyanov Physik-Institut, Universität Zürich, Zurich, Switzerland

Yan Sun Max Planck Institute for Chemical Physics of Solids, Dresden, Germany

A. Topp Max-Planck-Institut für Festkörperforschung, Stuttgart, Germany

J. Wiedenmann Physikalisches Institut (EP3), Institute for Topological Insulators, University of Würzburg, Würzburg, Germany

Chapter 1
Band Theory Without Any Hamiltonians or "The Way Band Theory Should Be Taught"

I. Robredo, B. A. Bernevig and Juan L. Mañes

Abstract In this chapter, we introduce the theory of Topological Quantum Chemistry. Within this formalism, we can predict the presence or absence of topological phases by studying the behavior of orbitals lying in some special positions of the crystal. Throughout the chapter, we analyze and study the main concepts of the theory following a well-known example, graphene.

1.1 Introduction

There are different approaches for studying the properties of crystals. An initial proposal may consist in solving the Schrödinger equation of the crystal in real space. Chemists usually follow this approach, since many physical properties can be described by localized orbitals. However, due to the nonvanishing overlap between orbitals, physicists prefer the description in terms of energy bands in reciprocal space, where the Schrödinger equation and the Hamiltonian become block diagonal.

These two approaches seem disjoint, especially when it comes down to the study of topological insulators (TIs) [1–3]. During his research, Zak et al. discovered a link between those two descriptions through the concept of band representation (BR) [4–7]. BRs are mathematical objects that link the real space orbital description to the momentum space description of energy bands in the Brillouin Zone (BZ). Shortly after, Zak realized that BRs can be decomposed into what he called "elementary band representations" (EBRs), i.e., a class of BRs that cannot be further decomposed.

I. Robredo
Donostia International Physics Center, 20018 Donostia-San Sebastián, Spain
e-mail: irobredo001@ikasle.ehu.es

I. Robredo · J. L. Mañes (✉)
Condensed Matter Physics Department, University of the Basque Country UPV/EHU, 48080 Bilbao, Spain
e-mail: juanl.manes@gmail.com

B. A. Bernevig
Department of Physics, Princeton University, Princeton, NJ 08544, USA
e-mail: bernevig@princeton.edu

© Springer Nature Switzerland AG 2018
D. Bercioux et al. (eds.), *Topological Matter*, Springer Series in Solid-State Sciences 190, https://doi.org/10.1007/978-3-319-76388-0_1

Besides, there are also band representations that satisfy an extra symmetry, Time Reversal Symmetry (TRS). These band representations are called Physical Band Representations (PBRs/PEBRs).

Later, Zak and Michel examined the connectivity of the (P)EBRs[1] and claimed that all of them were connected [8, 9]. However, it has been recently proven that this is not correct [10]. In fact, if a (P)EBR happens to be disconnected,[2] then *at least one of the disconnected sets is not a (P)EBR and cannot come from a set of localized orbitals*. This is precisely the condition for a set of bands to be *topological* [11–14]. In the course of this research, the theory of Topological Quantum Chemistry (TQC) was developed, which we present in this chapter. This formalism can predict whether a material can hold topological bands by just looking at how atoms are arranged in a lattice. Thus, it is a powerful tool for novel topological materials search.

The full discussion of the theory, results, and applications can be found in a recent series of papers [10, 14–18]. In this chapter, we intend to introduce the main concepts and results of the theory by solving two widely known examples, spinless and spinful graphene. The main mathematical tools used here are those of group theory. As the reader might not be familiar with the theory, we provide a brief, practical explanation of the tools and concepts before using them, such as the Bilbao Crystallographic Server [19–21], where all elementary band representations are tabulated.

The chapter is organized as follows: In Sect. 1.2, we review the basic aspects of the hexagonal lattice and present our convention. In Sect. 1.3, we explain what happens when we add p orbitals at carbon atom sites of graphene. In Sect. 1.4, we study how to induce a BR. In Sect. 1.5, we compute the little groups for some high-symmetry points in the BZ. In Sect. 1.6, we analyze the cases of graphene, both spinless and spinful: in the spinless case we find the Dirac cones, while in the spinful case they are gapped. In Sect. 1.7, we analyze the connectivity of the (P)EBRs arising in graphene. Finally, we discuss the results in Sect. 1.8.

1.2 Hexagonal Lattice

In order to understand what an EBR is, we will work out an example, graphene. In this section, we review some basic aspects of the hexagonal lattice and present our conventions.

Taking as origin the center of the tiles and the x, y-axes as shown in Fig. 1.1, the vectors describing the Bravais lattice are:

---

[1] The connectivity represents the number of energy bands that are connected together throughout the whole BZ and cannot be disconnected without breaking the crystal symmetry. In a more graphical sense, a set of connected bands is the one that can be drawn without lifting the pencil.

[2] A set of bands is disconnected if the bands are part of a (P)EBR, but there is a gap in the whole BZ that breaks them into different sets.

Fig. 1.1 Some tiles of the 17 wallpaper group

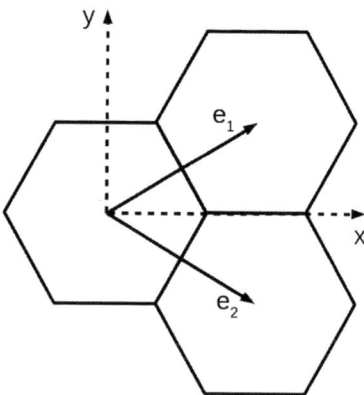

$$\begin{aligned}
\mathbf{e}_1 &= \frac{\sqrt{3}}{2}\hat{x} + \frac{1}{2}\hat{y} \\
\mathbf{e}_2 &= \frac{\sqrt{3}}{2}\hat{x} - \frac{1}{2}\hat{y}
\end{aligned} \quad (1.1)$$

where we have taken length units such that the norm of the vectors, i.e., the lattice constant, is 1. The generators[3] of the symmetry (point) group of the lattice are a 3-axis ($\{C_3|0\}$), a 2-axis ($\{C_2|0\}$), and a mirror plane ($\{m_{1\bar{1}}|0\}$[4]). Their effect on the basis vectors is the following (see Fig. 1.2).

$$\begin{aligned}
C_3 &: \quad (\mathbf{e}_1, \mathbf{e}_2) \to (-\mathbf{e}_2, \mathbf{e}_1 - \mathbf{e}_2) \\
C_2 &: \quad (\mathbf{e}_1, \mathbf{e}_2) \to (-\mathbf{e}_1, -\mathbf{e}_2) \\
m_{1\bar{1}} &: \quad (\mathbf{e}_1, \mathbf{e}_2) \to (\mathbf{e}_2, \mathbf{e}_1)
\end{aligned} \quad (1.2)$$

Before proceeding, let's define some important concepts[5]:

Definition 1.1 (*Orbit of q*) The *orbit of q* is the set of all positions in the same unit cell related to q by the elements of the symmetry group G, i.e., $Orb_q = \{gq | g \in G\}$.

Definition 1.2 (*Stabilizer group/Site-symmetry group*) The *stabilizer group* or *site-symmetry group* of a position q is the set of symmetry operations $g \in G$ that leave q fixed. It is denoted by $G_q = \{g | gq = q\} \subset G$. There are a couple of things to remark:

- G_q can include elements $\{R|\mathbf{r}\}$ with nonzero translations, $\mathbf{r} \neq 0$.

---

[3] Notice that we could have used the sixfold axis as the generator of all the rotations in this group. However, for reasons that will be clear later, we use a different set of generators.

[4] Here, by $1\bar{1}$ we refer to a mirror plane which is perpendicular to the direction $\mathbf{e}_1 - \mathbf{e}_2$, in this case, orthogonal to the y-axis.

[5] See Appendix A for a more complete set of definitions.

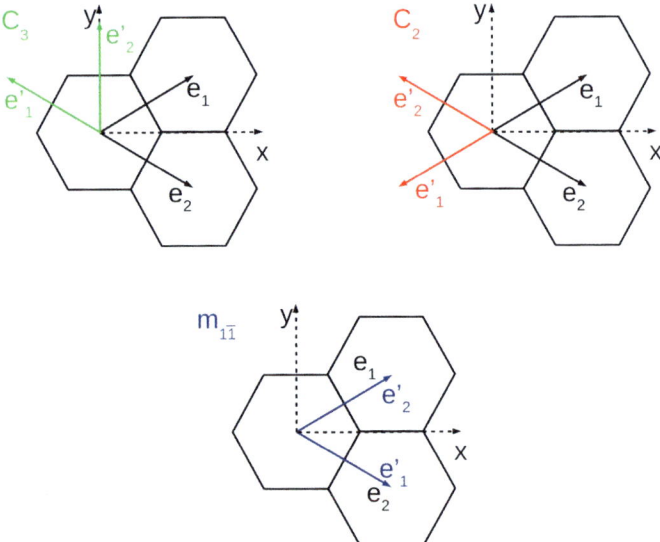

Fig. 1.2 Action of the symmetry operators on the basis vectors

- However, since any site-symmetry group leaves a point invariant, it is isomorphic to a crystallographic point group (there are 32 crystallographic point groups).

Definition 1.3 (*Coset representatives*) The *coset representatives* of a site-symmetry group can be defined as the set of elements that generate the orbit of a Wyckoff position.

Definition 1.4 (*Wyckoff position*) A *Wyckoff position* is any position in the unit cell of the crystal. Besides this general definition, there are *special* Wyckoff positions, which are positions that are left invariant by some symmetry operations, such as mirror planes and rotation axis.

Let q be a position in a unit cell. If there is an atom in that position q, to respect the symmetry of the crystal we must have an identical atom at every site in the *orbit* of q. In what follows, we compute some site-symmetry groups for special positions (see Fig. 1.3).

- $q = (\mathbf{e}_1 - \mathbf{e}_2)/2 = \mathbf{e}_y/2$ (*blue cross*)

$$\begin{aligned}
\{m_{1\bar{1}}|0\}: \quad & (\mathbf{e}_1, \mathbf{e}_2) \to (-\mathbf{e}_2, -\mathbf{e}_1) \\
& q = \frac{\mathbf{e}_1 - \mathbf{e}_2}{2} \to \frac{\mathbf{e}_1 - \mathbf{e}_2}{2} = q \\
\{C_2|1\bar{1}\}: \quad & (\mathbf{e}_1, \mathbf{e}_2) \to (-\mathbf{e}_1, -\mathbf{e}_2) + (\mathbf{e}_1 - \mathbf{e}_2) \\
& q = \frac{\mathbf{e}_1 - \mathbf{e}_2}{2} \to \frac{-\mathbf{e}_1 + \mathbf{e}_2}{2} + \mathbf{e}_1 - \mathbf{e}_2 = \frac{\mathbf{e}_1 - \mathbf{e}_2}{2} = q
\end{aligned} \tag{1.3}$$

Fig. 1.3 Maximal Wyckoff positions for the wallpaper group **17**

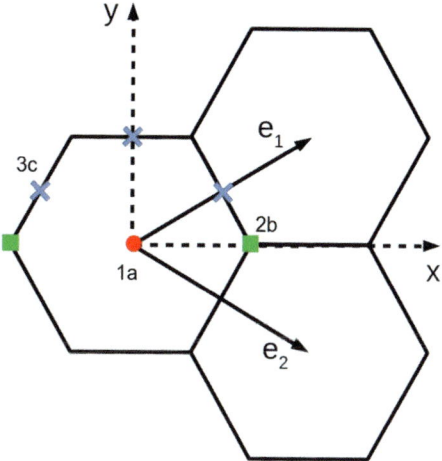

Thus the generators for the site-symmetry group at $q = \frac{e_1-e_2}{2}$ are

$$\{m_{11}|0\}, \{C_2|1\bar{1}\} \tag{1.4}$$

which generate a group isomorphic to C_{2v}.

- $q = (e_1 + e_2)/3$ (*green square*)

$$\begin{aligned}
\{m_{1\bar{1}}|0\}: \quad & (e_1, e_2) \to (e_2, e_1) \\
& q = \frac{(e_1+e_2)}{3} \to \frac{(e_1+e_2)}{3} = q \\
\{C_3|01\}: \quad & (e_1, e_2) \to (-e_2, e_1-e_2) + (e_2) \\
& q = \frac{(e_1+e_2)}{3} \to \frac{-e_2+e_1-e_2}{3} + e_2 = \frac{(e_1+e_2)}{3} = q
\end{aligned} \tag{1.5}$$

In this case, the site-symmetry group is isomorphic to C_{3v}.[6]

- $q = 0$ (*red dot*)
  In this case, all elements in the point group leave this point invariant, and the stabilizer group at $q = 0$ is not only isomorphic to the point group C_{6v} of the space group, but coincides with it.
- $q = x(e_1 + e_2)$, $x \in (0, \frac{1}{3})$ (*line connecting the dot and the square*)
  In this case, we consider the line that goes from the origin to one of the vertices of the hexagon. This set of points is left unchanged by the mirror plane $m_{1\bar{1}}$. Notice that $q = (x, x)$ interpolates between the origin and the corner of the lattice. Notice

---

[6] See Appendix B.

also that the element $m_{1\bar{1}}$ is common to the site-symmetry groups of the origin and the vertex of the hexagon.

1.2.1 Orbits for the Different q Points

Let's compute now the orbits of the points we have just discussed. To do this, we don't have to consider all the symmetry operations in the point group, only those that do not belong to the site-symmetry group. However, there is an ambiguity as to which of them to choose, because there are more than necessary.[7] The ones we choose are the *coset representatives*.[8]

- $q = (\mathbf{e}_1 - \mathbf{e}_2)/2 = \mathbf{e}_y/2$

  Since the site-symmetry group for this point contains as generators the mirror plane and the 2-axis, we will use the 3-axis to generate the orbit:

  $$\{C_3^+|0\}\left(\frac{1}{2}, -\frac{1}{2}\right) = \left(-\frac{1}{2}, 0\right)$$
  $$\{C_3^-|0\}\left(\frac{1}{2}, -\frac{1}{2}\right) = \left(0, \frac{1}{2}\right) \quad (1.6)$$

  So the orbit of q is composed by 3 points. We call these the 3c Wyckoff positions.

- $q = (\mathbf{e}_1 + \mathbf{e}_2)/3$

  In this case, the site-symmetry group contains the mirror plane and the 3-axis, so we need to consider the 2-axis:

  $$\{C_2|0\}\left(\frac{1}{3}, \frac{1}{3}\right) = \left(-\frac{1}{3}, -\frac{1}{3}\right) \quad (1.7)$$

  These positions are labeled as 2b Wyckoff positions.

- $q = 0$

  Since the site-symmetry group at this point is the whole point group, there are no other positions in its orbit. This position is denoted as 1a.

- $q = x(\mathbf{e}_1 + \mathbf{e}_2)$

---

[7] Imagine that our full group has 15 elements and that our site-symmetry group has 5. Then, the orbit of that point will have $15/5 = 3$ positions, i.e., we need 3 out of the $15 - 5 = 10$ remaining elements to generate the orbit. The other 7 elements will generate the same positions in the orbit; that is why we do not need to consider them.

[8] See Appendix A for a more formal definition.

The site-symmetry group for this set of points contains just the mirror plane, so any combination of the axes (the 2- and 3-axis) will generate a position in the orbit. In this case, there will be 6 positions in the orbit, which coincide with the ones generated by the 6-axis.

The site-symmetry groups at any two points in the orbit of q are conjugate to each other and, hence, isomorphic.[9] We are interested in the maximal Wyckoff positions, which are the positions whose site-symmetry group is a maximal subgroup of the full point group. Thus, as C_{6v} has only 3 maximal subgroups, C_{6v}, C_{3v}, and C_{2v}, the maximal Wyckoff positions will be the ones whose site-symmetry group is isomorphic to one of these. Actually, this is what we found; the site-symmetry groups for the positions 1a, 2b, and 3c are isomorphic to the maximal subgroups of C_{6v}. However, the last position for which we computed the site-symmetry group and the orbit is not a maximal Wyckoff position since its site-symmetry group is not maximal. The reason why we want to look only at maximal Wyckoff positions is that only they can give raise to elementary band representations, though not necessarily. We will discuss this point later on.

1.2.2 Adding Orbitals

We want to understand how bands arise from localized orbitals in real space. As mentioned above, we are interested only in orbitals that are localized at *maximal Wyckoff positions*. The reason for this will be clear once we construct the band representation, as we are interested in tabulating all *elementary* band representations, not all band representations in general. To illustrate how to construct these band representations, we solve an easy case, graphene.

Following what we have done so far, for us graphene consists in orbitals localized around the Wyckoff position 2b. In general, graphene has both s and p orbitals, but not all bands can be topological. So in this example, we will restrict ourselves to p_z orbitals, although we could consider *any* other set of orbitals.[10] To respect crystal symmetry, when we add an orbital at a Wyckoff position, we have to add the same orbital at every other position in the orbit. In the case of graphene, if we add a p_z orbital in the 2b Wyckoff position (where the atoms are centered) we have to add it in both positions of the orbit.

These orbitals will transform under a representation of the site-symmetry group. From the fact that the site-symmetry groups of different positions in the same orbit are conjugated to each other, it follows that once we know the representation under

---

[9]See Appendix B.

[10]Actually, saying that we consider only p_z orbitals is not entirely correct. What is true is that we are considering the *crystal orbitals* that transform under a certain irreducible representation of the site-symmetry group. In this case, we use the a_1 crystal orbital that transforms under the A_1 representation of C_{3v}. This a_1 orbital will be, in general, "contaminated" by pieces of higher atomic orbitals that transform under the same representation.

which the orbitals on *one* of the positions transform, we know how they transform on *any* other position of the orbit. This is easy to see:

Let ρ be the representation[11] under which a set of orbitals in position q_1 transform, h an element of the site-symmetry group G_{q_1}, and g_α one of the coset representatives. Then

$$hq_1 = q_1, \quad g_\alpha q_1 = q_\alpha \to q_1 = g_\alpha^{-1} q_\alpha$$
$$g_\alpha h g_\alpha^{-1} q_\alpha = g_\alpha h q_1 = g_\alpha q_1 = q_\alpha \qquad (1.8)$$

Thus $g_\alpha h g_\alpha^{-1}$ belongs to the site-symmetry group of q_α. Finally, if we know ρ we know also that $\rho_\alpha(h) = \rho(g_\alpha^{-1} h g_\alpha)$.[12] This makes sense because the site-symmetry groups of all positions in an orbit are conjugate to each other and, thus, isomorphic.

So far we have been talking about atomic orbitals but, as mentioned in a previous footnote, it is more accurate to speak of crystal orbitals. Atomic orbitals transform under irreducible representations of the full rotation group $SO(3)$, whereas crystal orbitals transform under irreducible representations of the site-symmetry group. Crystal orbitals can be described by a set of Wannier states, one per orbital,[13] that transform under that representation [11, 12].

So, let's actually introduce the orbitals and see what we get.

1.3 Adding *p* Orbitals at 2b Positions

In this section, we will study how *p* orbitals transform when placed in an environment with a reduced amount of symmetry: in this case, at Wyckoff 2b position of the space group P6mm (or, in the two-dimensional version, the wallpaper group **17**).

1.3.1 Spinless p Orbitals

Orbitals with angular momentum one or *p* orbitals are a basis for a three-dimensional ($l = 1$) or *vector* (V) representation of $O(3)$. However, the site-symmetry group for Wyckoff 2b is not the whole rotation group $O(3)$ but just one of its subgroups, namely C_{3v}. Thus, what matters physically is how *p* orbitals transform under the point group C_{3v}. Mathematically, we want to express the vector representation of $O(3)$ in terms of irreducible representations of C_{3v}. This process is known as *subduction*. In our case, if we denote by ρ the representation of $O(3)$ under which *p* orbitals transform,

---

[11] A representation assigns a square matrix or operator $\rho(g)$ to each element of the group, in such a way that when we compose two elements the product of the two matrices is equal to the matrix assigned to the resulting element, i.e., $\rho(g_1)\rho(g_2) = \rho(g_1 g_2)$.

[12] See Appendix C.

[13] If we have 2 orbitals per unit cell, and N cells, we will have 2N Wannier states.

Table 1.1 Effect of C_{3v} group elements on p orbitals

C_{3v}	E	C_3	$m_{1\bar{1}}$
p_x	p_x	$cp_x + sp_y$	p_x
p_y	p_y	$cp_y - sp_x$	$-p_y$
p_z	p_z	p_z	p_z

we want to find the corresponding subduced representation $\rho \downarrow G_{q^{2b}} \equiv \rho \downarrow C_{3v}$ of C_{3v}. Here, the down arrow represents the operation of *subduction*.

First of all, we have to determine how the three orbitals p_x, p_y, and p_z transform under the symmetries of the group C_{3v}. Actually, we only need to see how the generators act, since the rest of the elements can be obtained by matrix multiplication. The action of generators is given in Table 1.1, where c, s stand for $\cos(\frac{2\pi}{3})$, $\sin(\frac{2\pi}{3})$. Knowing this, we can construct the matrices of the representation[14]:

$$\rho^V(E) = \begin{pmatrix} 1 & 0 & 0 \\ 0 & 1 & 0 \\ 0 & 0 & 1 \end{pmatrix}, \quad \rho^V(m_{1\bar{1}}) = \begin{pmatrix} 1 & 0 & 0 \\ 0 & -1 & 0 \\ 0 & 0 & 1 \end{pmatrix}, \quad \rho^V(C_3) = \begin{pmatrix} \cos\left(\frac{2\pi}{3}\right) & \sin\left(\frac{2\pi}{3}\right) & 0 \\ -\sin\left(\frac{2\pi}{3}\right) & \cos\left(\frac{2\pi}{3}\right) & 0 \\ 0 & 0 & 1 \end{pmatrix}$$
(1.9)

We can see, just by inspection, that this representation is *reducible*, in the sense that we can decompose it into diagonal blocks. Formally, we need to compute the character[15] and compare it with the character table for the group C_{3v} (see Table 1.2).

It is important to note that *all* the symmetry elements of the group, not just the generators, must appear in the character table. However, to save space, the elements that have the same traces appear in the same column. These are called *classes*. In this case, our group has 6 elements (E, $C_3^{\pm}$ and three mirror planes m_i), but there are only 3 different classes. All the elements in a class must be "of the same type." In this case, the three classes correspond to the identity, the threefold rotations, and the mirror planes.

Although there is an algorithmic method[16] to compute how a group representation decomposes into irreducible representations (irreps) of one of its subgroups [22]; in this case, it is not hard to see by simple inspection that $V = A_1(p_z) + E(p_x, p_y)$. This is an example where an irreducible representation of $O(3)$ becomes reducible when restricted to a subgroup; i.e., when the symmetry is reduced.

So far we have been working with single-valued or spinless group representations, i.e., representations of subgroups of $O(3)$. But if we want to take spin into account, we should extend our methods to the so-called double-valued representations. The

---

[14]These matrices correspond to the basis of p_x, p_y, p_z. There is another convention where we change $(p_x, p_y) \to (p_x - ip_y, p_x + ip_y)$ so that the vectors in this new basis are eigenstates of L_z and $\rho(C_3)$ becomes diagonal, $\rho(C_3) = \text{diag}(e^{-2\pi i/3}, e^{2\pi i/3})$). On this basis, the matrix for the mirror plane is non-diagonal, $(\rho(m_{1\bar{1}}) = \begin{pmatrix} 0 & 1 \\ 1 & 0 \end{pmatrix}$.

[15]The character of a representation is the set the traces of its matrices.

[16]See the explanation around (1.17) below.

Table 1.2 Table of characters of the group C_{3v}. The first row gives the traces of the matrices for the vector representation, while the next two correspond to the blocks formed by p_z and p_x, p_y. Under the solid line, we have written the characters of the irreps (irreducible representations) of C_{3v}

C_{3v}	E	$C_3^{\pm}$	m_i
V	3	0	1
ρ^{p_z}	1	1	1
ρ^{p_x, p_y}	2	−1	0
A_1	1	1	1
A_2	1	1	−1
E	2	−1	0

reason is that spin transforms under $SU(2)$, which is the universal covering of the (proper) rotation group $SO(3)$.

1.3.2 Spinful p Orbitals

We will now focus on p_z orbitals with spin up and down, also at the Wyckoff 2b position. Angular momentum eigenstates rotate with the unitary operator $\exp(i\mathbf{n} \cdot \mathbf{J}\Omega)$, where $\mathbf{J}$ is the total angular momentum operator, $\mathbf{n}$ is a unit vector in the direction of the rotation axis, and Ω is the rotation angle. Representations of SO(3) correspond to integral total angular momentum. However, for half-integral angular momentum a rotation of 2π gives a minus sign instead of the identity. As mentioned earlier, this reflects the fact that half-integral angular momentum states transform under representations of $SU(2)$. Following what we did before, we construct the representation for spin up and down p_z orbitals.

For a $\frac{2\pi}{3}$ rotation about the C_3 axis in the z-direction, we have $e^{i\frac{\pi}{3}}$ for the spin-up state ($s_z = \frac{1}{2}$) and $e^{-i\frac{\pi}{3}}$ for spin down ($s_z = -\frac{1}{2}$).[17] We can write this representation in a more compact way as $\rho(C_3) = \exp(i\frac{\pi}{3}\sigma_z)$ where σ_z is the third Pauli matrix. The mirror plane is a bit trickier. We can think of a mirror plane as a 180° rotation around an axis orthogonal to the mirror plane followed by the inversion.[18] Spin rotates, but space inversion has no effect on it, so, for spin states, our mirror plane is just a 180° rotation about the y-direction (remember that the mirror is perpendicular to the y-axis). Thus the operator for the mirror plane is $\rho(m_{1\bar{1}}) = \exp(i\frac{\pi}{2}\sigma_y) = i\sigma_y$. The matrices for the generators will be

$$\rho(C_3) = \begin{pmatrix} e^{i\frac{\pi}{3}} & 0 \\ 0 & e^{-i\frac{\pi}{3}} \end{pmatrix}, \quad \rho(m_{1\bar{1}}) = \begin{pmatrix} 0 & 1 \\ -1 & 0 \end{pmatrix} \quad (1.10)$$

---

[17] Remember that p_z orbitals have $L_z = 0$.
[18] In this case, as the 180° rotation and the inversion commute, you can apply them in any order.

Table 1.3 Table of characters of the group C_{3v}

C_{3v}	E	$C_3^\pm$	m_i	$\bar{E}$	$\bar{C}_3^\pm$	$\bar{m}_i$
ρ	2	1	0	-2	-1	0
$\bar{\Gamma}_4$	1	-1	$-i$	-1	1	i
$\bar{\Gamma}_5$	1	-1	i	-1	1	$-i$
$\bar{\Gamma}_6$	2	1	0	-2	-1	0

Now we can write the character of this representation by computing the traces and compare them to the irreducible representations of the double group. Remember that for double groups a rotation of 2π is equal to minus the identity, so the number of elements doubles. For each element of the ordinary group, we have to include the result of composing the element with the 2π rotation $\bar{E}$. Note that the characters of "barred" and "unbarred" elements differ just by the sign (Table 1.3).

We can see that the representation under which our orbitals transform is an irreducible representation ($\bar{\Gamma}_6$) of the site-symmetry double group, and we may expect the emerging band representation to be elementary. The other two spinful irreps are total angular momentum $\frac{3}{2}$ representations. More precisely, the basis for $\bar{\Gamma}_5$ is $|\frac{3}{2}\rangle + i|-\frac{3}{2}\rangle$, while the basis for $\bar{\Gamma}_4$ is $|\frac{3}{2}\rangle - i|-\frac{3}{2}\rangle$, made from p_x, p_y spinful orbitals[19] [15, 17]. The remaining combinations of p_x, p_y spinful orbitals, like p_z, transform under $\bar{\Gamma}_6$.[20]

We have obtained the representations under which p_z orbitals, both spinless and spinful transform, under C_{3v} group, which is the site-symmetry group of the Wyckoff 2b position, the sites where carbon atoms lie in graphene. Now, we may ask:

- Q: What do orbitals at lattice sites, characterized by the representation, become?
- A: A band. These are the electronic bands.
- Q: Without a specific Hamiltonian, what can we say about the bands?
- A: Quite a lot. There are some properties that depend only on the symmetries of the crystal, such as the topological nature. The only thing we need for the Hamiltonians is to comply with the symmetries of the crystal, which is a property that needs to have in order to describe the system properly. What we do here is to understand bands not as sets of eigenvalues coming from a Hamiltonian, but as an abstract concept called *Band Representation*.

---

[19] $|\frac{3}{2}\rangle = (|p_x\rangle + i|p_y\rangle) \otimes |\uparrow\rangle$, $|-\frac{3}{2}\rangle = (|p_x\rangle - i|p_y\rangle) \otimes |\downarrow\rangle$.

[20] The combinations are is $(|p_x\rangle + i|p_y\rangle) \otimes |\downarrow\rangle$ and $(|p_x\rangle - i|p_y\rangle) \otimes |\uparrow\rangle$.

Next, we see what kind of band representations are induced from spinless and spinful orbital representations of the site-symmetry group. Mathematically speaking, we want to find $\rho_G = \rho_{G_q} \uparrow G \equiv \mathrm{Ind}_{G_q}^G \rho$.

1.4 Inducing a Band Representation

From the Wannier states $W_{i\alpha}(r - t_\mu)$,[21] we define Fourier transformed Wannier states:

$$a_{i\alpha}(\mathbf{k}, \mathbf{r}) = \frac{1}{\sqrt{N}} \sum_\mu e^{i\mathbf{k}\mathbf{t}_\mu} W_{i\alpha}(r - t_\mu) \qquad (1.11)$$

where $\mathbf{t}_\mu$ are all vectors in the Bravais lattice. We have gone from a $n \times n_q \times N$ (n positions in the orbit times n_q orbitals per site in the orbit times N cells in the crystal)-dimensional basis to a finite $n \times n_q$ basis for each $\mathbf{k}$. This corresponds to $n \times n_q$ bands. These functions are a set of Bloch wave functions that span bands in reciprocal space. What we want to know is how these functions transform in reciprocal space, since that will give us the *invariant subspace* (or irrep) to which they belong. And here is the key of this theory: Even if we don't know the actual form of the Hamiltonian, if it complies with the symmetries of the crystal, the eigenstates will transform under a certain representation of the space group, what is called a band representation [7]. So, if a set of eigenstates transform according to a certain irrep around some point in reciprocal space they belong to an invariant subspace of the Hilbert space. As the Hamiltonian commutes with the symmetries, this implies that *eigenstates that transform under an irrep will be degenerate*. Even if we are not able to predict the explicit form bands have, we can predict their crossings at high-symmetry $\mathbf{k}$ points, using group theory techniques. This will be more clear when we actually construct the Band Representation.

Our motivation being clear, we proceed to see how these $a_{i\alpha}(\mathbf{k}, \mathbf{r})$ functions transform under any element of the full group $h \in G$. We can derive it by knowing how Wannier functions transform[22] and then using (1.11). Wannier functions transform as

$$[\rho_G(h)]_{j\beta, i\alpha} W_{j\beta}(\mathbf{r} - \mathbf{t}_\mu) = [\rho(g_{\beta\alpha})]_{ji} W_{j\beta}(\mathbf{r} - R t_\mu - t_{\beta\alpha}) \qquad (1.12)$$

Thus

---

[21] Here, the i index labels the orbital (s, p, d...) while α labels the position on the orbit (1, 2, 3...). The last index is t_μ, which labels the cell of the crystal. This way we have labeled all orbitals in our crystal. We can see here how there is one Wannier function per orbital.

[22] See Appendix C for a complete derivation of the transformation properties of Wannier functions.

1 Band Theory Without Any Hamiltonians ...

$$[\rho_G(h)]_{j\beta,i\alpha}\, a_{j\beta}(\mathbf{k},\mathbf{r}) = \frac{1}{\sqrt{N}}\sum_\mu e^{i\mathbf{k}\mathbf{t}_\mu}[\rho_G(h)]_{j\beta,i\alpha} W_{j\beta}(r-t_\mu)$$

$$= \frac{1}{\sqrt{N}}\sum_\mu e^{i\mathbf{k}\mathbf{t}_\mu}[\rho(g_{\beta\alpha})]_{ji} W_{j\beta}(\mathbf{r}-R\mathbf{t}_\mu - t_{\beta\alpha})$$

$$= e^{-iR\mathbf{k}\mathbf{t}_{\beta\alpha}}[\rho(g_{\beta\alpha})]_{ji}\frac{1}{\sqrt{N}}\sum_\mu e^{i(R\mathbf{k})(R\mathbf{t}_\mu + \mathbf{t}_{\beta\alpha})} W_{j\beta}(\mathbf{r}-R\mathbf{t}_\mu - t_{\beta\alpha})$$

$$= e^{-iR\mathbf{k}\mathbf{t}_{\beta\alpha}}[\rho(g_{\beta\alpha})]_{ji} a_{j\beta}(R\mathbf{k},\mathbf{r})$$

(1.13)

where $g_{\beta\alpha} = g_\beta^{-1}\{E|-\mathbf{t}_{\beta\alpha}\}h g_\alpha \in G_q$.[23] For each value of $\mathbf{k}$ on the reciprocal space, this expression tells us how the Bloch wave functions transform. This is what we call a *Band Representation* or, more succinctly, bands.

The representation $\rho_G(h)$ is a $(n \times n_q)$-dimensional square matrix (orthogonal if the point group is single-valued, unitary otherwise) at each pair of $\mathbf{k}$, $\mathbf{k}'$ vectors. The only nonzero blocks are the $\mathbf{k}$, $\mathbf{k}' = R\mathbf{k}$, as can be seen from the above equation, since it relates the $a(\mathbf{k})$ state to the $a(R\mathbf{k})$ one. This block can be written as:

$$[\rho_G^{\mathbf{k}}(h)]_{j\beta,i\alpha} = e^{-iR\mathbf{k}\mathbf{t}_{\beta\alpha}}[\rho(g_\beta^{-1}\{E|-\mathbf{t}_{\beta\alpha}\}h g_\alpha)]_{ji}$$

(1.14)

Before actually building this representation for graphene in the case of spinless and spinful p_z orbitals, we can discuss the power of this approach. In what we have done so far we have not used the Hamiltonian in any way, apart from asking it to comply with the symmetries. But we claim that there is a lot of information about the bands in these representations at all $\mathbf{k}$ in the first Brillouin Zone (BZ). Lets see what kind of information we can get and how.

1.5 Little Groups at k Points in the First BZ

Let's denote by $G_\mathbf{k}$ the *little group* of a $\mathbf{k}$ point in the reciprocal space. We will see that the most interesting $\mathbf{k}$ points will be the ones with highest symmetry but, for the time being, $\mathbf{k}$ can have any value in the first BZ. The characters of the band representation are the traces of the $[\rho_G^{\mathbf{k}}(h)]_{j\beta,i\alpha}$ matrices for each h, i.e.,

$$\chi_G^{\mathbf{k}}(h) = [\rho_G^{\mathbf{k}}(h)]_{i\alpha,i\alpha} = \sum_{\alpha,i} e^{-iR\mathbf{k}\mathbf{t}_{\alpha\alpha}}[\rho(g)]_{ii} = \sum_\alpha e^{-iR\mathbf{k}\mathbf{t}_{\alpha\alpha}}\chi^{\mathbf{k}}(g_\alpha^{-1}\{E|-\mathbf{t}_{\alpha\alpha}\}h g_\alpha)$$

(1.15)

We know that Bloch wave functions transform under representations of the little group at different $\mathbf{k}$ points. Once we have the character of the representation under

---

[23] See Appendix C for further details.

which Bloch wave functions transform at any **k** point in the first BZ, we can *subduce* the full representation, from the full group G to the little group, $G_\mathbf{k}$.[24] Once we do this, we know the *small representation* under which Bloch wave functions transform at each **k** point. Now, we can ask ourselves if this subduced representation is reducible or not. In general, it will be reducible, and we will be able to express our representation as a sum of irreps of the little group.[25] So, strictly speaking, starting from the representation under which the Wannier functions transform, we want to construct:

$$(\rho \uparrow G) \downarrow G_\mathbf{k} \cong \bigoplus_i m_i \sigma_i^\mathbf{k} \tag{1.16}$$

where $\sigma_i^\mathbf{k}$ are irreps of the little group $G_\mathbf{k}$ and the m_i are the multiplicities of the representation, i.e., how many times that irrep appears. This number can be easily obtained using the so-called *magic formula* [22]:

$$m_i = \frac{1}{n} \sum_h \bar{\chi}_i(h) \chi^\mathbf{k}(h) \tag{1.17}$$

where n is the number of elements of the group, h are the elements of the group, the bar indicates complex conjugate, χ_i denote the characters of the irrep, and $\chi^\mathbf{k}$ is the character of the representation we want to decompose. So, we see how some properties of the bands are inherited from the way orbitals transform in real space. Using group theory, we can see how Bloch wave functions transform in the reciprocal space. In general, at **k** points with no symmetry, this will give us not very interesting information, since the irreps found at those points have no interest, in the sense that there are few of them, usually one-dimensional. However, at certain high-symmetry points, there will be irreps of dimension greater than 1 (up to a maximum of 8), coming from the argument given in the previous section. Notice that this degeneracy is not accidental; it comes from imposing the symmetries of the crystal. This means that these degeneracies are *protected by symmetry*, so we cannot break them by small perturbations of the Hamiltonian that respect the symmetry. This is one example of what this theory achieves: Using simple computations of group theory, one can predict the degeneracies of band crossings along the first BZ from the way orbitals transform in real space.

In the following section, we will compute the band representation arising from spinless and spinful p_z orbitals in graphene and see what we can predict within this theory.

---

[24] This is a rigorous mathematical procedure, common in group theory. In practice, it is like constructing the table of characters for the big group and removing the elements that do not belong to the little group.

[25] Here, the term "sum" has to be understood as sum of representations. For example, a one-dimensional representation "plus" a two-dimensional representation gives a three-dimensional one.

1.6 Example of Band Representation

In this section, we will explicitly compute the band representation arising from p_z orbitals, spinless and spinful in graphene. We will see how, even if we are not using any specific Hamiltonian, we can predict that spinless graphene has Dirac cones, while spinful does not, and that spinful graphene could be topological.[26]

1.6.1 Spinful Graphene

Let's start with the ingredients we need. First, remember that graphene consists in carbon atoms sitting at 2b Wyckoff positions, with coordinates $q_1 = \left(\frac{1}{3}, \frac{1}{3}\right)$ and $q_2 = \left(-\frac{1}{3}, -\frac{1}{3}\right)$. The site-symmetry group of this site is isomorph to C_{3v} and we choose the representation of the generators to be[27]:

$$\rho(C_3) = \exp\left(i\frac{\pi}{3}\sigma_z\right), \quad \rho(m_{1\bar{1}}) = i\sigma_x \qquad (1.18)$$

The coset representatives are chosen to be $g_1 = \{E|0\}$, $g_2 = \{C_2|0\}$. The first thing we need to compute are the $t_{\beta\alpha}$. In general the action of an element of the full group will have the following form:

$$hq_\alpha = \{E|\mathbf{t}_{\beta\alpha}\}q_\beta, \quad g_\beta^{-1}\{E|-\mathbf{t}_{\beta\alpha}\}hg_\alpha q_1 = q_1 \equiv gq_1 = q_1 \qquad (1.19)$$

where the vector $\mathbf{t}_{\beta\alpha}$ represents the possibility of an element to take some Wyckoff away to another cell. It can be shown in Appendix C:

$$\mathbf{t}_{\beta\alpha} = hq_\alpha - q_\beta \qquad (1.20)$$

We will use this last equation to compute the $\mathbf{t}_{\beta\alpha}$ for the different generators.

- $\{C_3|0\}$

$$\mathbf{t}_{\beta\alpha} = hq_\alpha - q_\beta$$
$$\mathbf{t}_{11} = \{C_3|0\}\left(\frac{1}{3}, \frac{1}{3}\right) - \left(\frac{1}{3}, \frac{1}{3}\right) = \left(\frac{1}{3}, -\frac{2}{3}\right) - \left(\frac{1}{3}, \frac{1}{3}\right) = (0, -1)$$
$$\mathbf{t}_{22} = \{C_3|0\}\left(-\frac{1}{3}, -\frac{1}{3}\right) - \left(-\frac{1}{3}, -\frac{1}{3}\right) = \left(-\frac{1}{3}, \frac{2}{3}\right) + \left(\frac{1}{3}, \frac{1}{3}\right) = (0, 1)$$
$$(1.21)$$

---

[26] This is not actually seen in real graphene, since the spin–orbit interaction is really small.
[27] As in Sect. 1.3.2.

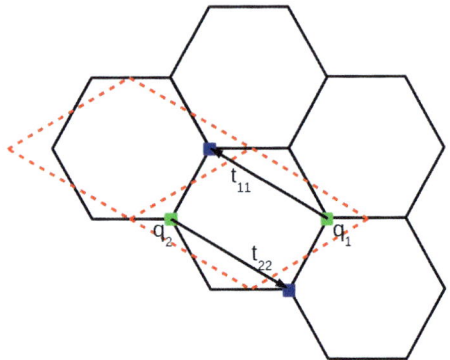

Fig. 1.4 Action of the 3-axis on Wyckoff positions. In green, the original positions. In blue, the new position after the action of the rotation. In red, the unit cells

Once determined that $t_{\beta 1}$ exists for a certain β, there is no other β for which $t_{\beta\alpha}$ makes sense. We can see in the Fig. 1.4 how $\mathbf{t}_{11}$ connects two q_1 positions in different cells, due to the action of the 3-axis. But this 3-axis *does not connect a position q_1 to a position q_2*, so $\mathbf{t}_{21}$ does not exist. What this really means is that when constructing the representation for the 3-axis with indices $(j\beta, i\alpha)$, the block with $\alpha = 1$, $\beta = 2$ will be full of zeroes, while the block with $\alpha = 1$, $\beta = 1$ will be the representation of some element of the site-symmetry group of position q_1. But this element can only be the one we obtained before, $g = g_\beta^{-1}\{E| - \mathbf{t}_{\beta\alpha}\}hg_\alpha$. We can use this equation to find this g element:

$$g_{11} = \{E|0\}^{-1}\{E| - \mathbf{t}_{11}\}\{C_3|0\}\{E|0\} = \{E| - 0\bar{1}\}\{C_3|0\} = \{C_3|0\bar{1}\}$$
$$g_{22} = \{C_2|0\}^{-1}\{E| - \mathbf{t}_{22}\}\{C_3|0\}\{C_2|0\} = \{C_3|0\bar{1}\} \quad (1.22)$$

So we can construct now the representation of this element. We will write it by blocks, each block defined by coordinates (β, α) being a matrix with indices (i, j). Using (1.14):

$$[\rho_G^k(\{C_3|0\})]_{j\beta,i\alpha} = \begin{pmatrix} e^{i(C_3\mathbf{k})\mathbf{e}_2} & 0 \\ 0 & e^{-i(C_3\mathbf{k})\mathbf{e}_2} \end{pmatrix} \otimes e^{i\frac{\pi}{3}\sigma_z} = \begin{pmatrix} e^{i(C_3\mathbf{k})\mathbf{e}_2}e^{i\frac{\pi}{3}\sigma_z} & 0 \\ 0 & e^{-i(C_3\mathbf{k})\mathbf{e}_2}e^{i\frac{\pi}{3}\sigma_z} \end{pmatrix} \quad (1.23)$$

where the product is a tensorial product.

- $\{m_{1\bar{1}}|0\}$

Following the same procedure as before:

$$\mathbf{t}_{11} = \{m_{1\bar{1}}|0\}\left(\frac{1}{3},\frac{1}{3}\right) - \left(\frac{1}{3},\frac{1}{3}\right) = (0,0)$$
$$\mathbf{t}_{22} = \{m_{1\bar{1}}|0\}\left(-\frac{1}{3},-\frac{1}{3}\right) - \left(-\frac{1}{3},-\frac{1}{3}\right) = (0,0) \quad (1.24)$$

The no-null blocks will be diagonal and, in this case, both $\mathbf{t}_{\alpha\beta}$ are zero. Let's find the elements g:

$$g_{11} = \{E|0\}^{-1}\{E| - \mathbf{t}_{11}\}\{m_{1\bar{1}}|0\}\{E|0\} = \{E|0\}\{m_{1\bar{1}}|0\} = \{m_{1\bar{1}}|0\}$$
$$g_{22} = \{C_2|0\}^{-1}\{E| - \mathbf{t}_{22}\}\{m_{1\bar{1}}|0\}\{C_2|0\} = \{C_2|0\}^{-1}\{E| - \mathbf{t}_{22}\}\{\bar{C}_2|0\}\{m_{1\bar{1}}|0\} = \{\bar{m}_{1\bar{1}}|0\} \quad (1.25)$$

where in the last step we have used the commutation relation of the 2-axis and the mirror plane, but notice that, in double groups, $C_2^2 = \bar{E} = -E \neq E$. We are able to build the representation for this element:

$$[\rho_G^{\mathbf{k}}(\{m_{1\bar{1}}|0\})]_{j\beta,i\alpha} = \begin{pmatrix} 1 & 0 \\ 0 & 1 \end{pmatrix} \otimes i\sigma_x = \begin{pmatrix} i\sigma_x & 0 \\ 0 & -i\sigma_x \end{pmatrix} \quad (1.26)$$

- $\{C_2|0\}$

So far, we have found that the induced representations are diagonal, giving us the feeling that it could be reducible. This has happened because the elements for which we have been constructing the representation were in the site-symmetry group, or differ by an integer lattice translation. This is a general remark. However, we will find now that, since the C_2 is not contained in the site-symmetry group, the representation will be off-diagonal and, thus, will make this representation irreducible. Let's compute the representation for the C_2.

$$\mathbf{t}_{21} = \{C_2|0\}\left(\frac{1}{3}, \frac{1}{3}\right) - \left(-\frac{1}{3}, -\frac{1}{3}\right) = (0, 0)$$
$$\mathbf{t}_{12} = \{C_2|0\}\left(-\frac{1}{3}, -\frac{1}{3}\right) - \left(\frac{1}{3}, \frac{1}{3}\right) = (0, 0) \quad (1.27)$$

We see that, in this case, the nonvanishing blocks are the ones with coordinates $\alpha = 1, \beta = 2$ and $\alpha = 2, \beta = 1$. So the representation will be off-diagonal. Let's compute the elements g:

$$g_{21} = \{C_2|0\}^{-1}\{E| - \mathbf{t}_{21}\}\{C_2|0\}\{E|0\} = \{E|0\}$$
$$g_{12} = \{E|0\}^{-1}\{E| - \mathbf{t}_{12}\}\{C_2|0\}\{C_2|0\} = \{\bar{E}|0\} \quad (1.28)$$

So, the representation for this element is:

$$[\rho_G^{\mathbf{k}}(\{C_2|0\})]_{j\beta,i\alpha} = \begin{pmatrix} 0 & -I \\ I & 0 \end{pmatrix} = -i\sigma_y \otimes \sigma_0 \quad (1.29)$$

where I is the 2×2 identity.

1.6.2 Spinless Graphene

Having determined the spinful representation, it is easy to see what the spinless representation is, by just getting rid of the spin degree of freedom. Now, the 4×4 matrices are 2×2 matrices:

$$[\rho_G^{\mathbf{k}}(\{C_3|0\})]_{\beta\alpha} = \begin{pmatrix} e^{i(C_3\mathbf{k})\mathbf{e}_2} & 0 \\ 0 & e^{-i(C_3\mathbf{k})\mathbf{e}_2} \end{pmatrix} \quad (1.30)$$

$$[\rho_G^{\mathbf{k}}(\{m_{1\bar{1}}|0\})]_{\beta\alpha} = \begin{pmatrix} 1 & 0 \\ 0 & 1 \end{pmatrix} \quad (1.31)$$

$$[\rho_G^{\mathbf{k}}(\{C_2|0\})]_{\beta\alpha} = \begin{pmatrix} 0 & -1 \\ 1 & 0 \end{pmatrix} \quad (1.32)$$

Now that we have the representation under which Bloch functions transform at any $\mathbf{k}$ point, we can see which degeneracies we will have due to the symmetry, by seeing how this representation *subduces* at different $\mathbf{k}$ points, since Bloch wave functions transform under representations of the little group of the $\mathbf{k}$ point. Let's see what happens at the $\mathbf{K} = \left(\frac{1}{3}, \frac{2}{3}\right)$ point.

The little group of the $\mathbf{K}$ point consists in the 3-axis and a mirror plane (isomorph to C_{3v}), but the mirror plane is not $m_{1\bar{1}}$ but m_{11}. We can compute explicitly the representation of this mirror plane to obtain the character to define its representation, but it is not necessary. Following the argument we gave before, the representation of an element that is not part of the site-symmetry group is off-diagonal when we induce it, since it mixes the two positions. So the trace of the representation of the mirror plane m_{11} will be 0. Because the 3-axis is in the little group, $(C_3\mathbf{K})\mathbf{e}_2 = \mathbf{K} \cdot \mathbf{e}_2$ and we take the trace of the 3-axis matrix to get the character: $[\rho_G^{\mathbf{K}}(\{C_3|0\})]_{\alpha\alpha} = 2\cos(\mathbf{K} \cdot \mathbf{e}_2) = 2\cos\left(\frac{4\pi}{3}\right) = -1$. So our representation has the following character in C_{3v} (Table 1.4):

We see that this representation is already an irreducible representation of C_{3v}; i.e., there will be a band crossing at the $\mathbf{K}$ point of the two bands that are doubly degenerate due to trivial spin degeneracy. And this crossing is protected by symmetry. We have found the famous Dirac cones of graphene. This is an example of how simple calculations using group theory lead to strong results.

Table 1.4 Table of characters of the group C_{3v}

C_{3v}	E	$C_3^{\pm}$	m_i
$\rho_G^{\mathbf{K}}$	2	-1	0
A_1	1	1	1
A_2	1	1	-1
E	2	-1	0

1.7 Subducing the Band Representation

We have by now obtained the band representation for spinless and spinful graphene. We will focus now on spinful graphene, since it is the one that can display topological properties. We will subduce now the representation at different, high-symmetry points in the first Brillouin Zone. In this case, we will study the points Γ, K and M (see Fig. 1.5).

We proceed as before; first, find the character of the representation for the elements of the little group. Then, see if the representation is irreducible or not, to see if bands cross at that point. Then, we will study how we can connect those bands.

1.7.1 Γ Point

The little group at this point is the full point group, C_{6v}. This a common property for all BZs of all space groups. This group contains 3-axis, 2-axis, planes, and 6-axis also. We haven't computed the representation for the 6-axis, but we can obtain it from the representations of 2- and 3-axis by combining them (e.g., a 6-axis is a 2-axis minus a 3-axis). Doing this way, we can write the character for this representation (Table 1.5):

Just by inspection of the table, we get that the representation is reducible, in fact (Table 1.6):

$$\rho_G^\Gamma = \bar{\Gamma}_7 \oplus \bar{\Gamma}_8 \tag{1.33}$$

Fig. 1.5 First Brillouin Zone for graphene

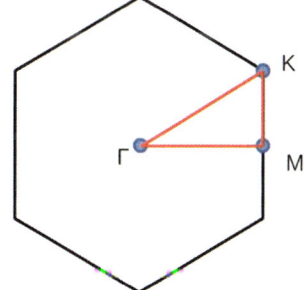

Table 1.5 Table of characters of the group C_{6v}

C_{6v}	E	$C_3^\pm$	$C_2, \bar{C}_2$	$C_6^\pm$	m_{11}	$m_{1\bar{1}}$	$\bar{E}$	$\bar{C}_3^\pm$	$\bar{C}_6^\pm$
ρ_G^Γ	4	2	0	0	0	0	−4	−2	0
$\bar{\Gamma}_7$	2	1	0	$-\sqrt{3}$	0	0	−2	−1	$\sqrt{3}$
$\bar{\Gamma}_8$	2	1	0	$\sqrt{3}$	0	0	−2	−1	$-\sqrt{3}$
$\bar{\Gamma}_9$	2	−2	0	0	0	0	−2	2	0

Table 1.6 Table of characters of the group C_{3v}

C_{3v}	E	$C_3^\pm$	m_i	$\bar{E}$	$\bar{C}_3^\pm$	$\bar{m}_i$
ρ_G^K	4	-1	0	-4	-1	0
$\bar{K}_4$	1	-1	$-i$	-1	1	i
$\bar{K}_5$	1	-1	i	-1	1	$-i$
$\bar{K}_6$	2	1	0	-2	-1	0

1.7.2 K Point

The little group of this point is C_{3v}. We compute the character of the representation as before:

In this case, we find that:

$$\rho_G^K = \bar{K}_4 \oplus \bar{K}_5 \oplus \bar{K}_6 \tag{1.34}$$

We see that if we have a spin–orbit the Dirac cones break since there is a splitting when considering double group representations, i.e., spin representations.

1.7.3 M Point

In this case, the little group is C_{2v}. In this case, we have no more work to do, since there is only one representation of the double group, $\bar{M}_5$, so the subduced representation will be a sum of two $\bar{M}_5$. However, we can do the math (Table 1.7):
which confirms what we knew. Explicitly:

$$\rho_G^M = \bar{M}_5 \oplus \bar{M}_5 \tag{1.35}$$

1.7.4 High-Symmetry Lines

We can use this machinery to see how these crossings will split when we get a bit away from the high-symmetry points. If we follow the high-symmetry lines depicted in Fig. 1.5, we see that the little group for all the points that lie in any of the lines is C_s, i.e., a mirror plane. So, let's see how our degeneracies break from the Γ point to the K point (Table 1.8):

Table 1.7 Table of characters of the group C_{2v}

C_{2v}	E	$C_2^\pm$	$m_{1\bar{1}}$	m_{11}	$\bar{E}$
ρ_G^K	4	0	0	0	-4
$\bar{M}_5$	2	0	0	0	-2

Table 1.8 Table of characters of the group C_s

C_s	E	m	$\bar{E}$	$\bar{m}$
$\bar{\Gamma}_7$	2	0	-2	0
$\bar{\Gamma}_8$	2	0	-2	0
$\bar{\Lambda}_3$	1	$-i$	-1	i
$\bar{\Lambda}_4$	1	i	-1	$-i$

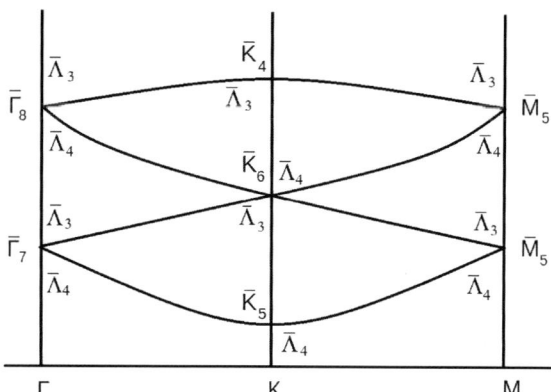

Fig. 1.6 Connected set of bands

So both representations split into two unidimensional ones. We can repeat this procedure for all irreps at high-symmetry points to obtain:

$$\begin{aligned}
\bar{\Gamma}_7 &\to \bar{\Lambda}_3 \oplus \bar{\Lambda}_4 \\
\bar{\Gamma}_8 &\to \bar{\Lambda}_3 \oplus \bar{\Lambda}_4 \\
\bar{K}_4 &\to \bar{\Lambda}_3 \\
\bar{K}_5 &\to \bar{\Lambda}_4 \\
\bar{K}_6 &\to \bar{\Lambda}_3 \oplus \bar{\Lambda}_4 \\
\bar{M}_5 &\to \bar{\Lambda}_3 \oplus \bar{\Lambda}_4
\end{aligned} \quad (1.36)$$

Now we can try to see how bands connect. The only restriction is that a band *cannot change the representation along a high-symmetry line*, so a band that comes from a $\bar{\Lambda}_3$ at the Γ point cannot arrive at a $\bar{\Lambda}_4$ at K point. Following this, we get the following picture:

We see in Fig. 1.6 how the four bands are *connected*, in the sense that we can draw them in a single trace. However, since group theory does not give us any prediction on the energetics (at which height the irreps go), we can have another figure:

The first one is an elementary band representation; i.e., it is not the sum of smaller dimensional band representations. But in Fig. 1.7, we see how the EBR is *disconnected* now. Since they both together form an EBR, it cannot be that both of them

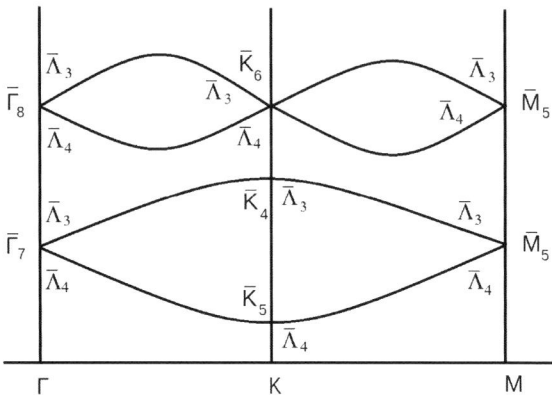

Fig. 1.7 Disconnected set of bands

are EBRs or even BRs. So it must happen that at least one of them is not a BR, i.e., not Wannier representable and, thus, *topological*.

1.8 Conclusion

We have seen how this theory, that combines group theory, Chemistry and Physics, provides a natural way to study topological phases of matter. Not only in a descriptive way, but also in a predictive one. With this framework, we transform the problem of identifying topological properties of a material into a Graph Theory problem. We follow the procedure described and analyze the different types of compatibility solutions we can have. These different compatibility solutions correspond to different types of band structures, such as insulators or semimetals. Consider the case where the Fermi level lies within one (P)EBR. Suppose now that by tuning an external parameter (SOC, electric field, strain, etc.) a gap is opened where the Fermi level sits. Then, at least one of the bands (conduction or valence) has to be topological, since it breaks a (P)EBR. This is the topological insulator case we have studied, graphene with SOC [13] in Sect. 1.6. This applies not only to topological insulators, but to semimetals and band theory in general.

Consider now the case where two (P)EBRs sit near the Fermi level. Suppose we have an insulator or semimetal, where both conduction and valence bands form a (P)EBR. By tuning of external parameters, a topological phase transition occurs when the gap closes and reopens, because a band inversion happens. After the transition, neither the valence nor the conduction bands form (P)EBRs, since the little groups at each **k** are not consistent with a (P)EBR [10]. In the case that the gap does not reopen fully, the material, after the topological phase transition, becomes a *protected semimetal*.

The way we classify different topological phase transitions within this framework is by a tuple of 2 numbers (n, m), where n denotes the number of EBR near the Fermi level in the trivial case (no SOC) and m the number in the topological phase. The tuning external parameter is chosen to be SOC, since it is the most natural parameter found in materials. Within this labeling the case of graphene will be a (1, 1). Another example of this type is Cu_3SbS_4, a well-known TI. With (1, 2) we have Bi^{1-} square nets and $Cu_2SnHgSe_4$, with (2, 2) Bi_2Se_3, KHgSb, and so on. More examples can be found on the main Topological Quantum Chemistry article in [10].

Acknowledgements I. Robredo wants to thank M.G. Vergniory for fruitful discussions and careful reading of the manuscript.

Appendix A: Definitions

Definition A.1 (*Bravais lattice*) A *Bravais lattice* is an infinite set of translations **t** generated by d linearly independent vectors $\mathbf{a}_i$, where d is the dimension of the crystal

$$\mathbf{t} = n_1\mathbf{a}_1 + \cdots + n_d\mathbf{a}_d \ , \ n_i \in Z \tag{A.1}$$

The Bravais lattice is thus isomorphic to Z^d.

Definition A.2 (*Crystal*) A *crystal* is a Bravais lattice arrangement of atoms, invariant under a space group G.

Definition A.3 (*Group of the crystal*) The *group of the crystal* is the space group G under which the crystal remains invariant. G is always an infinite group, as it includes all integer translations along the Bravais lattice. In Seitz notation, the elements of a space group G are denoted as

$$g = \{R|\mathbf{r}\} \tag{A.2}$$

where R is a point group element and $\mathbf{r}$ is a translation, which may or may not belong to the Bravais lattice. The action of $g \in G$ on a real space point $\mathbf{q}$ is given by

$$g\mathbf{q} = \{R|r\}\mathbf{q} = R\mathbf{q} + \mathbf{r} \tag{A.3}$$

The Bravais lattice is always a subgroup of the space group G. Its elements are of the form $\{E|\mathbf{t}\}$, where E is the identity operation.

Definition A.4 (*Stabilizer group/Site-symmetry group*) The *stabilizer group* or *site-symmetry group* of a position q is the set of symmetry operations $g \in G$ that leave q fixed. It is denoted by $G_q = \{g|gq = q\} \subset G$. There are a couple of things to remark:

- $g \in G_q$ may include a translation, $g = \{R|\mathbf{r}\}$, with $\mathbf{r} \neq 0$

- However, since any site-symmetry group leaves a point invariant, G_q is necessarily isomorphic to one of the 32 crystallographic point groups.

Definition A.5 (*Wyckoff position*) A general *Wyckoff position* is a position q in the unit cell of the crystal with a trivial site-symmetry group, i.e., the only element in G_q is the identity operation. A *special* Wyckoff position is a position q in the unit cell of the crystal with a non-trivial site-symmetry group; i.e., q is invariant under some symmetry operations, such as mirror planes and rotation axis.

Definition A.6 (*Orbit of q*) The *orbit of q* is the set of all positions which are related to q by elements of the symmetry group G; i.e., $Orb_q = \{gq|g \in G\}$ *and* belong to the same unit cell.

Definition A.7 (*Coset representatives*) The *coset representatives* of a site-symmetry group can be defined as the set of elements that generate the orbit of a Wyckoff position. Then each element q_α in the orbit of q may be written as $q_\alpha = g_\alpha q$.

Definition A.8 (*Coset decomposition*) The *coset decomposition* of the full space group is defined by

$$G = \bigcup_\alpha g_\alpha (G_q \ltimes Z^d) \tag{A.4}$$

where G_q is the site-symmetry group and g_α are the coset representatives. The piece multiplying the coset representatives is obtained as the semi-direct product of G_q and the translation group, that in d dimensions is isomorphic to Z^d. Each term $g_\alpha(G_q \ltimes Z^d)$ in (A.4) is a (left) *coset*.

This can be understood as follows. Let us take a position q with site-symmetry group G_q. Then G_q plus the translations in the Bravais lattice creates a replica of q at every primitive cell in the crystal. Acting with each coset representative g_α creates, throughout the crystal, replicas of every position in the orbit of q.

Definition A.9 (*Multiplicity of a Wyckoff position*) The *multiplicity of a Wyckoff position* is defined as the number of elements (positions) in the orbit of some Wyckoff position. It is obviously equal to the number of coset representatives.

This is what motivates the names for the different maximal Wyckoff positions 1a, 2b, 3c, etc. The number tells you the multiplicity of the position, while the letter labels the positions, from more to less symmetric.

Definition A.10 (*Maximal Wyckoff position*) A Wyckoff position q is said to be *non-maximal* if there exists a group H such that $G_q \subset H \subset G$. A Wyckoff position that is not non-maximal is *maximal*.

A sufficient (although not necessary) condition for a position q to be maximal is that q is the unique point fixed by every operation in G_q. As a particular case, in 2D, any site-symmetry group that contains rotations is maximal.

1 Band Theory Without Any Hamiltonians ...

Definition A.11 (*Little group*) Two reciprocal space vectors $\mathbf{k}_1$ and $\mathbf{k}_2$ are said to be equivalent, $\mathbf{k}_1 \equiv \mathbf{k}_2$, if there exists a reciprocal *lattice* vector $\mathbf{K}$ such that $\mathbf{k}_2 = \mathbf{k}_1 + \mathbf{K}$. Then the *little group* $G_\mathbf{k}$ of a vector $\mathbf{k}$ in reciprocal space is the set of elements $g \in G$ such that $g\mathbf{k} \equiv \mathbf{k}$. Note that the action of space group elements on reciprocal space vectors is defined by

$$g\mathbf{k} = \{R|t\}\mathbf{k} = R\mathbf{k} \tag{A.5}$$

Definition A.12 (*Small representation*) A *small representation* is a representation of the little group.

Appendix B: Proof That the Site-Symmetry Groups for the 3c Wyckoff Positions are Isomorphic to C_{3v}

In this appendix, we will prove two statements: First, that the site-symmetry group for the position $q = \left(\frac{1}{3}, \frac{1}{3}\right)$ is isomorphic to C_{3v}, and second, that the site-symmetry groups for positions in the same orbit are isomorphic to each other.

B.1 Site-Symmetry Group of $q = \left(\frac{1}{3}, \frac{1}{3}\right)$

First, we introduce the set of relations that define the group C_{3v}:

$C_3^3 = 1$:
$$(x, y) \to C_3 \to (y, -x - y) \to C_3 \to (-x - y, x) \to C_3 \to (x, y)$$

$C_3 m_{1\bar{1}} = m_{1\bar{1}} C_3^{-1}$:
$$(x, y) \to m_{1\bar{1}} \to (y, x) \to C_3 \to (x, -x - y)$$
$$(x, y) \to C_3^2 \to (-x - y, x) \to m_{1\bar{1}} \to (x, -x - y) \tag{B.1}$$

Now, let's see if the generators of the site-symmetry group follow the same relations:

$\{C_3|01\}^3 = 1$:
$$(x, y) \to \{C_3|01\} \to (y, -x - y + 1) \to \{C_3|01\}$$
$$\to (-x - y + 1, x) \to \{C_3|01\} \to (x, y)$$

$\{C_3|01\}\{m_{1\bar{1}}|0\} = \{m_{1\bar{1}}|0\}\{C_3|01\}^{-1}$:
$$(x, y) \to \{m_{1\bar{1}}|0\} \to (y, x) \to \{C_3|01\} \to (x, -x - y + 1)$$
$$(x, y) \to \{C_3|01\}^2 \to (-x - y + 1, x) \to \{m_{1\bar{1}}|0\} \to (x, -x - y + 1) \tag{B.2}$$

As we see, the group generators satisfy the same relations. Thus the groups are isomorphic.

B.2 Site-Symmetry Group of Positions in the Same Orbit

We know that the positions for the different elements in the same orbit are related to each other by

$$q_\alpha = g_\alpha q \tag{B.3}$$

for some q in the orbit and g_α a coset representative. Thus, for some $h \in G_q$,

$$hq = q \rightarrow g_\alpha h g_\alpha^{-1} q_\alpha = q_\alpha \tag{B.4}$$

and we see that $g_\alpha h g_\alpha^{-1} \in G_{q_\alpha}$. This is the definition of *conjugate group*. As two conjugate groups are isomorphic, it is enough to compute the site-symmetry group for one point in each orbit.

Appendix C: Wannier Function Transformation Properties

We will denote our Wannier functions on the unit cell by two indices: the orbital (latin) and site (greek). In the case of spinful p_z orbitals on 2b Wyckoff positions (graphene), the Wannier functions will be denoted as $W_{i\alpha}$, where i denotes spin up or down, and α denotes the site of the orbit. Wannier functions transform around each site as orbitals:

$$gW_{i1} = [\rho(g)]_{ji} W_{j1} \tag{C.1}$$

This follows from the Hamiltonian. If the Hamiltonian commutes with the symmetry operations, then its eigenstates[28] transform under representations of the symmetry group. In a unit cell, we have α positions in the orbit. The Wannier functions at those points are given, in terms of the functions around one position:

$$W_{i\alpha}(r) = g_\alpha W_{i1}(r) = W_{i1}(g_\alpha^{-1} r) \tag{C.2}$$

Let's see under which representation these transform:

$$hW_{i\alpha} = g_\alpha g g_\alpha^{-1} g_\alpha W_{i1} = g_\alpha g W_{i1} = g_\alpha [\rho(g)]_{ji} W_{j1} = [\rho(g_\alpha^{-1} h g_\alpha)]_{ji} W_{j\alpha} \tag{C.3}$$

where $h \in G_{q_\alpha}$ and $g \in G_{q_1}$.

---

[28] Or a set of states that generate the same Hilbert space.

Now, we can construct all Wannier functions on the full lattice by translating these functions along the lattice. $\{E|t_\mu\}W_{i\alpha}(r) = W_{i\alpha}(r - t_\mu)$, so we have a total of $n \times n_q \times N$ Wannier functions, where n_q is the number of orbitals per position in the orbit, n the multiplicity of the Wyckoff position and N the number of cells of our crystal. These functions form a *basis* for the representation of the space group induced from the representation of the site-symmetry group. Let the representation of the spatial group be ρ_G. Then, $\rho_G \equiv \rho \uparrow G$. This procedure is called *induction*. Let's proceed to see how Wannier states transform under an element $h = \{R|t\}$:

$$\begin{aligned}
hW_{i\alpha}(\mathbf{r} - \mathbf{t}_\mu) &= h\{E|\mathbf{t}_\mu\}W_{i\alpha}(\mathbf{r}) \\
&= \{E|R\mathbf{t}_\mu\}hW_{i\alpha}(\mathbf{r}) \\
&= \{E|R\mathbf{t}_\mu + \mathbf{t}_{\beta\alpha}\}g_\beta g g_\alpha^{-1} W_{i\alpha}(\mathbf{r}) \\
&= \{E|R\mathbf{t}_\mu + \mathbf{t}_{\beta\alpha}\}g_\beta g W_{i1}(\mathbf{r}) \\
&= \{E|R\mathbf{t}_\mu + \mathbf{t}_{\beta\alpha}\}g_\beta [\rho(g)]_{ji} W_{j1}(\mathbf{r}) \\
&= \{E|R\mathbf{t}_\mu + \mathbf{t}_{\beta\alpha}\}[\rho(g)]_{ji} W_{j\beta}(\mathbf{r}) \\
&= [\rho(g)]_{ji} W_{j\beta}(\mathbf{r} - R\mathbf{t}_\mu - \mathbf{t}_{\beta\alpha})
\end{aligned} \quad (C.4)$$

where in the third line we have used that the action of an element h on a Wyckoff position q_α is given by

$$hq_\alpha = \{E|\mathbf{t}_{\beta\alpha}\}q_\beta, \quad g_\beta^{-1}\{E|-\mathbf{t}_{\beta\alpha}\}hg_\alpha q_1 = q_1 \equiv gq_1 = q_1 \quad (C.5)$$

where the vector $\mathbf{t}_{\beta\alpha}$ represents the possibility of an element to take some Wyckoff away to another cell.[29] We see here that we can know how any Wannier in any position in any cell transform just by knowing how they transform around one of the positions of the orbit under an element $g \equiv g_\beta^{-1}\{E|-\mathbf{t}_{\beta\alpha}\}hg_\alpha \in G_{q_1}$. We can obtain from (1.19) that:

$$\mathbf{t}_{\beta\alpha} = hq_\alpha - q_\beta \quad (C.6)$$

Appendix D: Elementary Band Representation

In the main text, we have worked out an example of elementary band representation. We will give here some more general results about them. First, let's state some facts.

We say that two band representations ρ_G and σ_G are equivalent if and only if there exists a unitary matrix-valued function $S(\mathbf{k}, t, g)$ smooth in $\mathbf{k}$ and continuous in t such that, for all $g \in G$

---

[29] It can be easily seen from here that there is only one value of β for which α makes sense. As an example, let the element h take the Wyckoff position q_1 to q_3 in another cell, with a translation $\mathbf{a}$ being an integer Bravais lattice vector. In this notation, we will have that $\mathbf{t}_{31} = \mathbf{a}$, while the rest of $\mathbf{t}_{\beta 1}$ will not exist and, thus, the blocks of the full group representation that are not $\alpha = 1, \beta = 3$ will be 0.

- $S(\mathbf{k}, t, g)$ defines a band representation according to (1.14) for all $t \in [0, 1]$
- $S(\mathbf{k}, 0, g) = \rho_G^{\mathbf{k}}(g)$
- $S(\mathbf{k}, 1, g) = \sigma_G^{\mathbf{k}}(g)$

In the analyzed case of graphene, t would be the parameter of the line that connects two points.

A necessary condition is that both $\rho_G^{\mathbf{k}}(g)$ and $\sigma_G^{\mathbf{k}}(g)$ restrict to the same little group representations at all points in the Brillouin Zone. However, it is not sufficient: It may happen that both representations satisfy this condition but $S(\mathbf{k}, t, g)$ is not a band representation for all t. We need a sufficient condition for equivalence:

Given two sites q, q' (not necessarily in the same Wyckoff position) and representations of their site-symmetry groups (ρ of G_q and ρ' of $G_{q'}$), the band representations $\rho \uparrow G$ and $\rho' \uparrow G$ are equivalent if and only if there exists a site q_0 and representation σ of G_{q_0} such that $\rho = \sigma \uparrow G_q$ and $\rho' = \sigma \uparrow G_{q'}$.

Now let's discuss the compositeness of a band representation; i.e., if it is elementary or composite. We say that a band representation is composite if it can obtained as a sum of other band representations. A band representation that is not composite is called *elementary*.

Now that we know when a band is elementary, we will see what conditions must be met for these to exist.

All band representations admit a description in terms of localized Wannier functions. They are induced from the representations of some site-symmetry group with local orbitals. Notice that if we induce a band representation from a reducible representation of the site-symmetry group:

$$(\rho_1 \oplus \rho_2) \uparrow G = (\rho_1 \uparrow G) \oplus (\rho_2 \uparrow G) \qquad (C.7)$$

where we have used the distributive property of the direct sum. So, if we are interested in elementary band representations, we only need to take care of irreducible representations of the site-symmetry group. Moreover, since $(\rho \uparrow H) \uparrow G = \rho \uparrow G$, we only need to consider maximal subgroups of the space group.

We have determined that all elementary band representations can be induced from irreducible representations of the maximal site-symmetry groups. But this condition is not true in the opposite way; not all irreducible representation of the maximal site-symmetry groups *induce an elementary band representation*. These last cases, when what is induced is not an elementary band representation, are called *exceptions*. This may seem annoying, but they have already been tabulated in Topological Quantum Chemistry [10].

Hence (with some exceptions), band representations induced from irreducible representations of maximal site-symmetry groups give elementary band representations, whose bands are connected in the first BZ (they have no gap).

Band representations describe systems in the atomic limit, as they can be described by maximally localized Wannier orbitals. A trivial insulator is one whose bands can be obtained from maximally localized Wannier orbitals, so it does not have edge states.

So, a set of bands that is not a band representation cannot be described in terms of localized Wannier orbitals and is, hence, topological. We call this bands, that are a solution to compatibility relations, a *quasi-band representation*.

Let's analyze the following example, alike the graphene case. Suppose we have a Hamiltonian constructed from localized orbitals, whose EBR $\rho_G = \rho \uparrow G$, and that the energy bands of this system can be divided into two disconnected sets of bands overall **k** in the first BZ, separated by a spectral gap. This means that the action of every element in the symmetry group on one of the states of one the bands *does not take it out of it*. Formally, let P_i be the projector into the band i. Then:

$$[P_i, H] = 0, \quad [P_i, g] = 0 \tag{C.8}$$

for all $g \in G$. Now suppose that the bands of projector P_i transform under a band representation ρ_G^i. Then, the full ρ_G representation could be constructed as a direct sum of the band representations of the different bands. We reached a contradiction: Starting with an elementary band representation, we got a composite band representation. So, all bands that transform according to an elementary band representation *must be connected along the first BZ*, otherwise they are not a band representation and, thus, they are topological, in the sense that there is at least one of them that is topological.

Going back to the graphene case, we saw that the EBR we induced can be connected or disconnected. If it is connected, it describes a trivial insulator while, if disconnected, it describes a topological material.

References

1. R. Yu, X.L. Qi, A. Bernevig, Z. Fang, X. Dai, Equivalent expression of $\mathbb{Z}_2$ topological invariant for band insulators using the non-abelian berry connection. Phys. Rev. B **84**, 075119 (2011)
2. A. Alexandradinata, B. Andrei Bernevig, Berry-phase description of topological crystalline insulators. Phys. Rev. B **93**, 205104 (2016)
3. A. Alexandradinata, Z. Wang, B. Andrei Bernevig, Topological insulators from group cohomology. Phys. Rev. X **6**, 021008 (2016)
4. J. Zak, Band representations and symmetry types of bands in solids. Phys. Rev. B **23**, 2824–2835 (1981)
5. H. Bacry, L. Michel, J. Zak, Symmetry and classification of energy bands in crystals (Springer, Berlin, 1988), pp. 289–308
6. J. Zak, Symmetry specification of bands in solids. Phys. Rev. Lett. **45**, 1025–1028 (1980)
7. J. Zak, Band representations of space groups. Phys. Rev. B **26**(6), 3010–3023 (1982)
8. L. Michel, J. Zak, Connectivity of energy bands in crystals. Phys. Rev. B **59**, 5998–6001 (1999)
9. L. Michel, J. Zak, Elementary energy bands in crystals are connected. Phys. Rep. **341**(1), 377–395 (2001). Symmetry, invariants, topology
10. B. Bradlyn, L. Elcoro, J. Cano, M.G. Vergniory, Z. Wang, C. Felser, M.I. Aroyo, B. Andrei Bernevig, Topological quantum chemistry. Nature **547**, 298–305 (2017)
11. A.A. Soluyanov, D. Vanderbilt, Computing topological invariants without inversion symmetry. Phys. Rev. B - Condens. Matter Mater. Phys. **83**(23) (2011)
12. N. Marzari, A.A. Mostofi, J.R. Yates, I. Souza, D. Vanderbilt, Maximally localized wannier functions: theory and applications. Rev. Mod. Phys. **84**, 1419–1475 (2012)

13. C.L. Kane, E.J. Mele, Quantum spin hall effect in graphene. Phys. Rev. Lett. **95**(22), 226801 (2005)
14. B. Bradlyn, L. Elcoro, M.G. Vergniory, J. Cano, Z. Wang, C. Felser, M.I. Aroyo, B. Andrei Bernevig, Band connectivity for topological quantum chemistry: band structures as a graph theory problem. Phys. Rev. B **97**, 035138 (2018)
15. J. Cano, B. Bradlyn, Z. Wang, L. Elcoro, M.G. Vergniory, C. Felser, M.I. Aroyo, B. Andrei Bernevig, Building blocks of topological quantum chemistry: elementary band representations. Phys. Rev. B **97**, 035139 (2018)
16. B. Bradlyn, J. Cano, Z. Wang, M.G. Vergniory, C. Felser, R.J. Cava, B. Andrei Bernevig, Beyond Dirac and Weyl fermions: unconventional quasiparticles in conventional crystals. Science **353**(6299), aaf5037 (2016)
17. L. Elcoro, B. Bradlyn, Z. Wang, M.G. Vergniory, J. Cano, C. Felser, B. Andrei Bernevig, D. Orobengoa, G. de la Flor, M.I. Aroyo, Double crystallographic groups and their representations on the Bilbao crystallographic server. J. Appl. Crystallogr. **50**(5), 1457–1477 (2017)
18. M.G. Vergniory, L. Elcoro, Z. Wang, J. Cano, C. Felser, M.I. Aroyo, B. Andrei Bernevig, B. Bradlyn, Graph theory data for topological quantum chemistry. Phys. Rev. E **96**, 023310 (2017)
19. M.I. Aroyo, J.M. Perez-Mato, D. Orobengoa, E. Tasci, G. De La Flor, A. Kirov, Crystallography online: Bilbao crystallographic server. Bulg. Chem. Commun. **43**(2), 183–197 (2011). cited By 145
20. M. Aroyo, J. Perez-Mato, C. Capillas, Computing topological invariants without inversion symmetry. Z. fr Krist.- Cryst. Mater. **221**(1), 15–27 (2018)
21. M.I. Aroyo, A. Kirov, C. Capillas, J.M. Perez-Mato, H. Wondratschek, Bilbao crystallographic server. II. Representations of crystallographic point groups and space groups. Acta Crystallogr. Sect. A **62**(2), 115–128 (2006)
22. C.J. Bradley, A.P. Cracknell, *The Mathematical Theory of Symmetry in Solids: Representation Theory for Point Groups and Space Groups* (Clarendon Press, Oxford, 1972)

Chapter 2
Topological Crystalline Insulators

Titus Neupert and Frank Schindler

Abstract We give an introduction to topological crystalline insulators, that is, gapped ground states of quantum matter that are not adiabatically connected to an atomic limit without breaking symmetries that include spatial transformations, like mirror or rotational symmetries. To deduce the topological properties, we use non-Abelian Wilson loops. We also discuss in detail higher-order topological insulators with hinge and corner states, and in particular, present interacting bosonic models for the latter class of systems.

2.1 Wilson Loops and the Bulk-Boundary Correspondence

We first provide a unified picture of topological bulk-boundary correspondences in any dimension by making use of Brillouin zone Wilson loops.

2.1.1 Introduction and Motivation

In these notes, we are mostly concerned with the topological characterization of non-interacting electron Hamiltonians on a lattice in the presence of spatial symmetries. In general, an insulating topological phase of matter may be defined by the requirement that the many-body ground state of the corresponding Hamiltonian (given by a Slater determinant in the non-interacting case) cannot be adiabatically connected to the atomic limit of vanishing hopping between the sites of the lattice. Further requiring that certain symmetries such as time-reversal are not violated along any such

---

T. Neupert (✉) · F. Schindler
Department of Physics, University of Zurich, Wintherthurerstrasse 190,
8057 Zurich, Switzerland
e-mail: titus.neupert@uzh.ch

F. Schindler
e-mail: frank.schindler@uzh.ch

© Springer Nature Switzerland AG 2018
D. Bercioux et al. (eds.), *Topological Matter*, Springer Series in Solid-State Sciences 190, https://doi.org/10.1007/978-3-319-76388-0_2

adiabatic interpolation enriches the topological classification, in that phases which were classified as trivial in the previous sense now acquire a topological distinction which is protected by the respective symmetry.

To determine the topology of a given ground state, several topological invariants have been proposed, such as the Pfaffian invariant for two-dimensional (2D) time-reversal symmetric systems. However, they often require Bloch states to be provided in a smooth gauge across the whole Brillouin zone (BZ) for their evaluation, making them impractical for numerical calculations. In addition, most of them are specific to the dimension or symmetry class considered and thus do not generalize well.

Here, we employ non-Abelian Wilson loops as a generalization of the one-dimensional (1D) Berry phase to characterize topological properties in any dimension and any symmetry class. This provides a framework of topological invariants which makes direct contact with the protected boundary degrees of freedom of a given phase.

As a prerequisite, we assume a working knowledge of the (boundary) physics of non-crystalline topological phases and their topological invariants, as well as their classification by the tenfold way. Suitable introductions can be found in [1–5].

2.1.2 Definitions

We work in units where $\hbar = c = e = 1$ and denote by σ_i, $i = x, y, z$, the 2×2 Pauli matrices. We define $\sigma_0 = \mathbb{1}_{2 \times 2}$ for convenience. We express eigenfunctions of a translationally invariant single-particle Hamiltonian in the basis

$$\phi_{k,\alpha}(r) = \frac{1}{\sqrt{N}} \sum_R e^{ik \cdot (R + r_\alpha)} \varphi_{R,\alpha}(r - R - r_\alpha), \tag{2.1}$$

where $\varphi_{R,\alpha}$, $\alpha = 1, \ldots, N$, are the orbitals chosen as basis for the finite-dimensional Hilbert space in each unit cell, labeled by the lattice vector R, and r_α is the center of each of these orbitals relative to the origin of the unit cell. Including r_α in the exponential corresponds to a convenient choice of gauge when studying the response to external fields defined in continuous real space.

A general non-interacting Hamiltonian then has the Bloch matrix elements

$$\mathcal{H}_{\alpha,\beta}(k) = \int d^d r \, \phi^*_{k,\alpha}(r) \hat{H} \phi_{k,\beta}(r), \tag{2.2}$$

as well as energy eigenstates

$$\psi_{k,n}(r) = \sum_\alpha^N u_{k;n,\alpha} \phi_{k,\alpha}(r), \tag{2.3}$$

where

$$\sum_\beta \mathcal{H}_{\alpha,\beta}(\mathbf{k}) u_{\mathbf{k};n,\beta} = \varepsilon_n(\mathbf{k}) u_{\mathbf{k};n,\alpha}, \qquad n = 1, \ldots, N. \tag{2.4}$$

In the following, we are interested in situations where the system has an energy gap after the first $M < N$ bands, i.e., $\varepsilon_M(\mathbf{k}) < \varepsilon_{M+1}(\mathbf{k})$ for all $\mathbf{k}$.

2.1.3 Wilson Loop and Position Operator

Introduced in 1984 by Sir Michael Berry, the so-called Berry phase describes a phase factor which arises in addition to the dynamical evolution $e^{i \int E[\lambda(t)] dt}$ of a quantum mechanical state in an adiabatic interpolation of the corresponding Hamiltonian $\hat{H}[\lambda(t)]$ along a closed path $\lambda(t)$ in parameter space. It depends only on the geometry of the path chosen and can be expressed as a line integral of the Berry connection, which we define below for the case where the parameter λ is a single-particle momentum. If degeneracies between energy levels are encountered along the path, we have to consider the joint evolution of a set of eigenstates that may have degeneracies. If we consider M such states, the Berry phase generalizes to a $U(M)$ matrix, which may be expressed as the line integral of a non-Abelian Berry connection, and is called non-Abelian Wilson loop.

In the BZ, we may consider momentum $\mathbf{k}$ as a parameter of the Bloch Hamiltonian $\mathcal{H}(\mathbf{k})$. The corresponding non-Abelian Berry–Wilczek–Zee connection is then given by

$$\mathbf{A}_{m,n}(\mathbf{k}) = \langle u_{\mathbf{k},m} | \mathbf{\nabla}_\mathbf{k} | u_{\mathbf{k},n} \rangle, \qquad n, m = 1, \ldots, M. \tag{2.5}$$

Note that it is anti-Hermitian, that is, it satisfies $A^*_{n,m}(\mathbf{k}) = -A_{m,n}(\mathbf{k})$. Using matrix notation, we define the Wilson loop, a unitary operator, as

$$W[l] = \overline{\exp}\left[-\int_l d\mathbf{l} \cdot \mathbf{A}(\mathbf{k})\right], \tag{2.6}$$

where l is a loop in momentum space and the overline denotes path ordering of the exponential, where as usual operators at the beginning of the path occur to the right of operators at the end. This unitary operator acts on the occupied band manifold and can be numerically evaluated with the formula

$$W_{n_{R+1}, n_1}[l] = \lim_{R \to \infty} \sum_{n_2, \ldots n_R=1}^{M} \prod_{i=R}^{1} \left[\exp\left[-(\mathbf{k}_{i+1} - \mathbf{k}_i) \cdot \mathbf{A}(\mathbf{k}_{i+1})\right]\right]_{n_{i+1}, n_i}$$

$$= \lim_{R \to \infty} \sum_{n_2, \ldots n_R=1}^{M} \prod_{i=R}^{1} \left[\delta_{n_i, n_{i+1}} - (\mathbf{k}_{i+1} - \mathbf{k}_i) \cdot \mathbf{A}_{n_{i+1}, n_i}(\mathbf{k}_{i+1})\right]$$

$$= \lim_{R\to\infty} \sum_{n_2,\ldots n_R=1} \prod_{i=R}^{1} \Big[\langle u_{k_{i+1},n_i+1} | u_{k_{i+1},n_i} \rangle \quad (2.7)$$

$$- (k_{i+1} - k_i) \cdot \langle u_{k_{i+1},n_i+1} | \nabla_{k_{i+1}} | u_{k_{i+1},n_i} \rangle \Big]$$

$$= \lim_{R\to\infty} \sum_{n_2,\ldots n_R=1}^{M} \prod_{i=R}^{1} \langle u_{k_{i+1},n_i+1} | u_{k_i,n_i} \rangle$$

$$= \langle u_{k_1,n_1} | \lim_{R\to\infty} \prod_{i=R}^{2} \left(\sum_{n_i} |u_{k_i,n_i}\rangle \langle u_{k_i,n_i}| \right) |u_{k_1,n_1}\rangle, \quad (2.8)$$

where the path l is sampled into R momenta k_i, $i = 1, \ldots, R$, and the limit $R \to \infty$ is taken such that the distance between any two neighboring momentum points goes to zero. Further, $k_1 = k_{R+1}$ are the initial and final momenta along the loop, respectively, on which the Wilson loop matrix depends.

By the last line of (2.7), it becomes clear that $W[l]$ is gauge *covariant*, that is, transforms as an operator under a general gauge transformation $S(k) \in U(M)$ of the occupied subspace given by $|u_k\rangle \to S(k)|u_k\rangle$, only for a closed loop l (the case where l is non-contractible is also referred to as the Zak phase). However, the Wilson loop *spectrum* for a closed loop is gauge *invariant*, that is, the eigenvalues of $W[l]$ are not affected by gauge transformations (note that they also do not depend on the choice of $k_i = k_f$) and may therefore carry physical information. We will show in the following that this is indeed the case: The Wilson loop spectrum is related to the spectrum of the position operator projected into the space of occupied bands.

To proceed, we consider a geometry where l is parallel to the x coordinate axis and winds once around the BZ. Let $\vec{k}$ denote the $(d-1)$ dimensional vector of remaining good momentum quantum numbers. Then, $W(\vec{k})$ is labeled by these remaining momenta. Denote by $\exp(i\theta_{\alpha,\vec{k}})$, $\alpha = 1, \ldots, M$, the eigenvalues of $W(\vec{k})$. The set of phases $\{\theta_{\alpha,\vec{k}}\}$ forms a band structure in the $(d-1)$ dimensional BZ and is often equivalently referred to as the Wilson loop spectrum. Note that all $\theta_{\alpha,\vec{k}}$ are only defined modulo 2π, which makes the Wilson loop spectrum inherently different from the spectrum of a physical Bloch Hamiltonian.

The spectral equivalence we will show relates the eigenvalues of the operator $(-i/2\pi)\log[W(\vec{k})]$ with those of the projected position operator

$$P(\vec{k}) \hat{x} P(\vec{k}), \quad (2.9)$$

where the projector $P(\vec{k})$ onto all occupied band eigenstates along l (i.e., all states with wave vector $\vec{k}$) is given by

$$P(\vec{k}) = \sum_n^M \int_{-\pi}^{\pi} \frac{dk_x}{2\pi} |\psi_{k,n}\rangle \langle \psi_{k,n}|, \quad (2.10)$$

while the states $|\psi_{k,n}\rangle$ are given by (2.3). The eigenvalues of the projected position operator have the interpretation of the charge centers in the ground state of the Hamiltonian considered, while the eigenstates are known as hybrid Wannier states, which are localized in the x-direction and plane waves perpendicular to it [6].

To prove the equivalence, we start with the eigenfunctions of $P(\vec{k})\hat{x}P(\vec{k})$, which satisfy

$$\left[P(\vec{k})\hat{x}P(\vec{k}) - \frac{\tilde{\theta}_{\vec{k}}}{2\pi}\right]|\Psi_{\vec{k}}\rangle = 0. \tag{2.11}$$

Note that there are M eigenvectors, the form of the corresponding eigenvalues $\tilde{\theta}_{\alpha,\vec{k}}/(2\pi)$, $\alpha = 1, \ldots, M$ has been chosen for later convenience and in particular has not yet been logically connected to the $\theta_{\alpha,\vec{k}}$ making up the Wilson loop spectrum (however, we will do so shortly). An eigenfunction can be expanded as

$$|\Psi_{\vec{k}}\rangle = \sum_{n}^{M} \int dk_x \, f_{\vec{k},n}(k_x)|\psi_{k,n}\rangle, \tag{2.12}$$

where the coefficients $f_{\vec{k},n}$ satisfy the equation

$$\langle \psi_{k,n}|P(\vec{k})\hat{x}P(\vec{k})|\Psi_{\vec{k}}\rangle$$

$$= \sum_{m} \int d\tilde{k}_x \, \langle \psi_{k,n}|(i\partial_{\tilde{k}_x}) f_{\tilde{k},m}(\tilde{k}_x)|\psi_{\tilde{k},m}\rangle$$

$$= \sum_{m} \int d\tilde{k}_x \, i \frac{\partial f_{\tilde{k},m}(\tilde{k}_x)}{\partial \tilde{k}_x}(\delta_{m,n}\delta_{\tilde{k}_x,k_x}) \tag{2.13}$$

$$+ \sum_{m} \int d\tilde{k}_x \, f_{\tilde{k},m}(\tilde{k}_x) \int \frac{dx}{2\pi} \langle u_{k,n}|e^{-ik_xx}(i\partial_{\tilde{k}_x})e^{i\tilde{k}_xx}|u_{\tilde{k},m}\rangle$$

$$= i\frac{\partial f_{\vec{k},n}(k_x)}{\partial k_x} - f_{\vec{k},n}(k_x)\int \frac{dx}{2\pi}x + i\sum_m \int d\tilde{k}_x f_{\tilde{k},m}(\tilde{k}_x) \int \frac{dx}{2\pi} e^{-i(k_x-\tilde{k}_x)x}\langle u_{k,n}|\partial_{k_x}|u_{\tilde{k},m}\rangle$$

$$= i\frac{\partial f_{\vec{k},n}(k_x)}{\partial k_x} + i\sum_m f_{\vec{k},m}(k_x)\langle u_{k,n}|\partial_{k_x}|u_{k,m}\rangle$$

$$= i\frac{\partial f_{\vec{k},n}(k_x)}{\partial k_x} + i\sum_m^M A_{x;n,m}(\vec{k}) f_{\vec{k},m}(k_x). \tag{2.14}$$

(Note that we have to assume an appropriate regularization to make the term $\int dx\, x$ vanish in this continuum calculation, reflecting the ambiguity in choosing the origin of the coordinate system.) Then, integrating the resulting (2.11) for $f_{\vec{k},n}(k_x)$, we obtain

$$f_{\vec{k},n}(k_x) = e^{-i(k_x - k_x^0)\tilde{\theta}_{\vec{k}}/(2\pi)} \sum_{m}^{M} \overline{\exp}\left[-\int_{k_x^0}^{k_x} d\tilde{k}_x A_x(\tilde{k}_x, \vec{k})\right]_{n,m} f_{\vec{k},m}(k_x^0). \tag{2.15}$$

We now choose $k_x = k_x^0 + 2\pi$. Periodicity of $f_{\vec{k},m}(k_x^0)$ as $k_x^0 \to k_x^0 + 2\pi$ yields (choosing $k_x^0 = \pi$ without loss of generality)

$$\sum_m^M W(\vec{k})_{n,m} f_{\vec{k},m}(\pi) = e^{i\tilde{\theta}_{\vec{k}}} f_{\vec{k},n}(\pi), \tag{2.16}$$

showing that the expansion coefficients of an eigenstate of $P(\vec{k})\hat{x}P(\vec{k})$ with eigenvalue $\tilde{\theta}_{\vec{k}}/(2\pi)$ form eigenvectors of $W(\vec{k})$ with eigenvalues $e^{i\tilde{\theta}_{\vec{k}}}$. This establishes the spectral equivalence $\tilde{\theta}_{\vec{k}} = \theta_{\vec{k}}$.

Note that there are M eigenvalues of the Wilson loop, while the number of eigenvalues of $P(\vec{k})\hat{x}P(\vec{k})$ is extensive in the system size. Indeed, for each occupied band (i.e., every Wilson loop eigenvalue $\theta_{\alpha,\vec{k}}$, $\alpha = 1, \ldots, M$), there exists a ladder of eigenvalues of the projected position operator

$$\frac{\theta_{\alpha,\vec{k},X}}{2\pi} = \frac{\theta_{\alpha,\vec{k}}}{2\pi} + X, \quad X \in \mathbb{Z}, \quad \alpha = 1, \ldots, M. \tag{2.17}$$

Notice that we have set the lattice spacing in the x-direction to 1 for convenience here and in the following.

The eigenstates of the projected position operator are hybrid Wannier states which are maximally localized in x-direction, but take on plane wave form in the perpendicular directions. Note that since the eigenvalues of $W(\vec{k})$ along any non-contractible loop of $\vec{k}$ in the BZ define a map $S^1 \to U(1) \cong S^1$, their winding number, which is necessarily an integer, can, given additional crystalline symmetries, provide a topological invariant that cannot be changed by smooth deformations of the system's Hamiltonian. To familiarize the reader with the concepts introduced above, we now present the properties of Wilson loop spectra in the context of three simple models.

2.1.3.1 Example: Su–Schrieffer–Heeger Model

One of the simplest examples of a topological phase is exemplified by the Su–Schrieffer–Heeger (SSH) model, initially devised to model polyacetylene. It describes electrons hopping on a 1D dimerized lattice with two sites A and B in its unit cell (see Fig. 2.1a). In momentum space, the Bloch Hamiltonian reads

$$\mathcal{H}(k) = \begin{pmatrix} 0 & t + t'e^{ik} \\ t + t'e^{-ik} & 0 \end{pmatrix}. \tag{2.18}$$

The model has an inversion symmetry $I\mathcal{H}(k)I^{-1} = \mathcal{H}(-k)$, with $I = \sigma_x$. Since it does not couple sites A to A or B to B individually, it furthermore enjoys a chiral or sublattice symmetry $C\mathcal{H}(k)C^{-1} = -\mathcal{H}(k)$ with $C = \sigma_z$. [Notice some abuse of language here: The chiral symmetry is not a "symmetry" in the sense of a commuting operator on the level of the first quantized Bloch Hamiltonian. Still, as a mathemat-

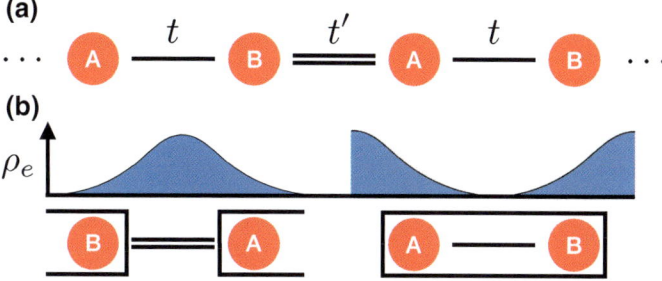

Fig. 2.1 Su–Schrieffer–Heeger model. **a** The model consists of electrons hopping on a dimerized chain with alternating hopping strengths t and t'. For the case of $t' > t$, the model is in its topological phase, which is adiabatically connected to the special case $t' \neq 0, t = 0$. In this limit, the presence of gapless edge modes is evident when the chain is cut after a full unit cell. **b** The polarization is a measure of where charges sit in the unit cell. Shown is the case $\mathsf{P} = 1/2$, where the charge center is displaced by exactly half a lattice spacing. When cutting the system after a full unit cell, the edge hosts a state of charge $1/2$. This is the simplest example of charge fractionalization in topological condensed matter systems

ical fact, this chiral symmetry can be helpful to infer and protect the existence of topological boundary modes.] While a standard discussion of the SSH model would focus on the chiral symmetry and its role in protecting topological phases, here we will first consider the implications of the crystalline inversion symmetry. It will be useful to note that the spectrum is given by $E = \pm\sqrt{t^2 + t'^2 + 2tt'\cos k}$ with a gap closing at $k = \pi$ for $t = t'$ and at $k = 0$ for $t = -t'$.

Let us start by calculating the Wilson loop for the case where $(t, t') = (0, 1)$. The eigenvectors of $\mathcal{H}(k)$ are then given by

$$|u_{k,1}\rangle = \frac{1}{\sqrt{2}}\begin{pmatrix}-e^{ik}\\1\end{pmatrix}, \quad |u_{k,2}\rangle = \frac{1}{\sqrt{2}}\begin{pmatrix}e^{ik}\\1\end{pmatrix}, \quad (2.19)$$

with energies -1 and $+1$, respectively. Since the occupied subspace is one-dimensional in this case, the Berry connection $A(k) = \langle u_{k,1}|\partial_k|u_{k,1}\rangle = i/2$ is Abelian and given by just a purely imaginary number (remember that it is anti-Hermitian in general). We thus obtain

$$\mathsf{P} := -\frac{i}{2\pi}\log W = -\frac{i}{2\pi}\int_0^{2\pi} A(k)dk = \frac{1}{2}. \quad (2.20)$$

The physical interpretation of P is given within the modern theory of polarization (see [7] for a pedagogical introduction) as that of a bulk electrical dipole moment or charge polarization, which is naturally only defined modulo 1 since the coordinate of a center of charge on the lattice is only defined up to a lattice translation (remember that we have chosen the lattice spacing $a = 1$). It is directly connected to the Wilson loop spectrum $\theta_{\alpha,\bar{k}}$ by a rescaling which makes sure that the periodicity of the charge centers defined in this way is that of the real-space lattice. See also Fig. 2.1b.

The result $\mathsf{P} = 1/2$ is by no means accidental: In fact, since the inversion symmetry reverses the path of integration in W, but leaves inner products such as $A(k)$ invariant, the Wilson loop eigenvalues of an inversion symmetric system satisfy $e^{i\theta} = e^{-i\theta}$ (see also Sect. 2.1.3.3 below). This requires that P be quantized to 0 ($\theta = 0$) or $1/2$ ($\theta = \pi$) in the Abelian case. This is a first example where a crystalline symmetry such as inversion, which acts non-locally in space, protects a topological phase by enforcing the quantization of a topological invariant to values that cannot be mapped into one another by an adiabatic evolution of the corresponding Hamiltonian. Note that since the eigenstates for the parameter choice $(t, t') = (1, 0)$ do not depend on k, we immediately obtain $\mathsf{P} = 0$ for this topologically trivial case.

By these considerations, it is clear that in fact the full parameter regime where $t < t'$ is topological, while the regime $t > t'$ is trivial. This is because it is possible to perform an adiabatic interpolation from the specific parameter choices $(t, t') \in \{(0, 1), (1, 0)\}$ considered above to all other values as long as there is no gap closing and no breaking of inversion symmetry, which is true provided that the line $t = t'$ is avoided in parameter space.

In general, a topological phase comes with topologically protected gapless boundary modes on boundaries which preserve the protecting symmetry. For inversion symmetry, however, there are no boundaries satisfying this requirement. Even though the model at $(t, t') = (0, 1)$ has zero-mode end states [since in this case, $\mathcal{H}(k)$ does not act at all on the A (B) site in the unit cell at the left (right) edge of the sample], these modes can be removed from zero energy by generic local perturbations even without a bulk gap closing. To protect the end modes, we need to invoke the chiral symmetry, which implies that an eigenstate at any energy E is paired up with an eigenstate at energy $-E$. Eigenstates of the chiral symmetry can then only appear at $E = 0$. A spatially and spectrally isolated boundary mode at $E = 0$ can thus not be removed by perturbations that retain the chiral symmetry. In conclusion, topological crystalline phases in 1D have no protected boundary degrees of freedom as long as we do not include further local symmetries.

In fact, in the presence of chiral symmetry, the above discussion can be generalized to arbitrary 1D models. In the eigenbasis of C, we can write any Hamiltonian with chiral symmetry in the form

$$\mathcal{H}(k) = \begin{pmatrix} 0 & q(k) \\ q^\dagger(k) & 0 \end{pmatrix}, \tag{2.21}$$

where for the SSH model the matrix $q(k)$ was given by just a complex number and in general we choose it to be a unitary matrix by an adiabatic deformation of the Hamiltonian. The chiral symmetry allows for the definition of a winding number

$$\nu = \frac{i}{2\pi} \int dk \, \text{Tr} \left[q(k) \partial_k q^\dagger(k) \right] \in \mathbb{Z}. \tag{2.22}$$

This winding number is one of the topological invariants alluded to in Sect. 2.1.1 and is only valid when chiral symmetry is present. We can make contact with the overarching concept of Wilson loops by calculating the connection

$$A = \frac{1}{2} q(k) \partial_k q^\dagger(k). \tag{2.23}$$

Thus, the Wilson loop eigenvalues $e^{i\theta_\alpha}$ satisfy

$$\frac{1}{2\pi} \sum_\alpha \theta_\alpha = \frac{\nu}{2} \bmod 1. \tag{2.24}$$

In particular, in the Abelian case, chiral symmetry thus implies the quantization of P to half-integer values, just as inversion symmetry did it above. An important distinction to be made is that with inversion symmetry, we have a $\mathbb{Z}_2$ topological classification (P can be either 0 or 1/2), while with chiral symmetry the winding number allows for a $\mathbb{Z}$ classification.

2.1.3.2 Example: Chern Insulator

Another paradigmatic example of a topologically protected phase is given by the (integer) quantum Hall effect of electrons subject to a perpendicular magnetic field in 2D continuous space. Here, we study its lattice realization, also called the quantum anomalous Hall effect or Chern insulator. We consider a 2D square lattice with open boundary conditions in x-direction and periodic boundary conditions in y-direction, retaining the momentum k_y as good quantum number.

To find an expression for the Hall conductivity for any Hamiltonian we could put on this lattice in terms of Wilson loops, let us perform a thought experiment where we roll up the y-direction to form a cylinder of circumference L (see Fig. 2.2a). Threading a magnetic flux ϕ along the x-direction through this cylinder amounts to the replacement $k_y \rightarrow k_y + \phi$ by a Peierls substitution. Note that in our units $\phi = 2\pi$ denotes a single flux quantum.

We now consider a Wilson loop along x-direction, labeled by k_y with eigenvalues $e^{i\theta_{\alpha,k_y}}$. The derivative $\partial_{k_y}\theta_{\alpha,k_y}$ of the αth Wilson loop eigenvalue is by the interpretation in terms of the modern theory of polarization explained in the previous section simply the 'velocity' in x-direction of the αth charge center at 'time' k_y. Integrating over k_y, i.e., adiabatically performing a flux insertion from $\phi = 0$ to $\phi = 2\pi$ (which brings the system back to its initial state), gives the full Hall conductivity as 2π (or, if e and $\hbar$ are reinstated, e^2/h) times

$$C = \sum_\alpha^M \int \frac{dk_y}{2\pi} \partial_{k_y}\theta_{\alpha,k_y}, \tag{2.25}$$

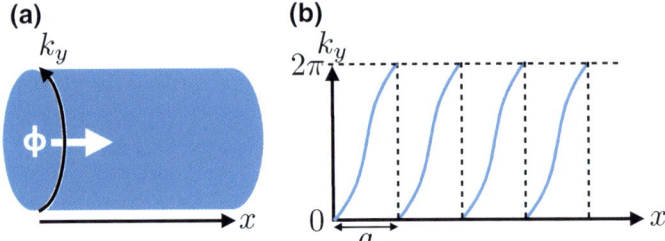

Fig. 2.2 Chern insulator geometry and charge center flow. **a** To calculate the Hall conductivity, we consider a Gedanken experiment where the y-direction of 2D space is compactified while we retain open boundary conditions in the x-direction. In particular, the translational symmetry along y allows for the introduction of the momentum k_y as a good quantum number to label blocks of the Hamiltonian and eigenstates. The Hall conductivity is then equal to the amount of charge transported along the y-direction in a single adiabatic cycle of flux insertion, where the inserted flux ϕ varies over time from 0 to 2π. **b** Charge center flow corresponding to a Chern number $C = 1$. Since one cycle of flux insertion corresponds to tuning k_y from 0 to 2π, we see that in one such cycle the charge center crosses exactly one unit cell

where C is known as the Chern number. To see how this formula works, note that the Hall conductivity is equal to the amount of charge transported in y-direction under the adiabatic insertion of a single flux quantum. Since we can only transport an integer number of charge around the cylinder in one such evolution (at least in the non-interacting systems we are considering here), C is necessarily quantized.

Making use of the relation $\sum_\alpha^M \theta_{\alpha,k_y} = \mathrm{i} \int dk_x \operatorname{Tr} A_x(k_x, k_y)$, which follows from (2.6), and requiring C to be gauge invariant, we can generalize (2.25) to

$$C = -\frac{\mathrm{i}}{2\pi} \int d^2 k \left[\partial_{k_x} \operatorname{Tr} A_y(\boldsymbol{k}) - \partial_{k_y} \operatorname{Tr} A_x(\boldsymbol{k}) \right]. \qquad (2.26)$$

The equality is directly seen in a gauge in which the integral of the first term $\partial_{k_x} \operatorname{Tr} A_y(\boldsymbol{k})$ does not contribute, which we have implicitly been working in (note that A here denotes the Berry connection, *not* the electromagnetic gauge field). The Chern number is thus the net number of charge centers crossing a given x position in the full k_y BZ. In the Wilson loop picture, it just corresponds to the winding number of the x-direction Wilson loop eigenvalues as k_y is varied along a non-contractible loop in the BZ, which is of course quantized (see Fig. 2.2b). While the Chern number is normally defined by employing the concept of Berry curvature, we have shown here that it may be equivalently expressed in terms of the spectral flow of Wilson loop eigenvalues as described at the end of Sect. 2.1.3.

2.1.3.3 Example: Time-Reversal Invariant Topological Insulator

Here, we explore the constraints imposed by time-reversal or inversion symmetries on Wilson loops. These symmetries protect topological insulators in two and three dimensions. In the presence of an anti-unitary time-reversal symmetry Θ, a Wilson loop $W_{2\pi \leftarrow 0}(\vec{k})$ along the x-direction, with k_x running from 0 to 2π, transforms as

$$\begin{aligned} \Theta W_{2\pi \leftarrow 0}(\vec{k}) \Theta^{-1} &= W^*_{0 \leftarrow 2\pi}(-\vec{k}) \\ &= W^T_{2\pi \leftarrow 0}(-\vec{k}) \\ \Rightarrow \quad \theta_\alpha(\vec{k}) &= \theta_\alpha(-\vec{k}). \end{aligned} \quad (2.27)$$

In particular, in a spinful system where $\Theta^2 = -1$, the representation of the time-reversal operation on the Wilson loop retains its property to square to -1, so that there is a Kramers degeneracy not only in the energy spectrum, but also in the Wilson loop spectrum. We thus recover the $\mathbb{Z}_2$ classification of 2D time-reversal invariant topological insulators from the spectral flow in the Wilson loop eigenvalues: Either the bands emerging from individual Kramers pairs connect back to the same pairs as $\vec{k}$ evolves along a non-contractible loop in the BZ, or they split up to connect to separate pairs.

Inversion I generates the following spectral pairing

$$\begin{aligned} I W_{2\pi \leftarrow 0}(\vec{k}) I^{-1} &= W_{0 \leftarrow 2\pi}(-\vec{k}) \\ &= W^\dagger_{2\pi \leftarrow 0}(-\vec{k}) \\ \Rightarrow \quad \theta_\alpha(\vec{k}) &= -\theta_\alpha(-\vec{k}). \end{aligned} \quad (2.28)$$

The combination of inversion I and time-reversal Θ then leads to a 'chiral symmetry' for the Wilson loop

$$\begin{aligned} I\Theta W_{2\pi \leftarrow 0}(\vec{k}) \Theta^{-1} I^{-1} &= W^*_{2\pi \leftarrow 0}(\vec{k}) \\ \Rightarrow \quad \theta_\alpha(\vec{k}) &= -\theta_\alpha(\vec{k}). \end{aligned} \quad (2.29)$$

Note that as the $\theta_\alpha(\vec{k})$ are only defined modulo 2π, we can have unidirectional flow in the Wilson loop spectrum: in the simplest case, in 2D we could have a single Wilson loop band which winds once along the θ-direction as k_y goes from 0 to 2π. This is in stark contrast to energy spectra, in which every unidirectionally dispersing band is paired up with a band going into the opposite direction so that the net chirality of the spectrum is always zero, a result which follows from the Nielsen–Ninomiya theorem under physically realistic circumstances such as locality [8].

2.1.4 Bulk-Boundary Correspondence

As alluded to in Sect. 2.1.1, Wilson loops not only provide a convenient formulation of many topological invariants, but are also in one-to-one correspondence with the boundary degrees of freedom of the system considered. We will now show that indeed the spectrum of a Hamiltonian in the presence of a boundary is smoothly connected to the spectrum of its Wilson loop along the direction perpendicular to the boundary. Note that since the Wilson loop is determined entirely by the bulk Bloch Hamiltonian, this relation provides an explicit realization of the bulk-boundary correspondence underlying all topological phases [9].

We consider a semi-infinite slab geometry with a single edge of the system at $x = 0$, while keeping $\vec{k}$ as good quantum numbers. From a topological viewpoint, the actual energetics of the band structure are irrelevant, and we can always deform the Hamiltonian for the sake of clarity to a spectrally flattened Hamiltonian where all bands above and below the gap are at energy $+1$ and -1, respectively, without closing the gap. It is therefore enough to work with

$$\mathcal{H}_{\text{flat}}(\vec{k}) = 1 - 2P(\vec{k}) \tag{2.30}$$

to model the bulk system. Here, $P(\vec{k})$ as defined in (2.10), repeated here for convenience,

$$P(\vec{k}) = \sum_{n}^{M} \int_{-\pi}^{\pi} \frac{dk_x}{2\pi} |\psi_{k,n}\rangle\langle\psi_{k,n}|, \tag{2.31}$$

is the projector onto the occupied subspace for a given $\vec{k}$. Note that $\mathcal{H}_{\text{flat}}(\vec{k})$ actually has the same eigenvectors as the original Hamiltonian. To model a system with boundary, we use

$$\mathcal{H}_{\text{bdr}}(\vec{k}) = P(\vec{k})V_0(\hat{x})P(\vec{k}) + 1 - P(\vec{k}), \tag{2.32}$$

with

$$V_0(x) = \begin{cases} 1 & x < 0 \\ -1 & x > 0 \end{cases} \tag{2.33}$$

so that we have $\mathcal{H}_{\text{bdr}}(\vec{k}) \to \mathcal{H}_{\text{flat}}(\vec{k})$ for $x \to +\infty$ and $\mathcal{H}_{\text{bdr}}(\vec{k}) \to 1$ for $x \to -\infty$ (see Fig. 2.3a). The latter limit corresponds to a description of the vacuum with the chemical potential chosen so that no electron states will be occupied, which we take to be the topologically trivial limit.

Since we take space to be infinitely extended away from the domain wall at $x = 0$, the spectrum of $\mathcal{H}_{\text{bdr}}(\vec{k})$ includes the spectrum of $\mathcal{H}_{\text{flat}}(\vec{k})$, given by ± 1 since $P^2(\vec{k}) = P(\vec{k})$, as well as that of the operator 1, trivially given by $+1$. The boundary region is of finite extent and can therefore contribute only a finite number of midgap states as the system has exponentially decaying correlations on either side of the boundary. There are therefore spectral accumulation points at ± 1, but otherwise we are left with a discrete spectrum (see Fig. 2.3b). We will focus on this part of the spectrum.

We will now deform the spectrum of $\mathcal{H}_{\text{bdr}}(\vec{k})$ to that of $(-i/2\pi)\log[W(\vec{k})]$ by considering an evolution that takes $P(\vec{k})V_0(\hat{x})P(\vec{k})$ to $P(\vec{k})\hat{x}P(\vec{k})$, the eigenvalues

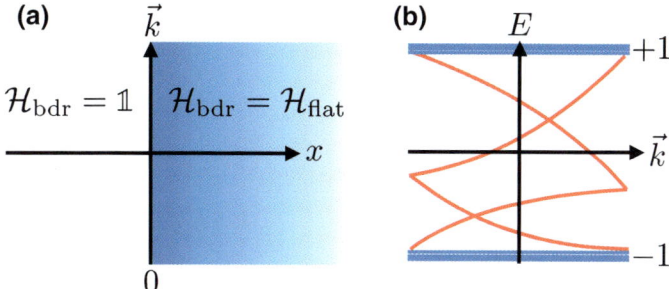

Fig. 2.3 Real-space setup and generic spectrum of $\mathcal{H}_{\text{bdr}}$. **a** For $V_0(x)$ as given by (2.33), $\mathcal{H}_{\text{bdr}}$ varies discontinuously from a trivial projector in the domain $x < 0$ to $\mathcal{H}_{\text{flat}}$ in the domain $x > 0$. Translational symmetry along x is thus broken, however it is preserved along all perpendicular directions, which still have good momentum quantum numbers $\vec{k}$. **b** The spectrum of $\mathcal{H}_{\text{bdr}}$ has accumulation points at ± 1, stemming from the semi-infinite regions to the left and right of the domain wall, and a discrete set of bands in between, coming from the finite domain wall region

of which were previously shown to be directly related to those of $(-i/2\pi)\log[W(\vec{k})]$. The deformation is continuous in $\vec{k}$ and therefore preserves both discreteness of the spectrum as well as its topological properties. An example for this interpolation is given by

$$V_t(x) = \begin{cases} -\frac{x}{t} & \text{for } |x| < t/(1-t) \\ -\frac{\text{sgn}(x)}{1-t} & \text{for } |x| \geq t/(1-t) \end{cases}, \quad 0 \leq t \leq 1. \quad (2.34)$$

Importantly, for any $t < 1$, $P(\vec{k})V_t(\hat{x})P(\vec{k})$ is a finite rank (finite support) perturbation of $(1-t)^{-1}P(\vec{k})V_0(\hat{x})P(\vec{k})$, so it will retain the property that the spectrum is discrete. However, the point $t = 1$ deserves closer inspection, as $P(\vec{k})\hat{x}P(\vec{k})$ is not a bounded operator. However, we can handle this subtlety by defining

$$h(r) = \begin{cases} r & \text{for } -w < r < w \\ \text{sgn}(r)w & \text{else} \end{cases} \quad (2.35)$$

and considering $h[P(\vec{k})V_t(\hat{x})P(\vec{k})]$ for some large w. The spectrum evolves uniformly continuously from $h[P(\vec{k})V_0(\hat{x})P(\vec{k})]$ to $h[P(\vec{k})V_1(\hat{x})P(\vec{k})]$ for any finite w [9].

The topology of the Wilson loop spectrum and the physical boundary spectrum is thus identical. Protected spectral flow in the former implies gapless boundary modes in the latter, as long as the form of the boundary [i.e., $V(x)$] does not break a symmetry that protects the bulk spectral flow.

2.1.4.1 Example: Chern Insulator Spectral Flow

We can obtain a simple Hamiltonian for a Chern insulator in 2D from that of the SSH model in 1D by tuning the latter from its topological to its trivial phase along a perpendicular direction k_y in the BZ. Along the way, we have to make sure that the whole system stays gapped. One Hamiltonian that does the job is given by [compare to (2.18)]

$$\mathscr{H}(\boldsymbol{k}) = \begin{pmatrix} \sin k_y & (1-\cos k_y) + e^{ik_x} \\ (1-\cos k_y) + e^{-ik_x} & -\sin k_y \end{pmatrix}. \tag{2.36}$$

Here, the part proportional to $\sin k_y$ is the term we added to keep the system gapped at all points in the new 2D BZ. The Wilson loop we considered in Sect. 2.1.3.1, and with it the polarization P, now becomes a function of k_y. We know that $\mathscr{H}(k_x, 0)$ corresponds to a topological SSH chain, while $\mathscr{H}(k_x, \pi)$ corresponds to a trivial one, implying $\mathsf{P}(0) = 1/2$ and $\mathsf{P}(\pi) = 0$. Remembering that P is only defined up to an integer, there are two possibilities for the Wilson loop spectral flow as k_y is varied from $-\pi$ to $+\pi$: Either the Wilson loop bands connect back to themselves trivially after the cycle has come to a close, or they do so only modulo an integer given by the Chern number C in (2.25). To infer which case applies to the model at hand, we can use the relation $\mathscr{H}(k_x, k_y) = -\sigma_3 \mathscr{H}(k_x, -k_y)\sigma_3$, which is a combination of chiral symmetry and y-mirror symmetry and must also hold in presence of a boundary (if the boundary potential is chosen such that it does not break this symmetry). It dictates that boundary spectra consist of *chiral* modes that connect the SSH spectra at $k_y = 0$ and $k_y = \pi$ as shown in Fig. 2.4. (Note that invoking the combination of chiral and mirror symmetry is only a convenient way to infer the boundary mode connectivity. No symmetry is needed to protect chiral boundary modes.) This is consistent with the Chern number, which for the model at hand evaluates to $C = 1$.

2.2 Topological Crystalline Insulators

Topological crystalline insulators [10] are protected by spatial symmetry transformations which act non-locally such as mirror or rotational symmetries. They are usually identified with two notions as follows: (i) their bulk ground state is not adiabatically connected to an atomic limit without breaking the protecting symmetry, (ii) they have gapless boundary modes which can only be gapped out by breaking the respective symmetry.

In fact, properties (i) and (ii) are not equivalent. We have already seen for the case of the SSH model protected by inversion symmetry that it is possible to have a model featuring (i) but not (ii). The reason was that although the model in its topological phase (as detectable by, e.g., the Wilson loop) is not adiabatically connected to any atomic limit, there is no boundary which is left invariant by inversion symmetry, and thus no protected edge modes (as long as we do not consider chiral symmetry, which is local and therefore non-crystalline). This is a general feature of topological

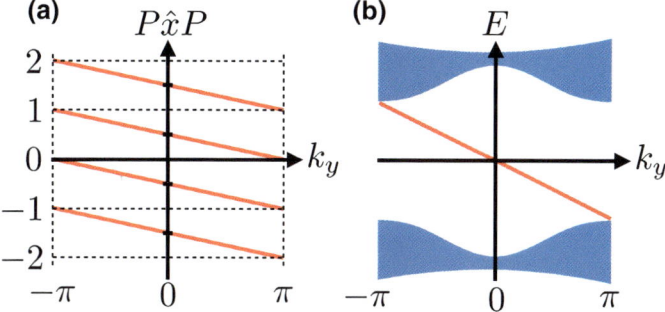

Fig. 2.4 Projected position operator spectral flow and its implications for the boundary Hamiltonian of a Chern insulator. **a** The model in (2.36) has two bands, therefore the Wilson loop is Abelian. Due to $C = 1$, by the spectral equivalence derived in Sect. 2.1.3, the eigenvalues of the projected position operator $P\hat{x}P$ flow from an integer n back to $n - 1 = n$ mod 1 exactly once as k_y is varied across a non-contractible loop in the BZ. **b** By the bulk-boundary correspondence derived in the present chapter, this implies a single chiral mode crossing the gapped bulk bands when the system is considered in the presence of a boundary termination

crystalline insulators: Gapless symmetry-protected boundary modes also require the boundary on which they are localized to preserve the corresponding symmetry. In the following, we discuss several pedagogical examples of topological crystalline phases and their invariants.

2.2.1   2D Topological Crystalline Insulator

Here we show how crystalline symmetries can enrich the topological classification of band structures. We begin with a model with chiral symmetry in 2D. A natural non-local symmetry in 2D we can add is a mirror symmetry, which leaves an edge invariant. While all 2D systems with just chiral symmetry (class AIII in the tenfold way) are topologically trivial, it will turn out that with mirror symmetry this is no longer the case when we require that mirror and chiral symmetry transformations commute. Note that in contrast, the Chern insulator model we considered in Sect. 2.1.4.1 breaks the chiral symmetry of the SSH models from which it was constructed by the gapping term proportional to $\sin k_y$. It therefore belongs to symmetry class A (no symmetries) and can be topological without crystalline symmetries.

The model we consider here is defined by the Bloch Hamiltonian

$$\mathcal{H}(k) = \begin{pmatrix} 0 & q(k) \\ q^\dagger(k) & 0 \end{pmatrix},$$

$$q(k) = \begin{pmatrix} (1 - \cos k_y) + e^{ik_x} + \lambda & \sin k_y \\ -\sin k_y & (1 - \cos k_y) + e^{-ik_x} - \lambda \end{pmatrix}. \quad (2.37)$$

The symmetry representations are

$$\mathcal{C}\mathcal{H}(\mathbf{k})\mathcal{C}^{-1} = -\mathcal{H}(\mathbf{k}), \quad M_y \mathcal{H}(k_x, k_y) M_y^{-1} = \mathcal{H}(k_x, -k_y),$$

$$C = \begin{pmatrix} \mathbb{1}_{2\times 2} & 0 \\ 0 & -\mathbb{1}_{2\times 2} \end{pmatrix}, \quad M_y = \begin{pmatrix} \sigma_z & 0 \\ 0 & \sigma_z \end{pmatrix}, \qquad (2.38)$$

where $\mathbb{1}_{2\times 2}$ denotes the 2×2 identity matrix and λ represents a numerically small perturbation that breaks M_x symmetry. When we calculate the winding number as defined in (2.22) along the path $k_x = 0 \to k_x = 2\pi$, $k_y = $ const. in the BZ, we find $\nu(k_y) = 0 \; \forall k_y$. We can most easily see this by evaluating $\nu(0) = 0$ and noting that as the spectrum is gapped throughout the BZ, and the model has chiral symmetry, the result holds for all k_y.

In the presence of mirror symmetry, however, we can refine the topological characterization. Since in our case mirror symmetry satisfies $M_y^2 = 1$, its representation has eigenvalues ± 1. Given any line l_{M_y} in the BZ which is left invariant under the action of M_y, the eigenstates $|u_{k,n}\rangle$ of $\mathcal{H}$ on l_{M_y} can be decomposed into two groups, $\{|u_{k,l}^+\rangle\}$ and $\{|u_{k,l'}^-\rangle\}$, with mirror eigenvalue ± 1, respectively. We can define the Wilson loop in each mirror subspace as

$$W^{\pm}[l_{M_y}] = \overline{\exp}\left[-\int_{l_{M_y}} \mathrm{d}\mathbf{l}_{M_y} \cdot \mathbf{A}^{\pm}(\mathbf{k})\right], \qquad (2.39)$$

where we have used the mirror-graded Berry connection

$$A_{m,n}^{\pm}(\mathbf{k}) = \langle u_{k,m}^{\pm}|\nabla_{\mathbf{k}}|u_{k,n}^{\pm}\rangle, \quad n, m = 1, \ldots, M. \qquad (2.40)$$

For the two mirror invariant paths $l_{M_y} : k_x = 0 \to k_x = 2\pi, k_y = 0, \pi$, the mirror-graded topological polarization invariants evaluate to

$$\mathsf{P}_{M_y}(k_y) = \frac{1}{2}\left[\left(-\frac{\mathrm{i}}{2\pi}\log W^+(k_y)\right) - \left(-\frac{\mathrm{i}}{2\pi}\log W^-(k_y)\right)\right] = \begin{cases} 1/2 & k_y = 0 \\ 0 & k_y = \pi \end{cases}, \qquad (2.41)$$

as can be directly seen from the relation of the model to two mirror-graded copies of the SSH model in the trivial ($k_y = \pi$) and nontrivial ($k_y = 0$) phase. This confirms that the 2D model is in a topologically nontrivial phase protected by mirror and chiral symmetry. With open boundary conditions, we will therefore find gapless states on both edges with normal to the x-direction (see Fig. 2.5a for such a geometry), because these are mapped onto themselves under M_y. Since the model corresponds to a topological-to-trivial tuning of two copies of the SSH model with opposite winding number, we expect two anti-propagating chiral edge states, which cannot gap out at their crossing at $k_y = 0$ since they belong to different mirror subspaces at this point (see Fig. 2.5b). A simple way to see this is that mirror symmetry maps $k_y \to -k_y$, while it does not change the energy E. Therefore it exchanges states pairwise at

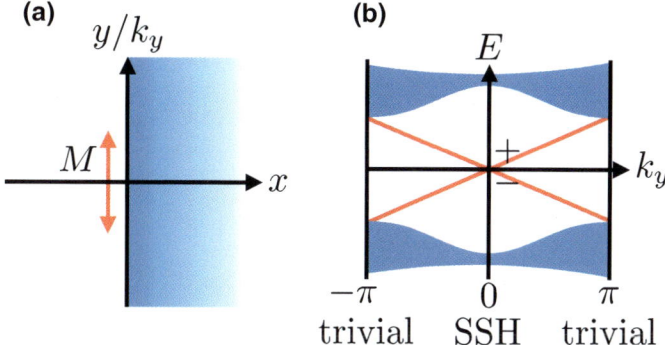

Fig. 2.5 Real-space geometry and spectrum of the mirror symmetric 2D model of a chiral symmetric topological crystalline insulator. **a** We consider a geometry where the system is terminated in x-direction but periodic in y-direction, retaining k_y as momentum quantum number. In particular, note that the surface in this semi-infinite slab geometry is mapped onto itself by the $M = M_y$ mirror symmetry and therefore hosts gapless modes stemming from the nontrivial topology of the bulk. **b** Schematic spectrum of the model given by (2.37) in the presence of the bulk termination in x-direction. There are two counter-propagating chiral modes which are necessarily crossing at $k_y = 0$ due to the chiral symmetry. At that point, they are also eigenstates of the mirror symmetry with eigenvalue ± 1, respectively. They are therefore protected from hybridization by the mirror symmetry

generic momenta k_y and $-k_y$ and we can form symmetric and anti-symmetric superpositions of them to get mirror eigenstates with eigenvalue $+1$ and -1, respectively. The trace of the representation of M_y on this two-dimensional subspace is therefore 0 at almost all momenta and in particular cannot change discontinuously at $k_y = 0$. Alternatively, direct inspection of the Hamiltonian (2.37) at $k_y = 0$ reveals that it is composed of two copies of the SSH model, and in view of the form of the mirror symmetry M_y, the two copies reside in opposite mirror subspaces. As a consequence, their end states (the edge modes at $k_y = 0$) also have opposite mirror eigenvalues and cannot hybridize.

Another 2D system which has two anti-propagating chiral edge modes is the quantum spin Hall effect protected by time-reversal symmetry, where the edge modes are localized on all boundaries. It corresponds to two Chern insulators, one for spin up and one for spin down. The present model may be viewed as a close relative, where the edge modes are protected by mirror and chiral symmetry as opposed to time reversal, and are only present on edges preserving the mirror symmetry.

2.2.2 Mirror Chern Number

In the previous section, we have witnessed an example of a general scheme to construct topological BZ invariants going beyond the tenfold way for systems protected

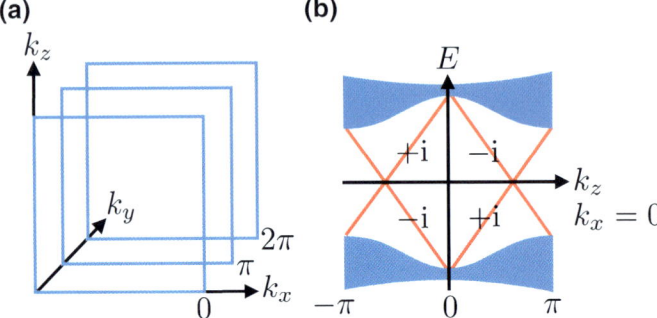

Fig. 2.6 Mirror Chern planes in the BZ and schematic surface spectrum for a time-reversal topological crystalline insulator with $C_m = 2$. *a* For the mirror symmetry M_y, there are two planes in the BZ which are left invariant by it and can therefore be used to define a mirror Chern number: The plane at $k_y = 0$ and the one at $k_y = \pi$. *b* A mirror Chern number $C_m = 2$ enforces the presence of two chiral left-movers and two chiral right-movers along mirror symmetric lines in the surface BZ of any surface mapped onto itself by the mirror symmetry. For M_y, this is, e.g., the case for the surface obtained by terminating the bulk in x-direction and retaining k_y and k_z as momentum quantum numbers. At $k_y = 0$, all bands are eigenstates of the mirror symmetry with eigenvalues as shown, and thus prevented from gapping out. At finite k_y, however, hybridization becomes possible and we are left with two Dirac cones in the surface BZ in the case at hand

by crystalline symmetries: Since a crystalline symmetry acts non-locally in space, it also maps different parts of the BZ onto each other. However, when there are submanifolds of the BZ which are left invariant by the action of the symmetry considered, we may evaluate a non-crystalline invariant on them, suited for the dimension and symmetry class of the corresponding submanifold, as long as we restrict ourselves to one of the symmetry's eigenspaces.

The most prominent example of this construction is the mirror Chern number C_m in three-dimensional systems. Since for a spinful system, mirror symmetry M squares to $M^2 = -1$, its representation in this case has eigenvalues $\pm i$. Let Σ be a surface in the BZ which is left invariant under the action of M, such as the surfaces shown in Fig. 2.6a for M_y. Then, the eigenstates $|u_{k,n}\rangle$ of the Hamiltonian on Σ can be decomposed into two groups, $\{|u^+_{k,l}\rangle\}$ and $\{|u^-_{k,l'}\rangle\}$, with mirror eigenvalues $+i$ and $-i$, respectively. Time-reversal symmetry maps one mirror subspace onto the other; if it is present, the two mirror eigenspaces are of the same dimension. We may define the Chern number in each mirror subspace as

$$C_\pm = -\frac{i}{2\pi} \int_\Sigma dk_x dk_z \operatorname{Tr}\left[\mathscr{F}^\pm_{xz}(\boldsymbol{k})\right]. \tag{2.42}$$

Here

$$\mathscr{F}^\pm_{ab}(\boldsymbol{k}) = \partial_a \mathscr{A}^+_b(\boldsymbol{k}) - \partial_b \mathscr{A}^+_a(\boldsymbol{k}) + \left[\mathscr{A}^+_a(\boldsymbol{k}), \mathscr{A}^+_b(\boldsymbol{k})\right] \tag{2.43}$$

is the non-Abelian Berry curvature field in the $\pm i$ mirror subspace, with $\mathscr{A}^{\pm}_{a;l,l'}(\bm{k}) = \langle u^{\pm}_{\bm{k},l}|\partial_a|u^{\pm}_{\bm{k},l'}\rangle$, and matrix multiplication is implied. Since $\text{Tr}\left[\mathscr{A}^{+}_a(\bm{k}),\mathscr{A}^{+}_b(\bm{k})\right]=0$, this corresponds to (2.26) restricted to a single mirror subspace. Note that in time-reversal symmetric systems we have $C_+ = -C_-$ and can thus define the mirror Chern number

$$C_{\text{m}} := (C_+ - C_-)/2. \tag{2.44}$$

A non-vanishing mirror Chern number implies that the Bloch Hamiltonian on Σ corresponds to a time-reversal pair of Chern insulators. Thus, the full model will host C_{m} Kramers pairs of gapless modes on an M-invariant line in any surface BZ corresponding to a real-space boundary which is mapped onto itself under the mirror symmetry M. These Kramers pairs of modes will be generically gapped out away from the lines in the surface BZ which are invariant under the mirror symmetry and therefore form surface Dirac cones. Indeed, when C_{m} is odd in a time-reversal symmetric system, this implies an odd number of Dirac cones in any surface BZ, since then the system realizes a conventional time-reversal invariant topological insulator with the Dirac cones located at time-reversal invariant surface momenta. When C_{m} is even, the surface Dirac cones exist only on mirror symmetric surfaces and are located at generic momenta along the mirror invariant lines of the surface BZ (see Fig. 2.6b). This inherently crystalline case is realized in the band structure of tin telluride, SnTe [11].

2.2.3 C_2T-Invariant Topological Crystalline Insulator

Here, we present another example of a topological crystalline insulator in 3D, introduced in [12], in order to show that surface Dirac cones protected by crystalline symmetries can also appear at generic, low-symmetry, momenta in the surface BZ. We consider a system that is invariant under the combination C_2T of a twofold rotation C_2 around the z-axis and time-reversal symmetry T. Note that we take both symmetries to be broken individually.

To understand how this symmetry can protect a topological phase, let us review how time-reversal protects a Dirac cone on the surface of a conventional topological insulator. The effective Hamiltonian on the boundary with surface normal along z of a 3D time-reversal symmetric topological insulator takes the form

$$\mathscr{H}(\bm{k}) = k_y\sigma_x - k_x\sigma_y. \tag{2.45}$$

The symmetries are realized as

$$\begin{aligned}T\mathscr{H}(\bm{k})T^{-1} &= \mathscr{H}(-\bm{k}), \quad T = i\sigma_y K,\\ C_2\mathscr{H}(\bm{k})C_2^{-1} &= \mathscr{H}(-\bm{k}), \quad C_2 = \sigma_z,\end{aligned} \tag{2.46}$$

where we denote by K complex conjugation. Now, the unique mass term for $\mathcal{H}(\boldsymbol{k})$ which gaps out the Dirac cone is $m\sigma_z$. This term is forbidden by time-reversal as expected, since it does not commute with T.

If we dispense with T symmetry and only require invariance under $C_2T = \sigma_x K$, the mass term is still forbidden. However, the addition of other constant terms to the Hamiltonian is now allowed. The freedom we have is to shift the Dirac cone away from the time-reversal symmetric point $\boldsymbol{k} = 0$ by changing the Hamiltonian to

$$\mathcal{H}(\boldsymbol{k}) = (k_y - a)\sigma_x - (k_x - b)\sigma_y, \tag{2.47}$$

with some arbitrary parameters a and b. Therefore, the phase stays topologically nontrivial, but has a different boundary spectrum from that of a normal topological insulator. On surfaces preserving $C_2^z T$ symmetry, any odd number of Dirac cones are stable but are in general shifted away from the time-reversal invariant surface momenta. On the surfaces that are not invariant under $C_2^z T$, the Dirac cones may be gapped out, since T is broken. This amounts to a $\mathbb{Z}_2$ topological classification of $C_2^z T$-invariant 3D topological crystalline insulators.

2.3 Higher-Order Topological Insulators

So far, when we discussed topological systems in d dimensions, we only considered $(d-1)$ dimensional boundaries which could host gapless states due to the nontrivial topology of the bulk. These systems belong to the class of first-order topological insulators, according to the nomenclature introduced in [13]. In the following, we will give an introduction to second-order topological insulators which have gapless modes on $(d-2)$ dimensional boundaries, that is, on corners in 2D and hinges in 3D, while the boundaries of dimension $(d-1)$ (i.e., the edges of a 2D system and the surfaces of a 3D system) are generically gapped. Higher-order topological insulators require spatial symmetries for their protection and thus constitute an extension of the notion of topological crystalline phases of matter.

2.3.1 2D Model with Corner Modes

A natural avenue of constructing a higher-order topological phase in 2D is to consider a 2D generalization of the SSH model with unit cell as shown in Fig. 2.7a (disregarding the colors in this figure for now) and alternating hoppings t and t' in both the x and y-directions. However, naively the bulk of the model defined this way with all hoppings of positive sign is gapless. This can be most easily seen in the fully atomic limit $t' = 0, t \neq 0$, where the Hamiltonian reduces to a sum over intra-unit cell Hamiltonians of the form

$$\mathcal{H} = t \begin{pmatrix} 0 & 1 & 0 & 1 \\ 1 & 0 & 1 & 0 \\ 0 & 1 & 0 & 1 \\ 1 & 0 & 1 & 0 \end{pmatrix}, \tag{2.48}$$

which has obviously zero determinant and therefore gapless modes.

This was amended in a model introduced in [14], which gave the first example of a higher-order topological insulator, by introducing a magnetic flux of π per plaquette. A specific gauge choice realizing this corresponds to reversing the sign of the hoppings along the blue lines in Fig. 2.7a. The model then has a gapped bulk, but gapless corner modes. This can be most easily seen in the fully dimerized limit $t = 0$, $t' \neq 0$, where one site in each corner unit cell is not acted upon by any term in the Hamiltonian. However, to protect the corner modes we have to include a spatial symmetry in addition to chiral symmetry, since we could otherwise perform an edge manipulation which leaves the bulk (and in particular, its gap) invariant but annihilates one corner mode with another. A natural candidate for this is the pair of diagonal mirror symmetries M_{xy} and $M_{x\bar{y}}$, which each leave a pair of corners invariant and therefore allow for protected gapless modes on them.

Note that we cannot arrive at the same phase by just combining two one-dimensional SSH models glued to the edges of a trivially gapped 2D system: By the mirror symmetry, the two SSH chains on edges that meet in a corner would have to be in the same topological phase. Thus, each would contribute one corner mode. At a single corner, we would therefore have a pair of modes which is not prevented by symmetry from being shifted to finite energies by a perturbation term. This consideration establishes the bulk model we introduced as an intrinsically 2D topological phase of matter. We will now present three alternative approaches to characterize the topology as well as the gapless corner modes of the model.

2.3.1.1 Elementary Mirror Subspace Analysis

The plaquettes along the $x\bar{y}$ diagonal are the only parts of the Hamiltonian mapped onto themselves by the $M_{x\bar{y}}$ mirror symmetry. In the fully dimerized limit $t' \neq 0$, $t = 0$, we may consider the Hamiltonian as well as the action of $M_{x\bar{y}}$ on a single inter-unit cell plaquette on the diagonal of the system as given by

$$\mathcal{H} = t' \begin{pmatrix} 0 & 1 & 0 & -1 \\ 1 & 0 & 1 & 0 \\ 0 & 1 & 0 & 1 \\ -1 & 0 & 1 & 0 \end{pmatrix}, \quad M_{x\bar{y}} = \begin{pmatrix} 0 & 0 & 1 & 0 \\ 0 & 1 & 0 & 0 \\ 1 & 0 & 0 & 0 \\ 0 & 0 & 0 & -1 \end{pmatrix}. \tag{2.49}$$

$M_{x\bar{y}}$ has eigenvectors

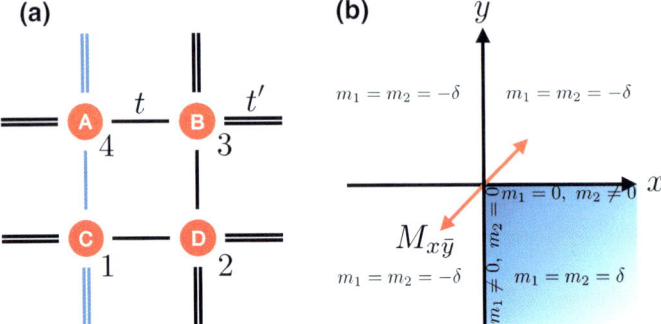

Fig. 2.7 Higher-order 2D SSH model. **a** The model features intra-unit cell hopping with strength t as well as inter-unit cell hopping with strength t'. For the topological phase we require $t' > t$. In particular, in the fully dimerized limit $t = 0, t' \neq 0$ it becomes evident that when we cut the system in two directions to create a corner, there is one dangling site which is not acted upon by any term in the Hamiltonian and therefore provides a zero-mode. The unit cell contains a π-flux per plaquette, which is realized by all blue hoppings being negative, while all black hoppings are positive. **b** Vortex geometry of the prefactors of the two masses in (2.56). We imply a smooth interpolation between the mass values given in the bulk, on the edges, and on the outside. At the corner, the masses vanish and they hence form a vortex-like structure around it

$$|+_1\rangle = \begin{pmatrix} 0 \\ 1 \\ 0 \\ 0 \end{pmatrix}, \quad |+_2\rangle = \frac{1}{\sqrt{2}} \begin{pmatrix} 1 \\ 0 \\ 1 \\ 0 \end{pmatrix}, \quad |-_1\rangle = \begin{pmatrix} 0 \\ 0 \\ 0 \\ 1 \end{pmatrix}, \quad |-_2\rangle = \frac{1}{\sqrt{2}} \begin{pmatrix} 1 \\ 0 \\ -1 \\ 0 \end{pmatrix} \tag{2.50}$$

with eigenvalues $+1, +1, -1, -1$, respectively. Since $[\mathcal{H}, M_{x\bar{y}}] = 0$ we know that the Hamiltonian block-diagonalizes into the two mirror subspaces, and we may calculate its form in each Block separately,

$$(\mathcal{H}_+)_{ij} = \langle +_i | \mathcal{H} | +_j \rangle \rightarrow \mathcal{H}_+ = \begin{pmatrix} 0 & \sqrt{2} \\ \sqrt{2} & 0 \end{pmatrix} = \mathcal{H}_-. \tag{2.51}$$

This, however, is exactly the form taken by a single SSH model in the fully dimerized, and topologically nontrivial, phase (this is because here we have focussed on a plaquette on the diagonal with t' hopping, an adjoining plaquette with t hopping would correspond to the weak bonds in the mirror subspace SSH model). We may therefore interpret our model along one diagonal as two nontrivial SSH models, one for each mirror subspace and protected by the chiral symmetry. Naively this would imply two end modes. However, this is not the case. In the upper left corner, for example, only a single A site is left from an inter-unit cell plaquette (see Fig. 2.7a), which happens to have mirror eigenvalue -1. Correspondingly, the lower right corner hosts a dangling D site, which has mirror eigenvalue $+1$. Thus, due to this modified bulk-boundary correspondence of the higher-order topological insulator, each of the two diagonal SSH chains has only one end state at opposite ends. These form the corner modes of the higher-order topological insulator.

2.3.1.2 Mirror-Graded Winding Number

In analogy to the mirror-graded Wilson loop introduced in Sect. 2.2.1, we can calculate the mirror-graded winding number suited for systems with chiral symmetry. For this we need the full Bloch Hamiltonian, which is given by

$$\mathcal{H}(\mathbf{k}) = (1 + \lambda \cos k_x)\tau_0\sigma_x + (1 + \lambda \cos k_y)\tau_y\sigma_y - \lambda \sin k_x \tau_z\sigma_y + \lambda \sin k_y \tau_x\sigma_y. \tag{2.52}$$

Note that by a term such as $\sigma_x \tau_0$ we really mean the tensor product $\sigma_x \otimes \tau_0$ of two Pauli matrices. Here, we have chosen $t = 1$ and $t' = \lambda$. The case where $\lambda > 1$ then corresponds to the topological phase. Along the diagonals of the BZ (and only there), the Hamiltonian may again be block-diagonalized by the mirror symmetries and thus decomposes into [choosing without loss of generality the $\mathbf{k} = (k, k)$ diagonal]

$$\mathcal{H}(k,k) = \begin{pmatrix} 0 & q_+(k) & 0 & 0 \\ q_+^\dagger(k) & 0 & 0 & 0 \\ 0 & 0 & 0 & q_-(k) \\ 0 & 0 & q_-^\dagger(k) & 0 \end{pmatrix}, \quad q_\pm(k) = \sqrt{2}(1 + \lambda e^{\mp ik}). \tag{2.53}$$

In this representation, the relevant mirror $M_{x\bar{y}}$ symmetry takes the block-diagonal form

$$M_{x\bar{y}} = \begin{pmatrix} 1 & 0 & 0 & 0 \\ 0 & -1 & 0 & 0 \\ 0 & 0 & 0 & 1 \\ 0 & 0 & 1 & 0 \end{pmatrix}. \tag{2.54}$$

We see that in the two mirror eigenspaces $\mathcal{H}(k, k)$ takes the form of an SSH model. Defining

$$\nu_\pm = \frac{i}{2\pi} \int dk \, \mathrm{Tr}\left[\tilde{q}_\pm(k) \partial_k \tilde{q}_\pm^\dagger(k)\right] \tag{2.55}$$

in analogy to (2.22), where we have appropriately normalized $\tilde{q}_\pm(k) = q_\pm(k)/\sqrt{2}$, we obtain $\nu_\pm = \pm 1$ and therefore $\nu_{M_{x\bar{y}}} = 1$ for the mirror-graded winding number $\nu_{M_{x\bar{y}}} = (\nu_+ - \nu_-)/2$. As long as the system obeys the mirror symmetry and the chiral symmetry, $\nu_{M_{x\bar{y}}}$ is a well-defined topological invariant that cannot be changed without closing the bulk gap of the 2D system.

2.3.1.3 Dirac Picture of Corner States

An alternative and very fruitful viewpoint of topological phases of matter arises from the study of continuum Dirac Hamiltonians corresponding to a given phase. For example, the band inversion of a first-order topological insulator can be efficiently captured by the Hamiltonian of a single gapped Dirac cone with mass m in the bulk of the material, and mass $(-m)$ in its exterior. One can then show that the domain wall in m binds exactly one gapless Dirac cone to the surface of the material. We want to develop an analogous understanding of higher-order topological phases as exemplified by the model studied in this section.

For the topological phase transition at $\lambda = 1$ in (2.52), there is a gap closing at $\boldsymbol{k}_0 = (\pi, \pi)$. Expanding $\mathscr{H}(\boldsymbol{k})$ around this point to first order and setting $\boldsymbol{k} = \boldsymbol{k}_0 + \boldsymbol{p}$, we obtain

$$\begin{aligned}\mathscr{H}(\boldsymbol{k}) &= (1-\lambda)\tau_0\sigma_x + (1-\lambda)\tau_y\sigma_y + \lambda p_x\,\tau_z\sigma_y - \lambda p_y\,\tau_x\sigma_y \\ &\approx \delta\tau_0\sigma_x + \delta\tau_y\sigma_y + p_x\,\tau_z\sigma_y - p_y\,\tau_x\sigma_y,\end{aligned} \quad (2.56)$$

where we have defined $\delta = (1-\lambda) \ll 1$ and $\lambda \approx 1$. Note that all matrices anticommute and that there are two mass terms, both proportional to δ, in accordance with the gap-closing phase transition at $\delta = 0$. When terminating the system, a boundary is modeled by a spatial dependence of these masses. We consider the geometry shown in Fig. 2.7b, where two edges meet in a corner. The mirror symmetry $M_{x\bar{y}}$ maps one edge to the other but leaves the corner invariant. As a result, the mirror symmetry does not pose any restrictions on the masses on one edge, but once their form is determined on one edge, they are also fixed on the other edge by $M_{x\bar{y}}$. In fact, since $M_{x\bar{y}}\tau_0\sigma_x M_{x\bar{y}}^{-1} = \tau_y\sigma_y$ with $M_{x\bar{y}} = (\tau_x\sigma_0 + \tau_z\sigma_0 + \tau_x\sigma_z - \tau_z\sigma_z)/2$ and vice versa, we may consider the particularly convenient choice of Fig. 2.7b for the mass configuration of the corner geometry.

From Fig. 2.7b, it becomes evident that the symmetries dictate that the masses, when considered as real and imaginary part of a complex number, wind once around the origin of the corresponding complex plane (at which the system becomes gapless) as we go once around the corner in real space. They are mathematically equivalent to a vortex in a p-wave superconductor, which is known to bind a single Majorana zero mode. We can therefore infer the presence of a single gapless corner state for the model considered in this section from its Dirac Hamiltonian.

To be more explicit, denoting by $m_1(x, y)$ and $m_2(x, y)$ the position-dependent prefactors of $\sigma_x\tau_0$ and $\sigma_y\tau_y$, respectively, we may adiabatically evolve the Hamiltonian to a form where the mass term vortex is realized in the particularly natural form $m_1(x, y) + i m_2(x, y) = x + i y = z$, where z denotes the complex number corresponding to the 2D real-space position (x, y). After performing a C_3 rotation about the (111)-axis in τ space, which effects the replacement $\tau_x \to \tau_y \to \tau_z \to \tau_x$, and exchanging the order of τ and σ in the tensor product, the resulting matrix takes on the particularly nice form

$$\mathcal{H}(\boldsymbol{k}) = \begin{pmatrix} 0 & q(\boldsymbol{k}) \\ q^\dagger(\boldsymbol{k}) & 0 \end{pmatrix},$$

$$q(\boldsymbol{k}) = \begin{pmatrix} m_1 - im_2 & -ip_x + p_y \\ -ip_x - p_y & m_1 + im_2 \end{pmatrix} = \begin{pmatrix} \bar{z} & -\partial_{\bar{z}} \\ -\partial_z & z \end{pmatrix}$$

$$\rightarrow \quad q^\dagger(\boldsymbol{k}) = \begin{pmatrix} z & \partial_{\bar{z}} \\ \partial_z & \bar{z} \end{pmatrix}. \tag{2.57}$$

While for $q^\dagger(\boldsymbol{k})$, there is one zero-energy solution $|\Psi\rangle = e^{-z\bar{z}}(1, 1)$, the corresponding solution $|\tilde{\Psi}\rangle = e^{\frac{z^2 + \bar{z}^2}{2}}(1, 1)$ for $q(\boldsymbol{k})$ is not normalizable. We thus conclude that there is one zero-mode with eigenfunction $(|\Psi\rangle, |0\rangle)$ localized at the corner of the sample.

2.3.2 3D Model with Hinge Modes

To construct a higher-order topological insulator in 3D, we start from a time-reversal invariant topological crystalline insulator with mirror Chern numbers in its bulk BZ. For the sake of simplicity we restrict to the case where only the mirror Chern number C_m belonging to the M_y symmetry is non-vanishing because the argument goes through for each mirror Chern number separately. We will now show that in an open geometry with surface normals along the xy and $x\bar{y}$ direction, $C_m = 2$ implies the presence of a single time-reversal pair of gapless chiral hinge modes on the intersection of the (110) and $(1\bar{1}0)$ surfaces (see Fig. 2.8).

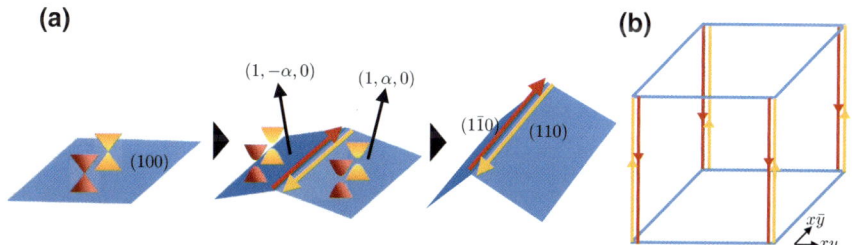

Fig. 2.8 Construction of a 3D second-order topological insulator. **a** We begin with a surface left invariant by M_y on which $C_m = 2$ implies two gapless Dirac cones. When slightly tilting the surface in opposite directions to form a kink, the Dirac cones on the new surfaces on either side of the kink may be gapped out with opposite masses, since the mirror symmetry maps one into the other and anti-commutes with the Dirac mass term. Since a domain wall in a Dirac mass binds a single zero-mode, and the two Dirac cones on each surface are mapped into each other by time-reversal, a Kramers pair of gapless hinge modes emerges on the intersection. When continuing to bend the surfaces to create a right angle, these modes cannot vanish since they are protected by the M_y mirror symmetry. **b** By this argument we can infer time-reversal paired hinge modes on each hinge along the x (and y, if we also take into account the mirror symmetry M_x along with $C_m = 2$) direction

As discussed in Sect. 2.2.2, for M_y symmetry, a nonzero $C_m = 2$ enforces two gapless Dirac cones in the surface BZ of the (100) termination, which is mapped onto itself by M_y. Note that on this surface, with normal in x-direction, k_y and k_z are still good momentum quantum numbers. The Hamiltonian for a single surface Dirac cone can be written as

$$\mathcal{H}(k_y, k_z) = v_1 \sigma_z k_y + v_z \sigma_x (k_z - k_z^0), \tag{2.58}$$

where the mirror symmetry is represented by $M_y = i\sigma_x$ and thus prevents a mass term of the form $+m\sigma_y$ from appearing. To arrive at a theory describing the intersection of the (110) and ($1\bar{1}0$) boundaries, we introduce a mirror-symmetric kink in the (100) surface (see Fig. 2.8a) and so first consider the intersection of two perturbatively small rotations of the (100) surface, one to a $(1, \alpha, 0)$ termination and the other to a $(1, -\alpha, 0)$ termination with $\alpha \ll 1$. The Hamiltonian on the $(1, \pm\alpha, 0)$ surface becomes

$$\begin{aligned}\mathcal{H}_\pm(k_y, k_z) &= v_1 \sigma_z (k_y \pm \rho) + v_z \sigma_x (k_z - k_z^0) \pm m\sigma_y \\ &\equiv v_1 \tilde{k}_y^\pm \sigma_z + v_z \tilde{k}_z \sigma_x \pm m\sigma_y,\end{aligned} \tag{2.59}$$

where m and ρ are small parameters of order α and we have omitted the irrelevant coordinate shifts in the last line. This Hamiltonian describes a gapped Dirac cone, the mass term is now allowed by mirror symmetry since the surfaces considered are no longer invariant under it. Instead, they are mapped onto each other and thus have to carry opposite mass. We note that by this consideration the hinge between the $(1, \alpha, 0)$ and $(1, -\alpha, 0)$ surfaces constitutes a domain wall in a Dirac mass extended in z-direction, which is known to host a single chiral mode [15].

We will now explicitly solve for this domain wall mode at $\tilde{k}_z = 0$ by going to real space in y-direction. Making the replacement $\tilde{k}_y^\pm \to -i\partial_y$, the Hamiltonian on either side of the hinge becomes

$$\mathcal{H}_\pm = \begin{pmatrix} -iv_1 \partial_y & \pm im \\ \mp im & iv_1 \partial_y \end{pmatrix}. \tag{2.60}$$

$\mathcal{H}_+$, for which $y > 0$, has one normalizable zero-energy solution given by $|\Psi_+\rangle = e^{-\kappa y}(1, 1)$ (where we assume $\kappa = m/v_1 > 0$ without loss of generality). $\mathcal{H}_-$, for which $y < 0$, has another normalizable zero-energy solution given by $|\Psi_-\rangle = e^{\kappa y}(1, 1)$. Since the spinor $(1, 1)$ of the solutions is the same on either side of the hinge, the two solutions can be matched up in a continuous wave function. We obtain a single normalizable zero-energy solution for the full system at $\tilde{k}_z = 0$, which is falling off exponentially away from the hinge with a real-space dependence given by $|\Psi\rangle = e^{-\kappa|y|}(1, 1)$. To determine its dispersion, we may calculate the energy shift for an infinitesimal $\tilde{k}_z$ in first-order perturbation theory to find

$$\Delta E(\tilde{k}_z) = \langle \Psi | v_z \tilde{k}_z \sigma_x | \Psi \rangle = +v_z \tilde{k}_z. \tag{2.61}$$

We have therefore established the presence of a single linearly dispersing chiral mode on the hinge between the $(1, \alpha, 0)$ and $(1, -\alpha, 0)$ surfaces by considering what happens to a single Dirac cone on the $(1, 0, 0)$ surface when a kink is introduced. The full model, which by $C_m = 2$ has two (100) surface Dirac cones paired by time-reversal symmetry, therefore hosts a Kramers pair of hinge modes on the intersection between the $(1, \alpha, 0)$ and $(1, -\alpha, 0)$ surfaces. These surfaces themselves are gapped. The two modes forming the hinge Kramers pair have opposite mirror eigenvalue. Increasing α non-perturbatively to 1 in a mirror-symmetric fashion cannot change the number of these hinge modes, since the chiral modes belong to different mirror subspaces and are thus stable to any perturbation preserving the mirror symmetry. By this reasoning, we end up with a pair of chiral modes at each hinge in the geometry of Fig. 2.8b.

2.3.3 Interacting Symmetry-Protected Topological Phases with Corner Modes

In this last section, we will switch gears and explore how one can construct interacting symmetry-protected topological (SPT) phases of bosons which share the phenomenology of higher-order topological insulators. Note that while non-interacting fermionic systems may have topologically nontrivial ground states, the same is not true for non-interacting bosonic systems whose ground state is a trivial Bose–Einstein condensate [16]. Therefore, for bosons we necessarily need interactions to stabilize a topological phase. We first give a lightning introduction to SPT phases via a very simple model in 1D. A topologically nontrivial SPT state is defined as the gapped ground state of a Hamiltonian, for which there exists no adiabatic interpolation to an atomic limit Hamiltonian without breaking the protecting symmetries or losing the locality of the Hamiltonian along the interpolation [17].

2.3.3.1 1D Model with Local Symmetry

Consider a chain of N spin-1/2 degrees of freedom with Hamiltonian

$$H = -\sum_{i=2}^{N-1} A_i, \quad A_i = \sigma^z_{i-1} \sigma^x_i \sigma^z_{i+1}, \tag{2.62}$$

which describes a system with open boundary conditions. All the A_i commute with each other and can therefore be simultaneously diagonalized.

The Hamiltonian H respects a time-reversal ($\mathbb{Z}_2^T$ in SPT lingo) symmetry $[T, H] = 0$ represented by the operator

$$T = K \prod_i \sigma_i^x. \qquad (2.63)$$

Note that $T^2 = +1$.

We now consider a set of operators

$$\Sigma^x = \sigma_1^x \sigma_2^z, \quad \Sigma^y = \sigma_1^y \sigma_2^z, \quad \Sigma^z = \sigma_1^z, \qquad (2.64)$$

which act locally on the left end of the chain and furnish a Pauli algebra. A similar set of operators can be defined for the other end of the chain.

Since $[T, \Sigma^a]_+ = 0$, where $[\cdot, \cdot]_+$ denotes the anti-commutator, these end operators cannot be added as a perturbation to the Hamiltonian without breaking the $\mathbb{Z}_2^T$ symmetry. However, they commute with all the A_i in H. This algebra can only be realized on a space with minimum dimension 2, imposing a twofold degeneracy on the eigenstates of H for each end of the chain. This degeneracy can be interpreted as one gapless spin-1/2 degree of freedom at each end of the chain. Note that a unitary version of T would commute with Σ^y rather than anti-commute and therefore not protect these edge degrees of freedom.

2.3.3.2 2D Model with Crystalline Symmetry

We can set up a very similar construction in 2D to arrive at a SPT model with gapless corner modes. Note however that we know from the classification of SPTs by group cohomology [17] that while in 1D the $\mathbb{Z}_2^T$ symmetry from before indeed protects a $\mathbb{Z}_2$ topological classification, in 2D there is no corresponding nontrivial phase. As was the case for non-interacting fermions, we therefore have to turn to spatial symmetries to protect corner states. Other than that, the construction is very similar to the 1D case.

Consider a square lattice of spin-1/2 degrees of freedom, again with Hamiltonian

$$H = -\sum_i A_i, \quad A_i = \sigma_i^x \prod_{j_i \in N(i)} \sigma_{j_i}^z, \qquad (2.65)$$

where the set $N(i)$ stands for the four next-to-nearest neighbor sites of site i, which are located along the xy and $x\bar{y}$ diagonals (see Fig. 2.9a). Again, verify that all A_i commute with each other and thus can be simultaneously diagonalized. We will be interested in open boundary conditions, in which case the sum over i runs only over the interior sites of the lattice, i.e., not the sites on the edges or corners.

Trivially, the model has the same symmetry as given by (2.63). However, we need to enrich it with a spatial transformation in order for it to protect topological features. We choose $T = K \prod_i \sigma_i^x$ as before and define

$$M_{x\bar{y}}: \quad (x, y) \to (-y, -x). \qquad (2.66)$$

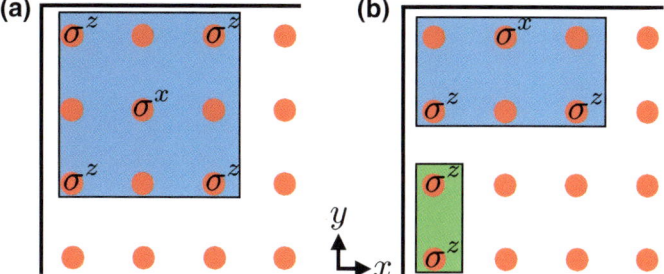

Fig. 2.9 Local operators of a higher-order SPT Hamiltonian with protected corner modes. Each site carries a spin-1/2 degree of freedom acted upon by the Pauli matrices σ^i, $i = 0, x, y, z$. **a** The bulk Hamiltonian consists of the sum over all sites of tensor products of σ_x acting on a given site with σ_z on all four adjoining sites along the two diagonals. **b** Two possible edge Hamiltonian elements which naively would both satisfy the symmetry $M_{xy}T$ when repeated over all edge sites. However, the ground state corresponding to the operator in green spontaneously breaks the symmetry and the operator is thus forbidden. Therefore, we may only terminate the edge with the operator in blue, leading to a twofold degeneracy in the resulting ground state for each corner

The model is then invariant under the symmetry $M_{x\bar{y}}T$. We will now show that this symmetry protects a pair of corner states along the $x\bar{y}$ diagonal. In order to also protect states at the other pair of corners we would have to perform the same analysis and require $M_{xy}T$ or C_4^z symmetry in addition.

Indeed, at each corner along the $x\bar{y}$ diagonal, we have a Pauli algebra generated by

$$\Sigma^x = \sigma_c^x \sigma_{j_c}^z, \quad \Sigma^y = \sigma_c^y \sigma_{j_c}^z, \quad \Sigma^z = \sigma_c^z, \quad (2.67)$$

where c denotes the corner site and j_c denotes the site which is the next-to-nearest neighbor of the corner along the diagonal. Crucially, there is a single next-to-nearest neighbor site for each corner, while in the bulk there are four next-to-nearest neighbors along the diagonals.

Since $[M_{x\bar{y}}T, \Sigma^a]_+ = 0$, these corner terms Σ^x, Σ^y, and Σ^z cannot be added as a perturbation to the Hamiltonian without breaking the symmetry. However, they commute with all the A_i in H, again imposing a twofold degeneracy on the eigenstates of H for each corner. We therefore have one gapless spin-1/2 degree of freedom at each corner lying along the diagonal corresponding to the respective mirror symmetry we require to hold.

Unlike in the 1D case, this is not the end of the story. We have merely shown that each corner provides a twofold degeneracy, but what about the edge degrees of freedom? In order to arrive at a higher-order phase, we need to gap them out. A natural way to do this is to include in the Hamiltonian not only the A_i terms with four next-to-nearest neighbors, but also the corresponding edge terms which only have two next-to-nearest neighbors. This is in fact symmetry-allowed for all the edge sites except the corners. The terms are sketched in Fig. 2.9b.

We may however also just put an Ising model on the edge by adding the Hamiltonian

$$H_{\text{edge}} = -\sum_{i \in E} \sigma_i^{\tilde{z}} \sigma_{N_E(i)}^{\tilde{z}}, \qquad (2.68)$$

where E denotes the set of all boundary sites, including the corners (see Fig. 2.9b), and $N_E(i)$ denotes one of the two nearest-neighbor sites of i on the edge chosen according to an arbitrary but globally fixed edge orientation. Hamiltonian (2.68) contains as many terms as there are edge and corner sites combined. The bulk Hamiltonian contains as many terms as there are bulk sites. We want to find a ground state that has simultaneously eigenvalue $+1$ with respect to all these commuting operators. At first sight, because there are as many terms as sites, these constraints fix the ground state completely. However, this is not true. The product of all terms in Hamiltonian (2.68) is the identity, because each site is acted upon by two $\sigma^{\tilde{z}}$ operators in this product. This means we have globally one less constraint than sites. This degeneracy corresponds to two magnetized ground states of the Ising model that is formed by the gapped edge. Luckily, either of these magnetized ground states of the quantum Ising model in 1D necessarily breaks the $M_{x\bar{y}}$ symmetry (remember that we are, as always in these notes, working at zero temperature). This spontaneous symmetry breaking preempts the definition of our topological case and renders the edge termination defined in (2.68) not permissible.

In conclusion, we have demonstrated how one can construct a 2D higher-order topological phase protected by mirror symmetries, where protection means that the symmetry may not be broken either explicitly or spontaneously in order for there to be gapless corner modes.

References

1. X.L. Qi, T.L. Hughes, S.C. Zhang, Phys. Rev. B **78**, 195424 (2008). https://doi.org/10.1103/PhysRevB.78.195424
2. M.Z. Hasan, C.L. Kane, Rev. Mod. Phys. **82**, 3045 (2010). https://doi.org/10.1103/RevModPhys.82.3045
3. B.A. Bernevig, *Topological Insulators and Topological Superconductors* (Princeton University Press, New Jercy, 2013)
4. J.K. Asbóth, L. Oroszlány, A. Pályi, A short course on topological insulators. Lect. Notes Phys. 919. https://link.springer.com/book/10.1007%2F978-3-319-25607-8
5. A. Bernevig, T. Neupert, ArXiv e-prints (2015). https://arxiv.org/abs/1506.05805
6. N. Marzari, A.A. Mostofi, J.R. Yates, I. Souza, D. Vanderbilt, Rev. Mod. Phys. **84**, 1419 (2012). https://doi.org/10.1103/RevModPhys.84.1419
7. N.A. Spaldin, J. Solid State Chem. Fr. **195**, 2 (2012). https://doi.org/10.1016/j.jssc.2012.05.010
8. H. Nielsen, M. Ninomiya, Phys. Lett. B **130**(6), 389 (1983). https://doi.org/10.1016/0370-2693(83)91529-0
9. L. Fidkowski, T.S. Jackson, I. Klich, Phys. Rev. Lett. **107**, 036601 (2011). https://doi.org/10.1103/PhysRevLett.107.036601
10. L. Fu, Phys. Rev. Lett. **106**, 106802 (2011). https://doi.org/10.1103/PhysRevLett.106.106802

11. T.H. Hsieh, H. Lin, J. Liu, W. Duan, A. Bansil, L. Fu, Nat. Commun. **3**, 982 (2012). https://doi.org/10.1038/ncomms1969
12. C. Fang, L. Fu, Phys. Rev. B **91**, 161105 (2015). https://doi.org/10.1103/PhysRevB.91.161105
13. F. Schindler, A.M. Cook, M.G. Vergniory, Z. Wang, S.S Parkin, B.A. Bernevig, T. Neupert, Higher-order topological insulators. Sci. Adv. **4**(6), eaat0346 (2018). http://advances.sciencemag.org/content/4/6/eaat0346
14. W.A. Benalcazar, B.A. Bernevig, T.L. Hughes, Science **357**(6346), 61 (2017). https://doi.org/10.1126/science.aah6442, http://science.sciencemag.org/content/357/6346/61
15. R. Jackiw, C. Rebbi, Phys. Rev. D **13**, 3398 (1976). https://doi.org/10.1103/PhysRevD.13.3398
16. Ann. Rev. Condens. Matter Phys. **6**(1), 299 (2015). https://doi.org/10.1146/annurev-conmatphys-031214-014740
17. X. Chen, Z.C. Gu, Z.X. Liu, X.G. Wen, Phys. Rev. B **87**, 155114 (2013). https://doi.org/10.1103/PhysRevB.87.155114

Chapter 3
Calculating Topological Invariants with Z2Pack

Dominik Gresch and Alexey Soluyanov

Abstract The topological phase of non-interacting electronic bandstructure can be classified by calculating integer invariants. In this chapter, we introduce the Chern invariant that classifies 2D materials in the absence of symmetry. We then show that this invariant can be used as the building block for the classification of topological insulators, semimetals, and symmetry-protected topological phases. We show how this classification is performed in practice by introducing Z2Pack, a tool which allows calculating topological invariants from $\mathbf{k} \cdot \mathbf{p}$ and tight-binding models, as well as first-principles calculations.

3.1 The Chern Number

In this section, we give a coarse introduction to topological invariants in the context of classifying crystalline solids. In the interest of brevity, we will skip many of the mathematical details, instead focusing on conveying an intuitive understanding as required to follow the rest of this chapter. For a more thorough description of the topics covered here, the reader is referred to [1, 2].

3.1.1 Topology in Non-interacting Materials

In this first section, we will introduce the notion of topological properties in the context of non-interacting materials. From their basic definition, we will see that topological phases must exhibit some interesting physical phenomena.

D. Gresch (✉)
ETH Zurich, Institut für Theoretische Physik, Wolfgang-Pauli-Str. 27,
8093 Zürich, Switzerland
e-mail: greschd@phys.ethz.ch

A. Soluyanov
Physik-Institut, Universität Zürich, Winterthurerstrasse 190,
8057 Zurich, Switzerland
e-mail: soluyanov@phys.ethz.ch

© Springer Nature Switzerland AG 2018
D. Bercioux et al. (eds.), *Topological Matter*, Springer Series in Solid-State Sciences 190, https://doi.org/10.1007/978-3-319-76388-0_3

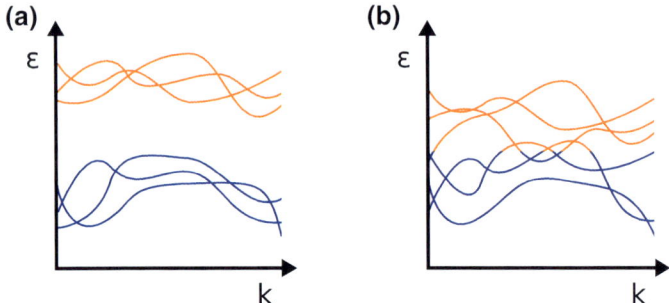

Fig. 3.1 **a** Bandstructure of an insulating material. Occupied (blue) and unoccupied (orange) states are separated by an energy gap for all **k**. **b** Bandstructure of a conducting material. Occupied and unoccupied states touch, and some bands are partially occupied

3.1.1.1 A Short Reminder on Band Theory

We will start with a short reminder about band theory. A more thorough description of the subject can be found in any solid-state physics textbook.

In the non-interacting limit, electronic states in crystalline materials can be described by a single-particle Hamiltonian $H(\mathbf{k})$, which is a smooth function of the crystal wave-vector **k**. The possible electronic states are given by the solutions of the time-independent Schrödinger equation

$$H(\mathbf{k}) \left|\psi_{n,\mathbf{k}}\right\rangle = \epsilon_{n,\mathbf{k}} \left|\psi_{n,\mathbf{k}}\right\rangle. \tag{3.1}$$

These so-called *Bloch states* $\left|\psi_{n,\mathbf{k}}\right\rangle$ are a superposition of plane waves with wave-vector **k**. As such, they can be written as

$$\left|\psi_{n,\mathbf{k}}\right\rangle = e^{i\mathbf{k}\cdot\mathbf{r}} \left|u_{n,\mathbf{k}}\right\rangle, \tag{3.2}$$

where $\left|u_{n,\mathbf{k}}\right\rangle$ is cell-periodic. This property is known as the Bloch theorem [3]. The energy eigenvalues $\epsilon_{n,\mathbf{k}}$ are called *energy bands*, with n being their band index.

The bulk properties of materials are determined largely by their bandstructure. For example, a material is insulating if there is an energy gap between the eigenstates which are occupied by electrons, and those that are empty, as shown in Fig. 3.1a. Conversely, a material is conducting if there is no such energy gap, such as when an energy band is only occupied for certain values of **k**.

It turns out, however, that characterizing materials by their bandstructure does not fully capture their physical properties. Instead, taking into account the shape of the Bloch states $\left|\psi_{n,\mathbf{k}}\right\rangle$ leads to a topological classification of materials.

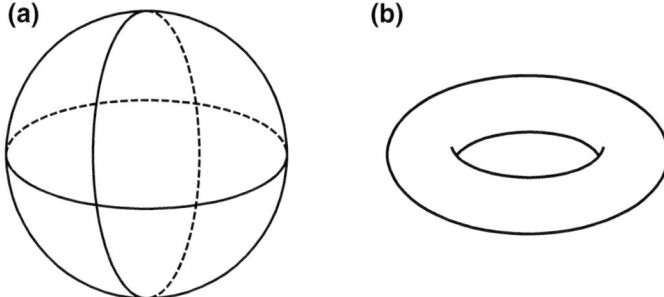

Fig. 3.2 Examples of closed orientable surfaces: **a** A sphere has no holes, and **b** A torus has one hole

3.1.1.2 Topological Properties

To motivate the concept of topological classification, we first show an example from its mathematical origins in geometry: Closed, orientable two-dimensional surfaces can be classified by their number of holes, called *genus*. A sphere, for example, has no holes, while a torus has exactly one (see Fig. 3.2). This property is conserved under smooth deformations of the surface. The only way to add or remove a hole is by tearing and gluing the surface. The genus is an example for a topological invariant – a quantized property that cannot be changed without changing the topological phase. For this reason, topological invariants are commonly used to *identify* topological phases.

In order to define topological phases for materials, we need a geometric object on which the topological properties can be defined. For this purpose, we pick a set of bands B. A very common choice for B is to pick the occupied subspace.[1] The set of states $\{|u_{n,\mathbf{k}}\rangle\}_{n \in B}$ span a vector space $V_\mathbf{k}$ (over $\mathbb{C}$) for each $\mathbf{k}$. If $V_\mathbf{k}$ is a smooth function of $\mathbf{k}$ and the space where $\mathbf{k}$ itself is defined is a manifold, this defines a so-called fiber bundle.

A simple geometrical example of a fiber bundle is given by a one-dimensional vector space defined on a circle. If the vector space is orthogonal to the plane described by the circle, the resulting object is a cylinder, as shown in Fig. 3.3a. If, however, the basis vector rotates by π as it goes around the circle, the resulting object is a Möbius strip. These two objects cannot be smoothly transformed into each other, making them topologically distinct.

---

[1] This is not always possible, for example in the case of semimetals where the occupation number changes with $\mathbf{k}$. In these cases, one often picks the N lowest energy bands instead.

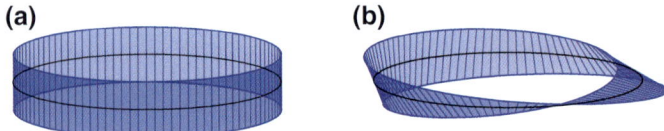

Fig. 3.3 a A cylinder, spanned by a vector which does not rotate as it goes around a circle. **b** A Möbius strip, spanned by a vector which rotates by π as it goes around a circle

3.1.1.3 Bulk-Edge Correspondence

In the previous section, the fact that the vector space $V_\mathbf{k}$ needs to be a smooth function of $\mathbf{k}$ was mentioned. This has a profound impact on the physical properties of topological states, as we shall now see.

Even though the Hamiltonian $H(\mathbf{k})$ is a smooth function of $\mathbf{k}$, the same is not necessarily true for $V_\mathbf{k}$. Consider the following one-dimensional example:

$$H(k) = -\cos(k)\,|a\rangle\langle a| + \cos(k)\,|b\rangle\langle b|, \tag{3.3}$$

where $|a\rangle$ and $|b\rangle$ are two arbitrary orthogonal states. For $k = 0$, the energy eigenvalues of $|a\rangle$ and $|b\rangle$ are -1 and 1, respectively. Consequently, $|a\rangle$ has band index 1, while $|b\rangle$ has index 2. As k changes, the energy eigenvalues shift until they are equal at $k = \pi/2$. At this point, the vector space $V_\mathbf{k} = \mathrm{span}\left(\{|u_{n,\mathbf{k}}\rangle\}_{n\in\{1\}}\right)$ switches discretely from being spanned by $|a\rangle$ to being spanned by $|b\rangle$. As a result, this space does not meet the criteria for topological categorization.

The smoothness of the vector space $V_\mathbf{k}$ can be broken if the order of energy eigenvalues between the states which are in the set B and those which are not changes. This can easily be avoided if we restrict our possible choice of bands B, such that they are always separated from the other bands by a direct energy gap. In other words, topological properties are defined for *isolated* sets of bands, which form smooth fiber bundles.

Another way to frame this is by looking at the possible transformations that can be done to a material without changing its topological properties. In addition to requiring that these transformations smoothly change the Hamiltonian, we impose that the band gap remains open. This definition leads to a remarkable physical property of topological phases: At the boundaries of topologically non-trivial insulating materials, stable conducting edge states must form. In going from the bulk of the topological material to vacuum, the system undergoes a smooth transition from a non-trivial to a trivial (vacuum) state. To allow for this, the aforementioned condition that the bands are separated in energy must be broken. This effect is known as the bulk-boundary correspondence, and variations of this effect govern the interesting transport phenomena to be found in many topological materials [4–6].

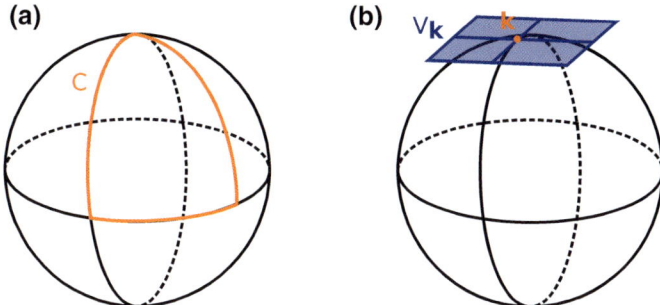

Fig. 3.4 **a** A closed path C on the surface of a sphere. **b** The tangential vector space $V_\mathbf{k}$ for a given point $\mathbf{k}$ on a sphere

Fig. 3.5 Parallel transport of a vector on a closed path on a sphere rotates the vector by an angle ϕ

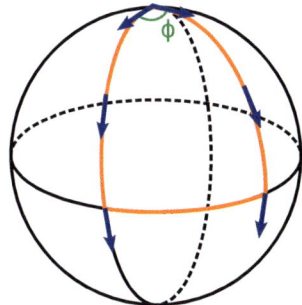

3.1.2 Defining the Chern Number

In the previous section, we have seen how topological properties in crystalline materials are defined on a conceptual level. Now, we will show an example for a topological invariant, which can be used to classify many topological phases in matter.

3.1.2.1 The Berry Phase and Chern Invariant

The basis for defining a topological invariant for electronic bands is the notion of a *geometric phase*. To illustrate this phase, imagine a closed loop C on a manifold. As an example, we choose a closed loop on a sphere, as shown in Fig. 3.4a. Adding the plane tangential to the sphere at each point gives us a fiber bundle (see Fig. 3.4b).

Now we choose a vector in the tangential space and move it along C in such a way that it remains locally parallel to itself, as shown in Fig. 3.5. This process is called *parallel transport*. We observe that the vector is rotated by some angle ϕ as it traverses the path C. Since this angle depends only on the geometry of the fiber bundle, it is called a geometric phase.

For electronic bands, such a phase, known as Berry's phase,[2] can be written as [7]

$$\gamma_C = i \oint_C \sum_{n \in B} \langle u_{n,\mathbf{k}} | \nabla_{\mathbf{k}} | u_{n,\mathbf{k}} \rangle . d\mathbf{k}, \quad (3.4)$$

where C is a closed loop in reciprocal space. Unlike the example above, the Berry's phase represents a rotation in the complex phase of a vector, not its real-space direction.[3] It is Gauge invariant up to multiples of 2π [7]. By defining the Berry potential

$$\mathscr{A}(\mathbf{k}) = i \sum_{n \in B} \langle u_{n,\mathbf{k}} | \nabla_{\mathbf{k}} | u_{n,\mathbf{k}} \rangle, \quad (3.5)$$

the Berry phase can be rewritten as

$$\gamma_C = \oint_C \mathscr{A}(\mathbf{k}) . d\mathbf{k}. \quad (3.6)$$

Note that unlike the Berry phase, the Berry potential is not a Gauge-invariant quantity. If the Berry potential is a smooth function of $\mathbf{k}$ (an important prerequisite, as we shall see soon), we can use Stokes' theorem to rewrite the Berry phase as a surface integral

$$\gamma_C = \int_S \nabla_{\mathbf{k}} \wedge \mathscr{A}(\mathbf{k}) . d\mathbf{k}, \quad (3.7)$$

where $C = \partial S$. Introducing the *Berry connection*

$$\mathscr{F} = \nabla_{\mathbf{k}} \wedge \mathscr{A}(\mathbf{k}), \quad (3.8)$$

which is again Gauge invariant, we can write this as

$$\gamma_C = \int_S \mathscr{F}(\mathbf{k}) . d\mathbf{S}. \quad (3.9)$$

For a closed, orientable two-dimensional surface S in reciprocal space, we can now define the Chern invariant as [8, 9]

$$C = \frac{1}{2\pi} \int_S \mathscr{F}(\mathbf{k}) . d\mathbf{S}. \quad (3.10)$$

Since the edge of a closed surface is a trivial path, (3.6) seems to suggest that the Chern number is always zero. However, we must now remember that (3.7) and (3.9)

---

[2] For simplicity, we consider the *total* Berry phase of all bands. The Berry phase can also be defined for a single band, in which case the sum over bands is dropped.

[3] To see this, try calculating the Berry phase for $|u_k\rangle = e^{ik/2} \begin{pmatrix} \cos(k) \\ \sin(k) \end{pmatrix}$, for $k \in [0, 2\pi]$.

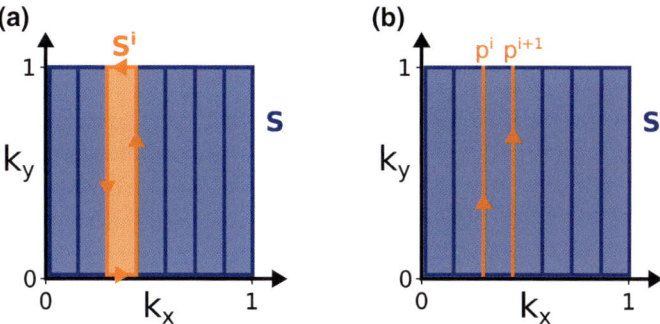

Fig. 3.6 **a** The surface S is divided into segments S^i. For each segment, the flux of Berry connection can be calculated from the Berry phase around its boundary. **b** The top and bottom paths of each boundary cancel, leaving paths p^i which cross the Brillouin zone at a constant k_x

are valid only if the Berry potential $\mathscr{A}(\mathbf{k})$ is smooth. Previously, we discussed that $V_{\mathbf{k}}$ spanned by $|u_{n,\mathbf{k}}\rangle$ must be a smooth function of $\mathbf{k}$ if we wish to define a topological classification. However, there can still be a winding in the phase of $|u_{n,\mathbf{k}}\rangle$ which makes the Berry potential non-smooth. As a result, the Chern number can take any integer value. In fact, the presence of a nonzero Chern number can be viewed as a topological obstruction to finding a globally smooth Gauge [10, 11].

3.1.2.2 The Chern Number as Change in Berry Phase

Having defined the Chern number in terms of the cell-periodic states $|u_{n,\mathbf{k}}\rangle$, we will now show an alternative form that is easier to calculate numerically and is used within the Z2Pack code [12]. For simplicity, we will look at the example where S is the Brillouin zone $\mathbf{k} \in [0, 1)^2$ of a two-dimensional material, in reduced coordinates. The results are equally applicable to other closed two-dimensional surfaces.

We divide the surface integral (3.10) for the Chern number into small segments S^i, as shown in Fig. 3.6a. The segments should be small enough that

$$C^i_{\text{part.}} = \frac{1}{2\pi} \int_{S^i} \mathscr{F}(\mathbf{k}) . \mathrm{d}\mathbf{S} \tag{3.11}$$

is much smaller than one. The Chern number is then given as the sum of all segment integrals,

$$C = \sum_i C^i_{\text{part.}} \tag{3.12}$$

Since $\mathscr{A}(\mathbf{k})$ can be made to be *locally* smooth [13, 14], we can use Stokes' theorem to obtain

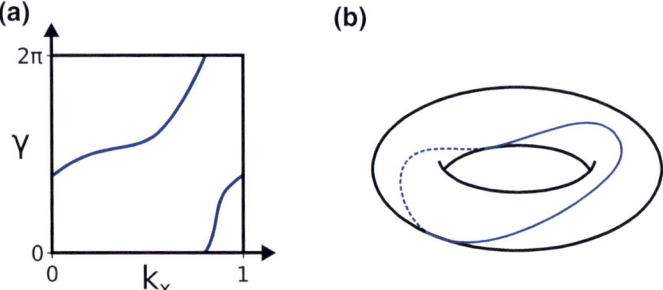

Fig. 3.7 a The Berry phase γ as a function k_x for an example system with $C = 1$. **b** Because both k_x and γ are periodic, the Chern number can be seen as the winding number of the Berry phase around a torus

$$C^i_{\text{part.}} \mod 1 = \frac{1}{2\pi} \int_{\partial S^i} \mathscr{A}(\mathbf{k}).d\mathbf{k} \mod 1 = \frac{\gamma_{\partial S^i}}{2\pi} \mod 1, \quad (3.13)$$

where the modulus comes from the fact that the Berry phase is defined only modulo 2π. Since we imposed that $C^i_{\text{part.}}$ must be much smaller than one, we can still uniquely determine its value from $\gamma_{\partial S^i}/2\pi$ by adding an integer that minimizes the absolute value. Since the top and bottom parts of ∂S^i cancel out due to periodicity, we can write the Berry phase as

$$\gamma_{\partial S^i} = \gamma_{p^{i+1}} - \gamma_{p^i}, \quad (3.14)$$

where p^i and p^{i+1} are the paths at either side of the segment S^i, as shown in Fig. 3.6b. The Berry phase can also be understood as a function of k_x, since each path p^i is given by a fixed k_x. Because both γ and k_x are periodic, the Berry phase describes a line on a torus, as shown in Fig. 3.7. The winding number of this line around the torus is exactly the Chern number [15]. In other words, the Chern number can be calculated by *continuously* tracking the Berry phase on lines of constant k_x as it goes across the Brillouin zone. In practice, enforcing this continuity is a difficult task and is the goal of the convergence options discussed in Sect. 3.2.4.

3.1.2.3 Wilson Loop and Hybrid Wannier Charge Centers

The problem of calculating the Chern number is now reduced to calculating the Berry phase for closed loops in the Brillouin zone. This can be done by calculating the so-called Wilson loop [16] $W(C)$. The Wilson loop can be understood as a matrix that maps the states at a starting point $\mathbf{k}_0$ along the loop onto their images after parallel transport along C. For a discretization $\{\mathbf{k}_0, \ldots, \mathbf{k}_{n-1}, \mathbf{k}_n = \mathbf{k}_0\}$ of the path C, the Wilson loop can be approximated as [12, 16]

$$W(C) = M^{\mathbf{k}_0,\mathbf{k}_1} \cdot \ldots \cdot M^{\mathbf{k}_{n-1},\mathbf{k}_n}, \quad (3.15)$$

where

$$M_{m,n}^{\mathbf{k}_i,\mathbf{k}_j} = \langle u_{m,\mathbf{k}_i} | u_{n,\mathbf{k}_j} \rangle \quad (3.16)$$

are the overlap matrices between Bloch functions at different $\mathbf{k}$. The eigenvalues λ_i of the Wilson loop are connected to the total Berry phase by [17]

$$\gamma_C = \sum_i \arg \lambda_i. \quad (3.17)$$

This reflects the fact that each λ_i is the rotation angle that is acquired by an eigenstate of the Wilson loop as it traverses the path C. Since the overlap matrices M can be readily computed, this gives a method for calculating the Chern number numerically. Of course, the convergence of the Wilson loop eigenvalues with respect to the discretization of C needs to be accounted for, which will be discussed in Sect. 3.2.4.

Another, equivalent, approach to calculating the Berry phase is by computing so-called hybrid Wannier charge centers [18, 19]. This method is based on the notion of Wannier orbitals, which are given by Fourier transforming the Bloch states:

$$|\mathbf{R}n\rangle = \frac{V}{(2\pi)^d} \int_{BZ} e^{-i\mathbf{k}\cdot\mathbf{R}} |\psi_{n,\mathbf{k}}\rangle \, d\mathbf{k}, \quad (3.18)$$

where d is the dimensionality of the system, and V is the unit cell volume. The resulting orbitals are localized, in contrast to the extended nature of the Bloch waves. Since the Bloch states that are used to construct Wannier orbitals can be changed by a Gauge transformation, the same is true for the Wannier orbitals. Their properties, in particular the localization and position in real space, depend sensitively on this choice of Gauge [20]. For the purposes of computing topological invariants, we introduce *hybrid* Wannier orbitals [19, 21], which are Fourier transformed only in one spatial direction and remain extended in the others:

$$|R_x, k_y, k_z; n\rangle = \frac{a_x}{2\pi} \int_{-\pi/a_x}^{\pi/a_x} e^{-ik_x R_x} |\psi_{n,\mathbf{k}}\rangle. \quad (3.19)$$

The average position of such an orbital can be thought of as a function of the remaining reciprocal space variables:

$$\bar{x}_n(k_y, k_z) = \langle 0, k_y, k_z; n | \hat{x} | 0, k_y, k_z; n \rangle. \quad (3.20)$$

This quantity known as the hybrid Wannier charge center (HWCC) is directly related to the Berry phase:

$$\gamma_C = \frac{2\pi}{a} \sum_n \bar{x}_n, \quad (3.21)$$

where C is the path along which the hybrid Wannier orbitals were Fourier transformed. Moreover, if the Gauge is chosen such that these hybrid Wannier orbitals are maximally localized, the individual HWCC corresponds to the eigenvalues of the Wilson loop [12]

$$\bar{x}_i = \frac{2\pi}{a} \arg(\lambda_i), \quad (3.22)$$

up to possible reordering.

This equivalence between hybrid Wannier charge centers and the Berry phase gives rise to a physical interpretation of the Chern number C. As the momentum (k_x, in the case of Fig. 3.7) is varied across the Brillouin zone, the average position of the electrons in the orthogonal direction can change. Due to the periodicity of k_x, it must come back to the same position within the unit cell, but it can change into a different unit cell. This represents a *charge pumping* process, where each cycle of k_x moves the charge by C unit cells.

3.2 The Z2Pack Code

Having defined the Chern number and how it can be calculated in theory, we will now see how this knowledge can be applied in practice. First, we will give a brief overview of the Z2Pack code, introducing the necessary components for calculating Chern numbers. Next, we will show two examples, the Haldane model of a Chern insulator and the Weyl semimetal. Finally, we conclude this section with a discussion of the convergence options available in Z2Pack.

3.2.1 Introduction to the Code

Z2Pack is a Python [22] library which provides functionality for computing topological invariants. A basic knowledge of the Python language is required for using the code. For this, the reader is referred to the many excellent Python tutorials available online, in particular the official Python tutorial [23]. In the following, we will give a short introduction to using Z2Pack. For a more detailed description of the classes and functions described here, the reader may wish to consult the online documentation at www.z2pack.ethz.ch/doc.

In order to calculate the Chern number with Z2Pack, two inputs are needed: a description of the material (system) and a parametrization of the surface on which the invariant should be calculated. These inputs are passed to a function which calculates the hybrid Wannier charge center evolution across the surface. Optionally, this function can regularly save its progress to a file, to allow restarting aborted calculations. The result of this calculation can then be used to evaluate the Chern number or other topological invariants and create plots. Figure 3.8 shows an overview of this process and the modules involved in each step. We will now describe these steps in a bit more detail and show some example code.

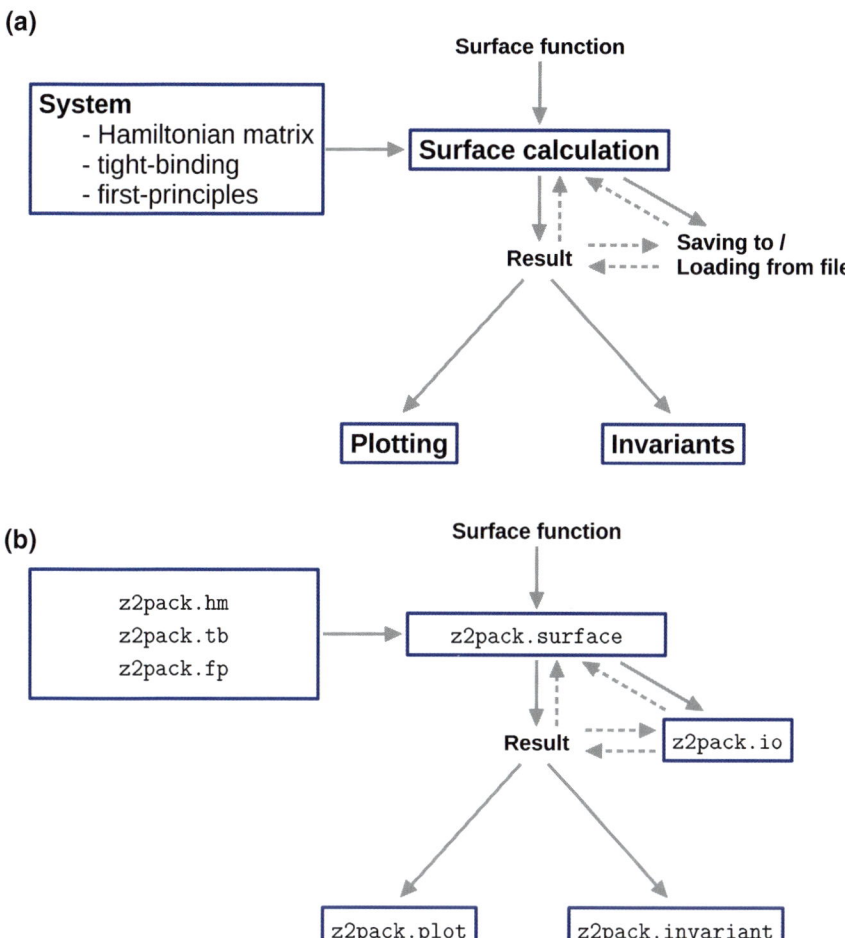

Fig. 3.8 a Overview of the process for calculating topological invariants for a reciprocal space surface of a given material. **b** Python modules corresponding to each of the steps in calculating topological invariants

The system for which topological invariants are to be calculated can be given in three different ways: First, it can be defined as an explicit function $H(\mathbf{k})$ describing the Hamiltonian matrix. This is useful for theoretical models, or when using the **k.p** approximation. Listing 3.1 shows how such a system can be created using the z2pack.hm.System class. The first, required, input is a function that takes **k** and returns the corresponding matrix $H(\mathbf{k})$. An optional keyword argument bands can be passed to the class, to describe which band indices the topological invariant should be calculated for. It can be given either as an integer N, such that the lowest N bands will be taken into account, or as an explicit list of band indices. As is customary in

Python, the lowest band has index 0. By default, the lower half of all bands are taken into account.

```
import z2pack

def hamilton(k):
    ...
    # return Hamiltonian matrix for k

system = z2pack.hm.System(hamilton)

# Choose which bands are taken into account
# by specifying the 'bands' keyword

# lowest 2 bands
system = z2pack.hm.System(hamilton, bands=2)

# first and third band
system = z2pack.hm.System(hamilton, bands=[0, 2])
```

Listing 3.1 Example code for creating a `System` class defined with an explicit Hamiltonian matrix.

Second, the system can be given as a tight-binding model. For this, the TBmodels package is used,[4] which allows for defining tight-binding models either manually or from the output of the Wannier90 [24, 25] code. This is shown in Listing 3.2. For more details about how to construct the tight-binding model, we refer to the TBmodels documentation: www.z2pack.ethz.ch/tbmodels.

```
import tbmodels

# Create tight-binding model
model = tbmodels.Model(...)

# Example: Model from Wannier90 output file
model = tbmodels.Model.from_hr_file('wannier90_hr.dat')

system = z2pack.tb.System(model)
```

Listing 3.2 Creating a tight-binding system by using the `tbmodels.Model` class.

Finally, the system can be given as a first-principles calculation. As we have seen in Sect. 3.1.2.3, the overlap matrices $M^{\mathbf{k}_i,\mathbf{k}_j}$ between states at different k-points along a path are needed to calculate the hybrid Wannier charge centers. Z2Pack makes use of the fact that the Wannier90 code [24, 25] also requires these as an input. As a result, Z2Pack is in principle compatible with all DFT codes which interface to Wannier90, and the user needs to create the same input files as for running Wannier90. Since Z2Pack needs to dynamically call the first-principles code for different k-points, a function with which the k-point input can be created also needs to be supplied. For

---

[4]TBmodels was initially developed as part of Z2Pack, but later separated because it can be used outside of the scope of calculating topological invariants.

3 Calculating Topological Invariants with Z2Pack

some codes, this is implemented in the z2pack.fp.kpoint module. Listing 3.3 shows how a first-principles system is defined to be used with VASP [26].

```
system = z2pack.fp.System(
    input_files=[
        'INCAR', 'POSCAR', 'POTCAR', 'wannier90.win'
    ],
    kpt_fct=z2pack.fp.kpoint.vasp,
    kpt_path='KPOINTS',
    command='mpirun $VASP >& log'
)
```

Listing 3.3 Defining a first-principles system for use with the VASP code.

Apart from the system, the only other input required for running a calculation is the surface on which the Chern number should be evaluated. This is simply given as a function

$$f : [0, 1]^2 \longrightarrow \mathbb{R}^d \qquad (3.23)$$
$$(s, t) \longmapsto \mathbf{k}$$

which parametrizes the surface. Listing 3.4 shows a simple example for a surface function. It is important to note here that $\mathbf{k}$ should be given in reduced coordinates $\mathbf{k} \in [0, 1)^d$. The reason for this is that it simplifies many things, for example, checking if the surface is a closed one.

```
# Defining an explicit function
def surface(s, t):
    return [s, t, 0]

# Equivalent expression using a lambda
surface = lambda s, t: [s, t, 0]
```

Listing 3.4 Two ways of defining a simple surface across the BZ at $k_z = 0$.

Given a system and surface, the hybrid Wannier charge centers can be calculated by calling the z2pack.surface.run function (see Listing 3.5). The return value of this function can then be passed to the z2pack.invariant.chern function to evaluate the Chern number. A simple plot of the sum of HWCC can be created by passing the result to the z2pack.plot.chern function. Since the plotting functionality is based on the popular matplotlib [27] library, the appearance of the plots can be fully customized using matplotlib commands.

```
result = z2pack.surface.run(
    system=system,
    surface=lambda s, t: [s, t, 0]
)

# Evaluate Chern number
z2pack.invariant.chern(result)
```

```
9   # Plot sum of HWCC
10  fig = z2pack.plot.chern(result)
11  fig.show()
```

Listing 3.5 Example code for calculating the hybrid Wannier charge centers, evaluating the Chern number and creating a simple plot.

The result object created by the `run` method can be saved into a file using the `z2pack.io.save` method (see Listing 3.6). To retrieve the stored object, the `z2pack.io.load` method can be used.

```
1   result = ...
2
3   # saving
4   z2pack.io.save(result, 'file_path.json')
5
6   # loading
7   result = z2pack.io.load('file_path.json')
```

Listing 3.6 Saving and loading Z2Pack results to a file.

Since the `run` calculation might take a while—especially for first-principles calculations—it is sometimes necessary to restart from an unfinished calculation. For this purpose, the `save_file` keyword can be specified when calling `run`, which means that the result will periodically be saved into the given file. If the `load` flag is set to `True`, the code will check for an existing result in that file before starting the calculation (see Listing 3.7). One needs to be careful, however, not to load old results the system or surface has changed. Another way to restart calculations is by explicitly passing a result, using the `init_result` keyword.

```
1   # Restart from file
2   result = z2pack.surface.run(
3       system=system,
4       surface=surface,
5       save_file='file_path.json',
6       load=True
7   )
8
9   # Restart from result
10  result2 = z2pack.surface.run(
11      system=system,
12      surface=surface,
13      init_result=result
14  )
```

Listing 3.7 The `run` method can be restarted either from a result saved in a file, or by explicitly passing a result.

Finally, during the `run` call, Z2Pack continuously writes information about the current status to the console. Depending on the use case, this might be an unwanted distraction. Since Z2Pack uses the Python standard module `logging` for this purpose, its verbosity can easily be changed by setting the so-called *level* of the messages that will be printed, as shown in Listing 3.8. The log level describes the severity of a

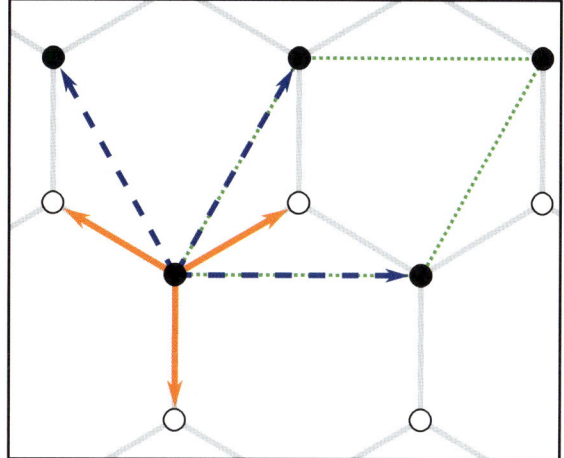

Fig. 3.9 A honeycomb lattice, with A and B sites marked with filled and empty circles, respectively. The unit cell is marked with a dotted green line. Nearest neighbors are indicated with a solid orange arrow, and next-nearest neighbors with a dashed blue arrow

given message. In Z2Pack, the two levels `logging.INFO` (for general messages) and `logging.WARNING` (for convergence issues) are used. Only messages which are at least as severe as the current level will be shown.

```
1  import logging
2
3  # Only show messages with at least 'WARNING' level importance
4  logging.getLogger('z2pack').setLevel(logging.WARNING)
```

Listing 3.8 By setting the log level, the messages printed by Z2Pack can be filtered by severity.

3.2.2 The Haldane Model [5]

The Haldane model [4] is a simple theoretical model for a Chern insulator. It describes two interleaved sub-lattices forming a honeycomb lattice, as shown in Fig. 3.9. The two sub-lattices have opposite on-site energies $\pm M$. Nearest- and next-nearest-neighbor hopping terms are included with strength t_1 and t_2, respectively. In order to break time-reversal symmetry, a microscopic magnetic field is introduced, adding a phase ϕ to the next-nearest-neighbor hopping. The full Hamiltonian of the system is given by

---

[5]Figures and Text in This Section Are Partly Copied from Previous Work of the Authors [28].

$$H(\mathbf{k}) = 2t_2 \cos\phi \left(\sum_i \cos(\mathbf{k}.\mathbf{b}_i)\right) \mathbb{I} + t_1 \sum_i \left[\cos(\mathbf{k}.\mathbf{a}_i)\sigma^x + \sin(\mathbf{k}.\mathbf{a}_i)\sigma^y\right] \quad (3.24)$$
$$+ \left[M - 2t_2 \left(\sum_i \sin(\mathbf{k}.\mathbf{b}_i)\right)\right] \sigma^z,$$

where $\mathbf{a}_i$ and $\mathbf{b}_i$ are the vectors connecting nearest- and next-nearest neighbors (solid orange / dashed blue arrows in Fig. 3.9), respectively, and σ^i are the Pauli matrices.

In this example, we will calculate the Chern number for the Haldane model for a particular value of the parameters M, t_1, t_2, and ϕ. First, we define a function describing the Hamiltonian, as a function of these parameters and $\mathbf{k}$, as shown in Listing 3.9.

```
# Define the Pauli matrices
IDENTITY = np.identity(2, dtype=complex)
PAULI_X = np.array([[0, 1], [1, 0]], dtype=complex)
PAULI_Y = np.array([[0, -1j], [1j, 0]], dtype=complex)
PAULI_Z = np.array([[1, 0], [0, -1]], dtype=complex)

# Define the function H(k)
def Hamilton(k, m, t1, t2, phi):
    kx, ky = k
    k_a = 2 * np.pi / 3. * np.array([
        kx + ky, -2 * kx + ky, kx - 2 * ky
    ])
    k_b = 2 * np.pi * np.array([kx, ky, ky - kx])
    H = (
        2 * t2 * np.cos(phi) *
        sum([np.cos(val) for val in k_b]) * IDENTITY
    )
    H += t1 * sum([np.cos(val) for val in k_a]) * PAULI_X
    H += t1 * sum([np.sin(val) for val in k_a]) * PAULI_Y
    H += m * PAULI_Z
    H -= (
        2 * t2 * np.sin(phi) *
        sum([np.sin(val) for val in k_b]) * PAULI_Z
    )
    return H
```

Listing 3.9 Defining a function that describes the Haldane Hamiltonian.

Next, we set some constants for the parameters M, t_1, t_2, ϕ and create a Z2Pack system from the Hamiltonian function as shown in Listing 3.10. We will take into account only the lower (occupied) band. Because the Hamilton function only takes a two-dimensional $\mathbf{k}$, we specify the dimension using the dim keyword. We can then run a surface calculation for this system.

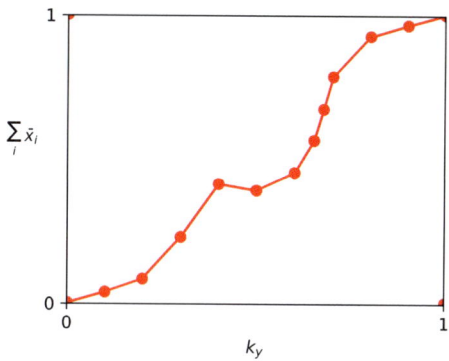

Fig. 3.10 The sum of HWCC as a function of k_x for the Haldane model with $M = 0.1, t_1 = 1, t_2 = 0.2$, and $\phi = \pi/2$. Since the HWCC winds around the torus once in positive direction, the Chern number is $C = 1$

```
# Set the constants for the Haldane model
M = 0.1
T1 = 1.
T2 = 0.2
PHI = 0.5 * np.pi

# Create a Z2Pack system
system = z2pack.hm.System(
    lambda k: Hamilton(k, m=M, t1=T1, t2=T2, phi=PHI),
    bands=1,
    dim=2
)
# Run the surface calculation
result = z2pack.surface.run(
    system=system,
    surface=lambda s, t: [t, s]
)
```

Listing 3.10 Defining a system and running the surface calculation for specific values of the Haldane parameters.

Finally, we evaluate the Chern number and create a figure that shows the HWCC evolution, as shown in Listing 3.11. This produces the image shown in Fig. 3.10. The complete Haldane model example can be seen in Listing 3.12.

```
# Evaluate the Chern number
print('Chern number:', z2pack.invariant.chern(result))

# Create a figure
fig, ax = plt.subplots(figsize=[4, 3])
z2pack.plot.chern(result, axis=ax)
ax.set_xlabel(r'$k_y$')
ax.set_ylabel(
    r'$\sum_i\,\bar{x}_i$', rotation='horizontal', ha='right'
)
ax.set_xticks([0, 1])
ax.set_yticks([0, 1])
```

```
13    fig.savefig('haldane.pdf', bbox_inches='tight')
```

Listing 3.11 Evaluating the Chern number and creating a plot from the calculation result.

```
1   import logging
2
3   import z2pack
4   import numpy as np
5   import matplotlib.pyplot as plt
6
7   logging.getLogger('z2pack').setLevel(logging.WARNING)
8
9   # Define the Pauli matrices
10  IDENTITY = np.identity(2, dtype=complex)
11  PAULI_X = np.array([[0, 1], [1, 0]], dtype=complex)
12  PAULI_Y = np.array([[0, -1j], [1j, 0]], dtype=complex)
13  PAULI_Z = np.array([[1, 0], [0, -1]], dtype=complex)
14
15  # Define the function H(k)
16  def Hamilton(k, m, t1, t2, phi):
17      kx, ky = k
18      k_a = 2 * np.pi / 3. * np.array([
19          kx + ky, -2 * kx + ky, kx - 2 * ky
20      ])
21      k_b = 2 * np.pi * np.array([kx, ky, ky - kx])
22      H = (
23          2 * t2 * np.cos(phi) *
24          sum([np.cos(val) for val in k_b]) * IDENTITY
25      )
26      H += t1 * sum([np.cos(val) for val in k_a]) * PAULI_X
27      H += t1 * sum([np.sin(val) for val in k_a]) * PAULI_Y
28      H += m * PAULI_Z
29      H -= (
30          2 * t2 * np.sin(phi) *
31          sum([np.sin(val) for val in k_b]) * PAULI_Z
32      )
33      return H
34
35  # Set the constants for the Haldane model
36  M = 0.1
37  T1 = 1.
38  T2 = 0.2
39  PHI = 0.5 * np.pi
40
41  # Create a Z2Pack system
42  system = z2pack.hm.System(
43      lambda k: Hamilton(k, m=M, t1=T1, t2=T2, phi=PHI),
44      bands=1,
45      dim=2
46  )
47  # Run the surface calculation
48  result = z2pack.surface.run(
49      system=system,
```

3 Calculating Topological Invariants with Z2Pack   81

```
50          surface=lambda s, t: [t, s]
51      )
52
53      # Evaluate the Chern number
54      print('Chern number:', z2pack.invariant.chern(result))
55
56      # Create a figure
57      fig, ax = plt.subplots(figsize=[4, 3])
58      z2pack.plot.chern(result, axis=ax)
59      ax.set_xlabel(r'$k_y$')
60      ax.set_ylabel(
61          r'$\sum_i\,\bar{x}_i$', rotation='horizontal', ha='right'
62      )
63      ax.set_xticks([0, 1])
64      ax.set_yticks([0, 1])
65      fig.savefig('haldane.pdf', bbox_inches='tight')
```

Listing 3.12 The complete Haldane model example.

3.2.3 Identifying Weyl Semimetals

So far, we have considered the Chern number in the context of insulating materials. From the discussion in Sect. 3.1.1.3, we know that a Chern number can be defined on any closed two-dimensional surface in the Brillouin zone where the bands are gapped. However, in three-dimensional materials, the band gap can still close outside of that specific surface. This can be used to classify topological semimetals, in particular to identify so-called Weyl nodes.

Weyl nodes are linear touching points of two bands in a single point. Their Hamiltonian can locally be described as [29]

$$H(\mathbf{k}) = \sum_{\substack{i \in \{x,y,z\} \\ j \in \{0,x,y,z\}}} A_{i,j}\, k_i\, \sigma^j, \qquad (3.25)$$

where σ^j are the Pauli matrices, and $A_{i,j}$ characterizes the Weyl node. Topologically, Weyl nodes are remarkable because they are a quantized source or sink of Berry connection, depending on their chirality [30]. Since the Chern number measures the flux of Berry connection through a surface, we can determine the chirality of a Weyl node by calculating the Chern number on a sphere enclosing it [12, 31].

Listing 3.13 shows how the Chern number can be calculated for a simple symmetric Weyl node $H(\mathbf{k}) = \sum_i k_i \sigma^i$. The techniques used are the same as for the Haldane example. For defining the surface—a sphere of radius $r = 0.01$—Z2Pack provides a helper function `z2pack.shape.Sphere`, with which a sphere can be defined through its center and radius. The plot created in this example is shown in Fig. 3.11.

```
 1  import numpy as np
 2  import matplotlib.pyplot as plt
 3
 4  import z2pack
 5
 6  # Define Pauli vector
 7  PAULI_X = np.array([[0, 1], [1, 0]], dtype=complex)
 8  PAULI_Y = np.array([[0, -1j], [1j, 0]], dtype=complex)
 9  PAULI_Z = np.array([[1, 0], [0, -1]], dtype=complex)
10  PAULI_VECTOR = list([PAULI_X, PAULI_Y, PAULI_Z])
11
12  def Hamilton(k):
13      """simple 2-band hamiltonian k.sigma with a Weyl point at k=0"""
14      res = np.zeros((2, 2), dtype=complex)
15      for kval, p_mat in zip(k, PAULI_VECTOR):
16          res += kval * p_mat
17      return res
18
19  # Create the System
20  system = z2pack.hm.System(Hamilton)
21
22  # the surface is a sphere around the Weyl point
23  result = z2pack.surface.run(
24      system=system,
25      surface=z2pack.shape.Sphere([0., 0., 0.], 0.01)
26  )
27  print('Chern number:', z2pack.invariant.chern(result))
28
29  # Create plot
30  fig, ax = plt.subplots(figsize=[4, 3])
31  z2pack.plot.chern(result, axis=ax)
32
33  ax.set_xlabel(r'$\theta$')
34  ax.set_xticks([0, 1])
35  ax.set_xticklabels([r'$0$', r'$\pi$'])
36  ax.set_ylabel(r'$\bar{\phi}$', rotation='horizontal')
37  ax.set_yticks([0, 1])
38  ax.set_yticklabels([r'$0$', r'$2\pi$'])
39  ax.set_title(r'$\vec{k}.\vec{\sigma}$')
40  plt.savefig('weyl.pdf', bbox_inches='tight')
```

Listing 3.13 Calculating the Chern number for a simple **k.p** model of a Weyl node.

3.2.4 Convergence Options

In the previous examples, we have used the `surface.run` function without specifying any convergence options. This means that we relied on the default values defined in Z2Pack. While these should work for a wide range of potential applica-

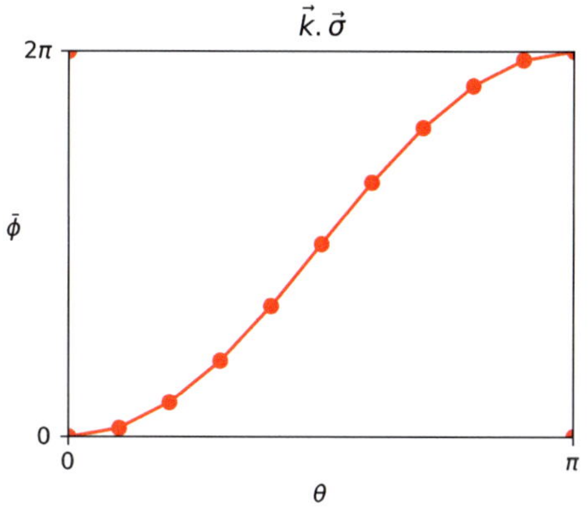

Fig. 3.11 The HWCC on loops around a sphere enclosing a Weyl node, as a function of the altitude angle θ

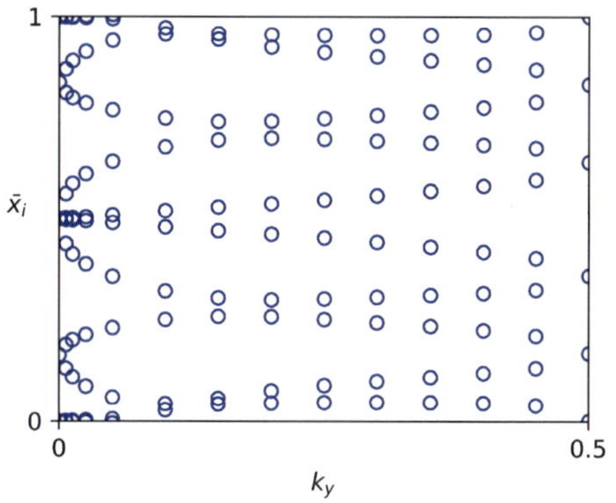

Fig. 3.12 Hybrid Wannier charge center evolution along half of the $k_z = 0$ plane for Bismuth

tions, it is still important to understand the convergence mechanisms and how they can be tuned.

First, we should note that for the convergence criteria described here, the individual hybrid Wannier charge centers are taken into account, not only their sum. Figure 3.12 shows a typical evolution of HWCC $\bar{x}_i$ evaluated at discrete values of k_y. Note that the HWCC is not connected across different k_y, since it is not possible to uniquely identify them.

The first convergence option defined in Z2Pack is convergence with respect to the discretization along the line for which the HWCC is calculated. In rough terms, the number of k-points along the line is increased until the change in HWCC positions is less than a certain threshold, `pos_tol`. How many k-points are used in each step is defined by the `iterator` keyword. This input can be any iterable object (for example a list) of integers. By default, it is set to `range(8, 27, 2)`, meaning that the code will start with eight k-points and then increase in steps of two until 26. If convergence is still not reached after this point, a warning will be generated. In this case, the best course of action is usually to increase the maximum number of k-points and restart the calculation from the previous result.

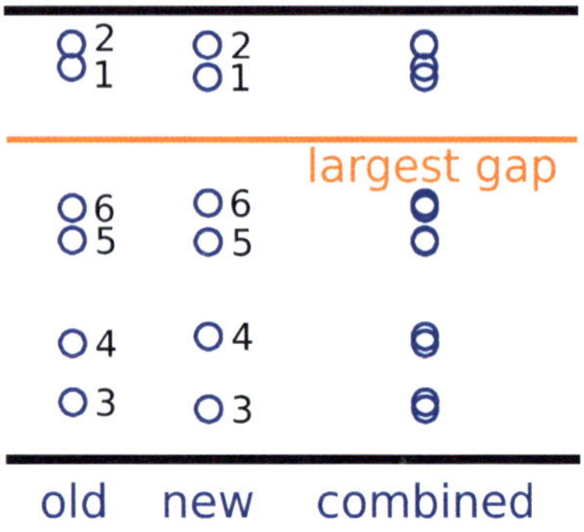

Fig. 3.13 For comparing their positions, HWCC is indexed starting from the largest gap between any two charge centers

old	new	combined
8^2_1	8^2_1	8
$\begin{matrix}6\\5\end{matrix}$	$\begin{matrix}6\\5\end{matrix}$	largest gap $\begin{matrix}\circ\\\circ\end{matrix}$
4	4	$\ominus$
3	3	$\ominus$

One detail that might be worth noting is how the movement of HWCC is calculated. Since, as mentioned above, the HWCC cannot be uniquely identified, we can not simply calculate the movement for each charge center individually. Because a HWCC can cross from 1 to 0 or vice versa, indexing them by their position from zero also does not work. Instead, the HWCC is indexed starting from the *largest gap*[6] between any two HWCCs (when considering both the old and new charge centers), as shown in Fig. 3.13. The positions of HWCC with the same index are then compared, and the maximum of these differences is computed.

In addition to convergence along a single k-point line, convergence in the orthogonal direction needs to be taken into account. This corresponds to the discretization shown in Sect. 3.1.2.2. Using the same technique as before, the movement of HWCC between two neighboring lines is calculated. If it is larger the threshold value `move_tol`, an additional line is added between the two neighbors. To avoid calculations running indefinitely, which could occur if there is a discontinuity in the HWCC spectrum due to a band gap closure, a minimum allowed distance between neighboring lines `min_neighbor_dist` is defined. Again, a warning is issued if convergence cannot be reached. It is important to note that, due to the way the movement is calculated, two HWCCs that exactly exchange places cannot be detected. This can happen in cases where the band gap becomes very small at some point in the Brillouin zone, and the character of the bands changes very rapidly. For such systems, it is important to increase the initial number of lines through the `num_lines` keyword.

Finally, Z2Pack also monitors the distance between the middle of the largest gap in each line and the HWCC positions in neighboring lines (see Fig. 3.14). If the distance is smaller than `gap_tol` times the size of the largest gap, an additional line is again placed between the two neighboring lines. The reason for this additional test should become obvious in the next section, as it is related to how the $\mathbb{Z}_2$ invariant is calculated. Scaling the tolerance with the size of the largest gap is necessary in this case because otherwise the condition can be impossible to fulfill, especially when there are many evenly spaced HWCC.

---

[6]Note that this gap in the HWCC spectrum is not related to the band gap of the energy spectrum.

Fig. 3.14 The minimum distance between the middle of the largest gap (orange diamond) and HWCC (blue circle) in neighboring lines determines whether the `gap_tol` criterion is met

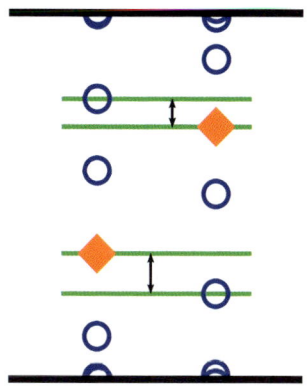

3.3 Time-Reversal Symmetry: $\mathbb{Z}_2$ Classification[7]

In the previous sections, we have seen how an isolated set of bands can be classified topologically according to their Chern number. Now, we will show how this classification can be enriched in the presence of symmetries. In particular, we will show that time-reversal invariant materials can be classified according to a $\mathbb{Z}_2$ index. After a theoretical introduction, we describe how the $\mathbb{Z}_2$ index is computed with Z2Pack. Finally, the example of a tight-binding model in the non-trivial $\mathbb{Z}_2$ phase is shown.

3.3.1 Individual Chern Numbers

In Sect. 3.1.1.2, we have seen that topological phases can be defined on manifolds in reciprocal space, if we choose a set of Bloch functions $\{|u_{n,\mathbf{k}}\rangle\}$ such that they span a smooth vector space $V_\mathbf{k}$. The most convenient way of achieving this smoothness, which we have used so far, is by choosing an isolated set of bands. This leads to a classification into topological states which can only be adiabatically changed by closing the band gap and are characterized by the Chern number. However, choosing isolated bands is by no means the only possible way to create a smooth $V_\mathbf{k}$. For the Hamiltonian of (3.3) for example, we could just pick state $|a\rangle$ everywhere.

Here, we aim to find a more complex topological classification by subdividing the occupied states into smooth parts. In general, if the Hilbert space $\mathcal{H}$ of a given problem can be written as a sum of smooth Hilbert spaces,

$$\mathcal{H} = \bigoplus_i \mathcal{H}_i, \qquad (3.26)$$

---

[7]Figures and text in this section were partly copied from previous work of the authors [12, 28].

then each of the Hilbert spaces has a well-defined Chern number C_i. These *individual Chern numbers* [19] sum together to the Chern number of the full Hilbert space:

$$C = \sum_i C_i. \tag{3.27}$$

However, in general, these individual Chern numbers do not carry much meaning, since the choice how to split up the Hilbert space is arbitrary. In the presence of a symmetry S, however, the Hilbert space can be split up according to the symmetry eigenvalues. For example, consider a mirror symmetry with eigenvalues $\pm i$. On the mirror-symmetric surface, S and $H(\mathbf{k})$ commute. Therefore, the Bloch functions $|u_{n,\mathbf{k}}\rangle$ can be separated into $+i$ and $-i$ eigenstates. Both eigenspaces have a well-defined Chern number:

$$C = C_i + C_{-i}. \tag{3.28}$$

This gives rise to a *symmetry-protected* [32–34] topological classification. Materials can have a zero total Chern number, but nonzero individual Chern numbers. Such a topological phase is protected as long as both the band gap remain open and the symmetry is respected. If the symmetry is broken, a mixing of the two eigenspaces can change the topological phase.

Time-reversal symmetry θ leads to a particularly interesting and well-known topological classification. Unlike spatial symmetries, it is an anti-unitary symmetry and squares to -1 in the spinful case. As a result, the Bloch functions come in so-called Kramers pairs [5, 6]

$$\theta |u_{m,\mathbf{k}}^I\rangle = |u_{m,\mathbf{k}}^{II}\rangle \tag{3.29}$$
$$\theta |u_{m,\mathbf{k}}^{II}\rangle = -|u_{m,\mathbf{k}}^I\rangle.$$

There is a Gauge in which these states are smooth [14, 19], and thus, they have well-defined, opposite [18] individual Chern numbers

$$C_m^I = -C_m^{II}. \tag{3.30}$$

Furthermore, the hybrid Wannier charge centers are related by [18]

$$\bar{x}_m^I(k_y) = \bar{x}_m^{II}(-k_y), \tag{3.31}$$

meaning that they are degenerate for the time-reversal invariant lines $k_y = 0, \pi$. In order to define a topological invariant, we group the states by their pair indices I, II. The two groups then have individual Chern numbers

$$C^I = -C^{II}. \tag{3.32}$$

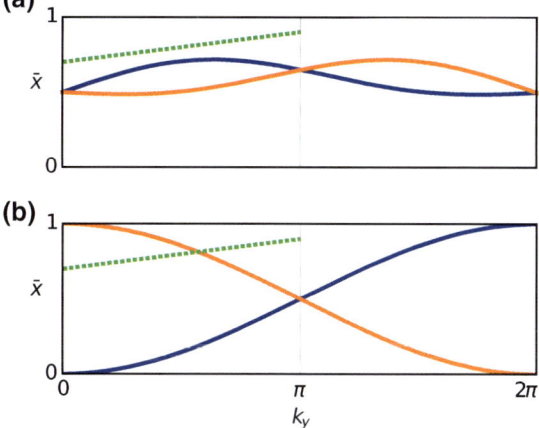

Fig. 3.15 Hybrid Wannier charge centers for a two-band time-reversal invariant system. **a** Trivial phase. The two bands each have a zero individual Chern number. **b** Non-trivial phase. The two bands have individual Chern numbers ±1

However, these Chern numbers are not Gauge invariant. This can be seen by changing the sign of one of the two states:

$$|\tilde{u}_m^{II}\rangle = |u_m^{I}\rangle \qquad (3.33)$$
$$|\tilde{u}_m^{I}\rangle = -|u_m^{II}\rangle$$

These states still obey equation 3.29, and the individual Chern number of each state remains the same. Yet the two states have switched their pair indices. As a result, the Chern number C^{I} is changed by $C_m^{II} - C_m^{I} = 2C_m^{II}$. Since this re-labeling of Kramers pairs can only ever change the Chern numbers by an even number, a topological invariant can be defined as

$$\mathbb{Z}_2 = C_m^{I} \mod 2. \qquad (3.34)$$

In practice, the states do not need to be split by their pair indices to calculate the $\mathbb{Z}_2$ invariant. Instead, we can use the fact that the hybrid Wannier charge centers must be doubly degenerate at the time-reversal invariant momenta. An arbitrary line between zero and π (dotted green line in Fig. 3.15) will cross an even number of HWCC in the topologically trivial case and an odd number in the non-trivial case [18]. This principle is used in Z2Pack to calculate the $\mathbb{Z}_2$ invariant.

When computing the $\mathbb{Z}_2$ invariant numerically, the challenge in using the approach described above is that we cannot uniquely identify hybrid Wannier charge centers. In other words, we do not know how the HWCC connects between two discrete values of k_y. We can get around this issue, however, by cleverly choosing the line $x_{cut}(k_y)$ for which the number of crossings is counted. Since we want a crossing to be as obvious as possible, we choose it to always be in the middle of the largest gap between any two HWCCs, as shown in Fig. 3.16. The number of crossings is then

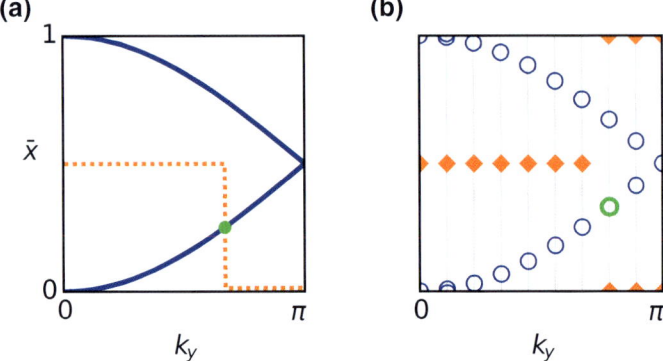

Fig. 3.16 Sketch showing the $\mathbb{Z}_2$ calculation. **a** Continuous case. The HWCC (solid blue line) is crossed exactly once by x_{cut} (dashed orange line), at the green point. **b** Discrete case. The HWCC (blue circles) and middle of the largest gap (orange diamonds) are known only for discrete k_y. Crossings are counted when the HWCC value lies between the largest gaps of the current and previous lines (green circle)

counted by summing up the HWCCs which lie between the current and previous value of the largest gap.

The interface for calculating the $\mathbb{Z}_2$ invariant in Z2Pack is very similar to that for calculating the Chern number. Given the result of a surface calculation, it can be evaluated with the `z2pack.invariant.z2` function. Note that the surface which is used to calculate the HWCC should cover only half the Brillouin zone.

3.3.2 Tight-Binding Example

For the final example in this chapter, we will consider a system of two interpenetrating square lattices A and B each carrying one electron per unit cell, as shown in Fig. 3.17. Let us take into account nearest and next-nearest-neighbor hopping terms, with strength t_1 and t_2, respectively. Each lattice site has two possible states (spin up / down), both carrying equal on-site energies $+1$ for sub-lattice A, and -1 for sub-lattice B.

Including only hopping terms between orbitals of the same spin direction, let the nearest-neighbor hopping terms from sub-lattice A to B have phases $\{1, i, -i, -1\}$ (counter-clockwise) for the spin-up case and its conjugate for the spin down case. Next-nearest-neighbor hopping terms do not carry a phase, but are positive for sub-lattice A and negative for sub-lattice B. The resulting Hamiltonian is

Fig. 3.17 Two inter-penetrating square lattices. The unit cell is shown in green (dotted line). Solid orange arrows connect nearest neighbors, and dashed blue arrows connect next-nearest neighbors

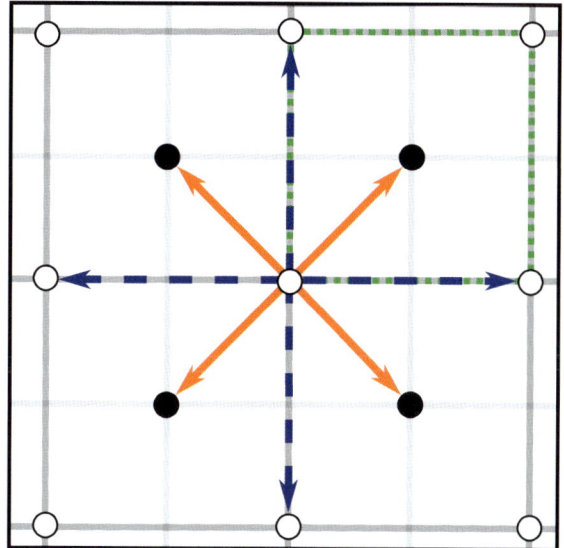

$$H(k_x, k_y) = \left(1 + 2t_2 \left[\cos k_x + \cos k_y\right]\right) \sigma_z \otimes \sigma_0 \quad (3.35)$$
$$- 2t_1 \left[\sin\left(\frac{k_x + k_y}{2}\right) \sigma_y \otimes \sigma_0 + \sin\left(\frac{k_x - k_y}{2}\right) \sigma_x \otimes \sigma_z\right].$$

The tight-binding model is built using the TBmodels code, as shown in listing 3.14. In the constructor of the tmodels.Model, the positions, on-site energies, and occupation number are set. Then, the add_hop method is used to add all hopping terms. Note that the inverse hopping terms (e.g., nearest-neighbor hopping from sub-lattice B to A) are added automatically. The surface calculation is performed in exactly the same way as for the previous examples, except that the surface now only covers half the Brillouin zone. Finally, the z2pack.plot.wcc method is used to plot the HWCC, and z2pack.invariant.z2 is used to calculate the $\mathbb{Z}_2$ invariant. The resulting plot can be seen in Fig. 3.18, showing the non-trivial $\mathbb{Z}_2$ phase.

```
import itertools

import z2pack
import tbmodels
import matplotlib.pyplot as plt

T1, T2 = (0.2, 0.3)

# Create a ''bare'' tight-binding model, with only
# on-site energies.
model = tbmodels.Model(
    on_site=(1, 1, -1, -1),
```

```
        pos=[[0., 0.], [0., 0.], [0.5, 0.5], [0.5, 0.5]],
        occ=2,
    )

    # Add nearest neighbor hopping terms
    for phase, R in zip(
        [1, 1j, -1j, -1],
        itertools.product([0, -1], [0, -1])
    ):
        model.add_hop(
            overlap=phase * T1,
            orbital_1=0,
            orbital_2=2,
            R=R
        )
        model.add_hop(
            overlap=phase.conjugate() * T1,
            orbital_1=1,
            orbital_2=3,
            R=R
        )

    # Add next-nearest neighbor hopping terms
    for R in (
        (r[0], r[1]) for r in itertools.permutations([0, 1])
    ):
        model.add_hop(T2, 0, 0, R)
        model.add_hop(T2, 1, 1, R)
        model.add_hop(-T2, 2, 2, R)
        model.add_hop(-T2, 3, 3, R)

    # Create System instance
    tb_system = z2pack.tb.System(model, dim=2)

    # Run HWCC calculation
    result = z2pack.surface.run(
        system=tb_system, surface=lambda s, t: [t, s / 2.]
    )

    # Create figure
    fig, ax = plt.subplots(figsize=[4, 3])
    z2pack.plot.wcc(result, axis=ax)
    ax.set_xlabel(r'$k_y$')
    ax.set_xticks([0, 1])
    ax.set_xticklabels([r'$0$', r'$\pi$'])
    ax.set_ylabel(r'$\bar{x}_i$', rotation='horizontal')
    ax.set_yticks([0, 1])
    plt.savefig('tb_wcc.pdf', bbox_inches='tight')

    # Calculate Z2 invariant
    print("Z2 invariant:", z2pack.invariant.z2(result))
```

Listing 3.14 Calculating the $\mathbb{Z}_2$ invariant for a tight-binding model of two inter-penetrating square lattices.

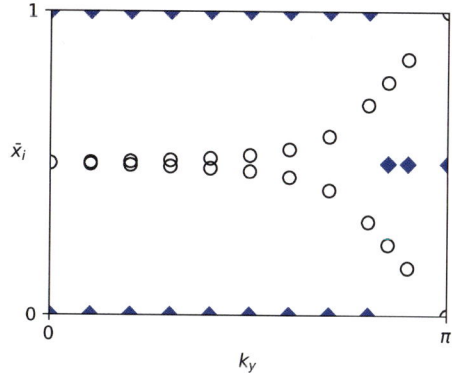

Fig. 3.18 Hybrid Wannier charge center evolution (black circles) and their largest gap (blue diamonds) for the system of two inter-penetrating square lattices, with $t_1 = 0.2$ and $t_2 = 0.3$

Acknowledgements The authors were supported by Microsoft Research, the Swiss National Science Foundation through the National Competence Centers in Research MARVEL and QSIT, and the ERC Advanced Grant SIMCOFE.

References

1. B.A. Bernevig, T.L. Hughes, *Topological Insulators and Topological Superconductors* (Princeton University Press, New Jercy, 2013)
2. M. Nakahara, *Geometry, Topology and Physics* (Taylor and Francis Group, London, 2003)
3. F. Bloch, Über die quantenmechanik der elektronen in kristallgittern. Z. Phys. **52**(7), 555–600 (1929)
4. F.D.M. Haldane, Model for a quantum Hall effect without Landau levels: condensed-matter realization of the parity anomaly. Phys. Rev. Lett. **61**(18), 2015–2018 (1988)
5. C.L. Kane, E.J. Mele, $\mathbb{Z}_2$ topological order and the quantum spin Hall effect. Phys. Rev. Lett. **95**, 146802 (2005)
6. C.L. Kane, E.J. Mele, Quantum spin Hall effect in graphene. Phys. Rev. Lett. **95**, 226801 (2005)
7. J. Zak, Berry's phase for energy bands in solids. Phys. Rev. Lett. **62**(23), 2747–2750 (1989)
8. S.-S. Chern, Characteristic classes of Hermitian manifolds. Ann. Math. 85–121 (1946)
9. D.J. Thouless, M. Kohmoto, M.P. Nightingale, M. den Nijs, Quantized Hall conductance in a two-dimensional periodic potential. Phys. Rev. Lett. **49**(6), 405–408 (1982)
10. T. Thonhauser, D. Vanderbilt, Insulator/Chern-insulator transition in the Haldane model. Phys. Rev. B **74**(23), 235111 (2006)
11. D.J. Thouless, Wannier functions for magnetic sub-bands. J. Phys. C **17**(12), L325 (1984)
12. D. Gresch, G. Autès, O.V. Yazyev, M. Troyer, D. Vanderbilt, B.A. Bernevig, A.A. Soluyanov, Z2Pack: numerical implementation of hybrid Wannier centers for identifying topological materials. Phys. Rev. B **95**, 075146 (2017)
13. A.A. Soluyanov, D. Vanderbilt, Smooth gauge for topological insulators. Phys. Rev. B **85**(11), 115415 (2012)
14. G.W. Winkler, A.A. Soluyanov, M. Troyer, Smooth gauge and Wannier functions for topological band structures in arbitrary dimensions. Phys. Rev. B **93**(3), 035453 (2016)
15. A.A. Soluyanov, Topological Aspects of Band Theory. Ph.D. thesis, Rutgers University-Graduate School-New Brunswick, 2012
16. A. Alexandradinata, X. Dai, B.A. Bernevig, Wilson-loop characterization of inversion-symmetric topological insulators. Phys. Rev. B **89**(15), 155114 (2014)

17. R. Leone, The geometry of (non)-Abelian adiabatic pumping. J. Phys. A Math. Theor. **44**(29), 295301 (2011)
18. A.A. Soluyanov, D. Vanderbilt, Computing topological invariants without inversion symmetry. Phys. Rev. B **83**, 235401 (2011)
19. A.A. Soluyanov, D. Vanderbilt, Wannier representation of $\mathbb{Z}_2$ topological insulators. Phys. Rev. B **83**(3), 035108 (2011)
20. N. Marzari, D. Vanderbilt, Maximally localized generalized Wannier functions for composite energy bands. Phys. Rev. B **56**(20), 12847–12865 (1997)
21. C. Sgiarovello, M. Peressi, R. Resta, Electron localization in the insulating state: application to crystalline semiconductors. Phys. Rev. B **64**(11), 115202 (2001)
22. G. Van Rossum et al., Python programming language. in *USENIX Annual Technical Conference*, vol. 41 (2007)
23. The Python Tutorial, Accessed 14 Jan 2018
24. A.A. Mostofi, J.R. Yates, Y.-S. Lee, I. Souza, D. Vanderbilt, N. Marzari, wannier90: A tool for obtaining maximally-localised Wannier functions. Comput. Phys. Commun. **178**(9), 685–699 (2008)
25. A.A. Mostofi, J.R. Yates, G. Pizzi, Y.-S. Lee, I. Souza, D. Vanderbilt, N. Marzari, An updated version of wannier90: a tool for obtaining maximally-localised Wannier functions. Comput. Phys. Commun. **185**(8), 2309–2310 (2014)
26. G. Kresse, J. Furthmüller, Efficiency of ab-initio total energy calculations for metals and semiconductors using a plane-wave basis set. Comput. Mater. Sci. **6**(1), 15–50 (1996)
27. J.D. Hunter, Matplotlib: a 2D graphics environment. Comput. Sci. Eng. **9**(3), 90–95 (2007)
28. D. Gresch, Identifying Topological States in Matter, Master's thesis, ETH Zurich, Zurich, 2015
29. H.B. Nielsen, M. Ninomiya, The Adler–Bell–Jackiw anomaly and Weyl fermions in a crystal. Phys. Lett. B **130**(6), 389–396 (1983)
30. G. Volovik, Zeros in the Fermion spectrum in superfluid systems as diabolical points. JETP Lett. **46**(2) (1987)
31. A.A. Soluyanov, D. Gresch, Z. Wang, Q. Wu, M. Troyer, X. Dai, B.A. Bernevig, Type-II Weyl semimetals. Nature **527**(7579), 495–498 (2015)
32. A. Alexandradinata, B.A. Bernevig, Berry-phase description of topological crystalline insulators. Phys. Rev. B **93**(20), 205104 (2016)
33. L. Fu, Topological crystalline insulators. Phys. Rev. Lett. **106**(10), 106802 (2011)
34. J.C.Y. Teo, L. Fu, C.L. Kane, Surface states and topological invariants in three-dimensional topological insulators: application to $Bi_{1-x}Sb_x$. Phys. Rev. B **78**, 045426 (2008)

Chapter 4
Transport in Topological Insulator Nanowires

Jens H. Bardarson and Roni Ilan

Abstract In this chapter, we review our work on the theory of quantum transport in topological insulator nanowires. We discuss both normal state properties and superconducting proximity effects, including the effects of magnetic fields and disorder. Throughout we assume that the bulk is insulating and inert, and work with a surface-only theory. The essential transport properties are understood in terms of three special modes: in the normal state, half a flux quantum along the length of the wire induces a perfectly transmitted mode protected by an effective time-reversal symmetry; a transverse magnetic field induces chiral modes at the sides of the wire, with different chiralities residing on different sides protecting them from backscattering; and finally, Majorana zero modes are obtained at the ends of a wire in a proximity to a superconductor, when combined with a flux along the wire. Some parts of our discussion have a small overlap with the discussion in the review [1]. We do not aim to give a complete review of the published literature, instead the focus is mainly on our own and directly related work.

4.1 Overview and General Considerations

Topological insulators (TI's) [2–4] are characterized by their bulk–boundary correspondence: the bulk has a gap that is inverted in comparison with the atomic insulator (vacuum), resulting in a robust metallic state at the surface. The quantum Hall effect,

---

J. H. Bardarson (✉)
Department of Physics, KTH Royal Institute of Technology,
SE-106 91 Stockholm, Sweden
e-mail: bardarson@kth.se

R. Ilan
Raymond and Beverly Sackler School of Physics and Astronomy,
Tel-Aviv University, 69978 Tel-Aviv, Israel
e-mail: ronilan@tauex.tau.ac.il

© Springer Nature Switzerland AG 2018
D. Bercioux et al. (eds.), *Topological Matter*, Springer Series in Solid-State
Sciences 190, https://doi.org/10.1007/978-3-319-76388-0_4

with its Landau levels and chiral edge states, is a good example. In this case, the Hamiltonian has no symmetries (apart from charge conservation) and quantum Hall states are realized in even spatial dimensions. The presence of symmetries allows for symmetry-protected topological phases [5], as long as the symmetry is not broken. The quantum spin Hall effect in 2D is the time-reversal invariant version of the quantum Hall effect, and the metallic surface consists of two counter-propagating helical edge states that are Kramers pairs and therefore not coupled by time-reversal preserving disorder [6–8]. Particle-hole symmetry allows for topological superconductivity in which case the surface states are particle-hole symmetric Majorana zero modes [9, 10].

In this review, we focus on 3D topological insulators protected by time-reversal symmetry [4]. In this case, the surface is 2D and the low energy degrees of freedom comprise an odd number (which we take to be one) of Dirac fermions. A defining feature of topological insulators is reflected in the fact that a single Dirac fermion cannot be localized, no matter how strong the disorder [11, 12]. Instead, disorder always drives the surface in the thermodynamic limit into a metallic phase referred to as the symplectic metal [13, 14]. Interference in the symplectic metal gives rise to weak antilocalization [15], the phenomena that the lowest-order quantum correction to the classical Drude conductivity is positive, leading to an enhanced conductance. While one can understand this as being destructive interference of time-reversed loops due to the Berry phase picked up by the Dirac fermion as it loops around, enforced by spin–momentum locking, it is not a signature of topology—any 2D strongly spin–orbit-coupled metal is, ignoring interactions, symplectic.

This Berry phase and the time-reversal symmetry strongly affect transport properties and are the key effects in the physics we discuss in this chapter. In the presence of a time-reversal symmetry $\mathscr{T}$ that satisfies $\mathscr{T}^2 = -1$, the scattering matrix S that relates incoming modes to outgoing modes in a two-terminal scattering set-up is antisymmetric: $S^T = -S$ [16]. As a consequence, backscattering is forbidden (the diagonal elements of the scattering matrix are zero) and in the presence of an odd number of modes, a perfectly transmitted mode [17] with transmission unity is obtained. In the presence of a perfectly transmitted mode, the conductance, via the Landauer formula, $G \geq e^2/h$, and localization cannot take place. In the field theory of diffusion, this is encoded in the presence of a topological term [13, 14]; all topological insulators and superconductors can in fact be classified according to the presence or absence of a topological term in the corresponding nonlinear sigma model describing diffusion [18, 19].

In the limit of large number of modes, and conductance $G \gg e^2/h$, the distinction between and odd and even number of modes is not important. This observation has been used to argue for the absence of localization in weak topological insulators [20, 21]. Here, it means that in the thermodynamic limit, transport cannot distinguish a 3D topological insulator surface from any regular spin–orbit-coupled metal, and no direct signatures of topology are to be obtained. This is the main motivation for exploring the transport properties of topological insulator nanowires. By reducing the size of the surfaces, the distinction between an even and odd number of modes becomes important and a direct signature of topology can be obtained in the presence of a perfectly transmitted mode at the Dirac point, which results in a quantized conductance of e^2/h [22]. The perfectly transmitted mode requires magnetic flux along the length of the wire [22–25]. A transverse magnetic field induces quantum Hall phases at the top and bottom of a wire, with chiral modes at the sides [26]. These modes and their essential transport properties are discussed in Sect. 4.2.

The odd number of modes is especially important when it comes to superconducting proximity effect: an s-wave superconductor coupled to a Dirac fermion with an odd number of modes can result in topological superconductivity with Majorana zero modes [10, 27]. Such a topological superconducting wire, when coupled with the perfectly transmitted or chiral mode of the normal state, has distinct transport signatures. Due to Béri degeneracy [28], the NS conductance of a normal metal–superconductor interface, in the single mode limit, is either 0 or e^2/h [29]. A magnetic flux along the wire allows to turn the topological superconductivity on and off. These and other related superconducting transport phenomena are discussed in Sect. 4.3.

4.2 Topological Insulator Nanowires: Normal State Properties

Topological insulator nanowires come in many shapes and sizes [30–35]. Their cross sections are commonly rectangular, and the wires look like ribbons. Their bulk is frequently inescapably doped during synthesis and is far from being an ideal insulator. Nevertheless, the surface is a significant contributor to transport, and often the most characteristic features of experimental data are surface features [36]. That, in addition to rapid improvements in the material science and synthesis of wires with more and more insulating bulks, motives us to make the simplifying theoretical assumptions of inert insulating bulk. The metallic surface state is modelled by a single Dirac fermion, which is sharply localized at the surface. By taking into account the effects of disorder, doping and magnetic field on transport of such a surface state, this theory has proven to be sufficiently detailed to describe the essential features of most experiments.

What are the defining properties of a wire? The most important feature is the aspect ratio of circumference P and the length L, which we usually take to be of the order of unity or smaller. In the limit $P/L \gg 1$, transport is independent of boundary conditions [37], and therefore magnetic flux, and the conductance flows into the symplectic metal [11, 12]. A typical circumference is of order 100 nm, which is a couple of orders of magnitude larger than a typical carbon nanotube. This has two important consequences: first, the magnetic field strengths needed to thread a flux quantum through the wire are easily realized experimentally, and second, the energy scale of confinement in the transverse direction $\Delta_P = \hbar v_F 2\pi/P$ is small enough that the number of modes can be tuned by gating, while at the same time temperature can be lowered such that individual modes can be resolved. Similarly, disorder broadening of transverse modes $\Gamma < \Delta_P$ [38]. The transport regime is therefore quasi-one-dimensional, with typically multiple modes taking part in transport but separate transverse modes being resolvable.

In this quasi-1D limit, signatures of topology are visible in transport properties. These features, relying on topology, are insensitive to the detailed geometry of the wires. Most of the time, no significant qualitative changes are observed in the results of transport calculations if one assumes the wires to be perfectly cylindrical instead of having the more realistic rectangular shapes. This assumption simplifies notation and some calculations, and we therefore often make it. A magnetic field transverse to the length of the wire, however, breaks rotational symmetry; a cylindrical shape no longer leads to simplifications and we revert to rectangular shapes.

4.2.1 Band Structure of a Clean Wire

In a compact geometry, such as that of a nanowire, the Berry phase due to spin–momentum locking leaves its hallmarks on the electronic band structure. The spin of a Dirac fermion is locked to the momentum direction and therefore rotates as the momentum goes in a loop. This is what happens when the Dirac fermion encircles the circumference of a nanowire. As a result, with a 2π rotation of a spin giving a minus sign, the wave function is antiperiodic [24, 39]. Alternatively, one can keep the wave function periodic and include a spin connection term in the Hamiltonian [23, 40]; these descriptions are equivalent up to a gauge transformation. Antiperiodicity requires nonzero transverse momentum, and therefore energy, necessitating a gap in the energy spectrum.

We demonstrate this for a simple model of a topological insulator surface state, with the effective Hamiltonian of a single Dirac fermion living on the surface of a cylindrical wire of circumference P. The surface Hamiltonian is given by (we set $\hbar = 1$)

$$H = -i v_F \left[\sigma_x \partial_x + 2\pi/P \sigma_y \partial_s \right] \quad (4.1)$$

which is equivalent to the Hamiltonian of a Dirac fermion in a flat surface, with the crucial difference of antiperiodic boundary condition on the wave function in the compact coordinate $0 \leq s \leq P$: $\psi(s + P) = -\psi(s)$. v_F is the Fermi velocity, σ_x and σ_y the Pauli spin matrices. The wave functions on the cylinder take the form

$$\psi_{k,n} = e^{ikx + i\ell_n s} \chi_{k,n},$$

with $\ell_n = n - 1/2$ and $n \in \mathbb{Z}$. The spinor $\chi_{k,n}$ satisfies $\hat{p} \cdot \sigma \chi_{k,n} = \pm \chi_{k,n}$, with $\hat{p}$ the unit vector in the direction of the momentum. The band structure

$$E_{k,n} = \pm v_F \sqrt{k^2 + (2\pi/P)^2 \ell_n^2}. \quad (4.2)$$

Fig. 4.1 A cylindrical topological insulator nanowire threaded by a coaxial magnetic field, resulting in total flux ϕ through the wire's cross section

is gapped with a finite gap of magnitude $\Delta_P = 2\pi v_F/P$ at $k = 0$. All energy bands are doubly degenerate.

We emphasize that the Hamiltonian given in (4.1) is in fact suitable to account for the physics of the surface states of wires with any (constant) cross section, provided they are strictly two dimensional and uniform. Nevertheless, as already mentioned, realistic systems do not necessarily meet these requirements. For example, they may have slightly different effective Dirac Hamiltonian depending on the surface termination. The band structure of rectangular wires has been studied both analytically and numerically taking into consideration corrections due to such details [41]. The result is qualitatively the same as obtained with the purely two-dimensional surface theory. Therefore, for the rest of this chapter we will rely mostly on the effective Hamiltonian equation (4.1).

4.2.2 Aharonov–Bohm Effect and Magnetoconductance Oscillations

A magnetic flux ϕ threading the wire's cross section, as in Fig. 4.1, results in an Aharonov–Bohm phase for the surface electrons. The flux is included in the Hamiltonian via minimal substitution as an azimuthal vector potential

$$H = v_F \left[-i\sigma_x \partial_x + \sigma_y(-i\partial_s + \eta) 2\pi/P \right], \quad (4.3)$$

where $\eta = \phi/\phi_0$ is the number of flux quanta $\phi_0 = h/e$ through the cross section. By a gauge transformation, the flux can alternatively be absorbed into the boundary conditions as an Aharonov–Bohm phase: $\psi(s+P) = -e^{2\pi\eta}\psi(s)$. The spectrum becomes η dependent

$$E_{k,n}(\eta) = \pm v_F \sqrt{k^2 + (2\pi/P)^2(\ell_n + \eta)^2}, \quad (4.4)$$

and is shown in Fig. 4.2 for different values of η. By construction, the spectrum is periodic in η and repeats whenever η changes by an integer.

For generic values of η all bands are non-degenerate. At integer and half-integer values, instead, all nonlinear bands are degenerate; in the integer case, crucially, there is a single additional non-degenerate linearly dispersing band, corresponding to the

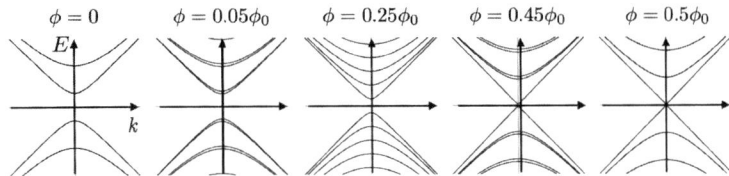

Fig. 4.2 Schematic band structure of a topological insulator nanowire threaded by flux. The spectrum is composed of discrete energy bands with a given angular momentum ℓ_n and varying as a function of the momentum k along the wire. At zero flux, the spectrum is gapped, and all bands are doubly degenerate. At a finite non-integer or half-integer flux, there is no band degeneracy and time reversal is broken at the surface. At half-integer flux, time reversal is effectively restored, and all bands, expect the linearly dispersing one, are doubly degenerate

value of ℓ_n for which $\ell_n + \eta = 0$. At a fixed chemical potential μ, the number of modes at the fermi energy can therefore be modified by tuning the flux. At integer values of η, this number is always even, while at half-integer values it is always odd. The difference $\Delta N = N(\eta = 0) - N(\eta = 1/2)$ in number of modes is $\Delta N = \pm 1$, with the sign depending on the value of the chemical potential; at the Dirac point $\Delta N = -1$.

In a perfectly ballistic wire, the two terminal conductance, according to the Landauer equation, is proportional to the number of modes, $G = (e^2/h)N(\mu, \eta)$. The above considerations then suggest that one should observe Aharonov–Bohm oscillations in the conductance with a period $\Delta \eta = 1$, corresponding to a flux periodicity of $\Delta \phi = \phi_0$, and amplitude $\Delta G = \pm e^2/h$, with a chemical potential dependent sign. Real wires are never perfectly ballistic as there is always some amount of disorder present. However, as long as the disorder-induced level broadening Γ is small compared with the level spacing, $\Gamma < \Delta_P$, the above expectation should hold, with the only modification a reduced amplitude ΔG of the oscillations. This is borne out in numerical calculations [22], which modelled disorder by including a scalar potential $V(\mathbf{x})$ in the Hamiltonian (4.3) and solving for the scattering matrix using the transfer matrix technique described in Sect. 4.4, the results of which are shown in Fig. 4.3. At chemical potentials away from the Dirac point and at weak disorder ($K_0 = 0.2$), a clear ϕ_0-periodic oscillations with a chemical potential dependent sign—determined by wether the blue dotted curve of $\eta = 0.5$ or green solid curve at $\eta = 0$ is higher—are clearly seen. The transport at the Dirac point is dominated by a perfectly transmitted mode discussed in the next section.

In the opposite limit, $\Gamma \gg \Delta_P$, of strongly disordered wires, the discrete structure of the number of modes is replaced by a smoothly increasing density of states. The conductance is no longer given by the simple mode counting argument. To understand the flux dependence of the conductance, we need to consider the symmetries of the Hamiltonian (4.3). Away from integer and half-integer values of η, the time-reversal symmetry $T = i\sigma_y K$, with K complex conjugation, is manifestly broken by the η term. However, time reversal reemerges at half-integral and integral values of η. This is best seen in the representation where we have gauged the flux into

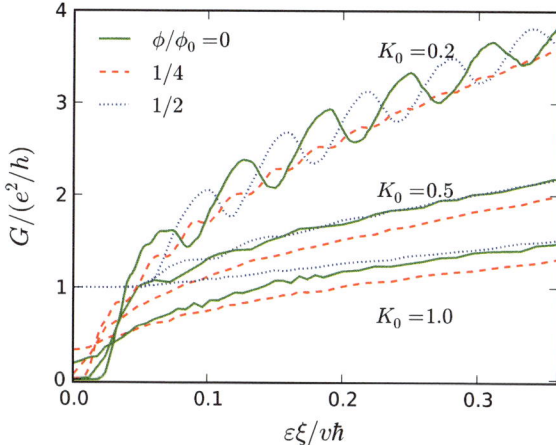

Fig. 4.3 Conductance of a topological insulator nanowire as a function of chemical potential (here denoted by ϵ) for three different values of flux ϕ and three values of the disorder strength K_0. The disorder is Gaussian distributed $\langle V(\mathbf{x})V(\mathbf{x}')\rangle = K_0(\hbar v_F)^2/(2\pi\xi^2)\exp(-|\mathbf{x}-\mathbf{x}'|^2/2\xi^2)$, with ξ the disorder correlation length. The circumference of the wire was taken to be $P = 100\xi$ and the length $L = 200\xi$. Figure taken from [22]

the boundary condition $\psi(s+P) = e^{i(2\pi\eta+\pi)}\psi(s)$; the corresponding Hamiltonian is η independent. The boundary conditions break time-reversal symmetry, except when $\psi(s+P) = \pm\psi(s)$, corresponding to integer or half-integer values of η. In particular, at $\eta = 1/2$, the boundary conditions are periodic allowing for a solution with zero angular momentum and no gap—the linearly dispersing mode.

With strongly overlapping modes, $\Gamma \gg \Delta_P$, and large enough chemical potential such that $G > e^2/h$, the flux dependence of the conductance is determined by weak antilocalization. The conductance in the presence of time-reversal symmetry is enhanced compared with that in the absence of time-reversal symmetry, due to destructive interference between time-reversed loops. Since both half-integer and integer values of η result in time-reversal symmetry, the period of the flux dependent conductance is $\Delta\eta = 1/2$ corresponding to flux period of $\phi_0/2$ (see $K_0 = 0.5$ and $K = 1.0$ curves in Fig. 4.3)—the period is half as large as in the regime of weakly coupled modes.

There is ample experimental evidence for the weakly coupled-mode regime being realized in current topological insulator nanowires [30–36]. The magnetoconductance is found to oscillate with a period of ϕ_0 with an amplitude whose sign can be changed by gating. This remains true even when there is a significant bulk contribution to the conductance. The same period is found in the flux dependence of conductance fluctuations [38].

We have assumed in our discussion uniformly doped wires such that the chemical potential μ is constant. In addition to random variations, the chemical potential can have smooth variations due to the experimental set-up. For example, the top and bottom parts of the wire may have different charge density due to the presence of a substrate. This was studied theoretically and experimentally in the case of HgTe nanowires in [42].

4.2.3 Perfectly Transmitted Mode

The combination of time-reversal symmetry and an odd number of modes, obtained at a half-integral flux η, implies the existence of a perfectly transmitted mode. The two-terminal scattering matrix is antisymmetric $S^T = -S$, and the eigenvalues of the transmission matrix come in degenerate pairs [16]; in the case of an odd number of modes, one eigenvalue is exactly unity—the perfectly transmitted mode. At the Dirac point, the conductance at half-integral flux is therefore quantized at e^2/h, irrespective of the strength or type of disorder, as long as it respects time-reversal symmetry. Away from half-integral flux, the conductance in the ballistic limit drops to zero as a Lorentzian with a peak width $\delta\eta = P/\pi L$ [37], while disorder enhances the conductance [11, 22]. This is evident in the numerical data of Fig. 4.3 where the conductance at $\eta = 0.5$ goes to e^2/h at the Dirac point for all values of disorder.

These transport signatures of the perfectly transmitted mode are unique to topological insulator nanowires. The first theoretical realization of a perfectly transmitted mode was in carbon nanotubes, where an effective symplectic time-reversal symmetry is obtained in the absence of intervalley scattering and trigonal warping [43]. Due to fermion doubling [44, 45], however, the perfectly transmitted modes always come in pairs, and the conductance therefore would be quantized at multiples of $2e^2/h$ instead of the e^2/h that characterizes the topological insulators. Furthermore, the emergent symmetry in the carbon nanotubes is easily broken and the magnetic field strengths needed to obtain a flux of $\eta = 0.5$ are huge due to the small radius of carbon nanotubes.

The effective time reversal at $\eta = 0.5$ requires a constant flux through the wire. Variations in the wire circumference lead to local variations in the flux, breaking the time-reversal symmetry. Random surface ripples combined with disorder result in a reduction of the conductance at the Dirac point that is no longer generally quantized [46]. At larger chemical potentials, the amplitude of the Aharonov–Bohm oscillations of the conductance reduce with increasing magnetic flux, and can, in the case of large surface ripples, completely wash out the oscillations. Experimentally realized wires can be made with a uniform enough surfaces that this effect is small.

In the absence of disorder, the Hamiltonian (4.3) at $\mu = 0$ has a chiral symmetry $\sigma_z H \sigma_z = -H$, which is not broken by the random ripples in the surface. This symmetry places the wire in the AIII symmetry class, which has a $\mathbb{Z}$ topological classification in one dimension. The flux ϕ tunes the wire between topologically distinct insulating states. At the transition between two such phases, the sign of one reflection eigenvalue changes sign, requiring a perfect transmission and quantized conductance [46, 47].

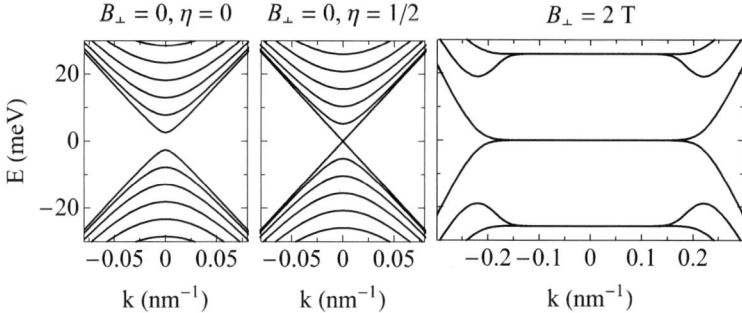

Fig. 4.4 The spectrum of a wire of cross section 40 nm by 160 nm with and without perpendicular field $B_\perp$, with and without a vortex along the wire, as a function of the momentum along the wire. Note that in the last case when $B_\perp$ is nonzero, the spectrum does not depend on the presence of absence of the vortex. Figure taken from [48]

4.2.4 Wires in a Perpendicular Field: Chiral Transport

Apart from the perfectly transmitted mode, there exist other ways in which topologically protected chiral transport can emerge in topological insulator wires or films, which do not rely on time-reversal symmetry, but rather on breaking it. This can be achieved either by coupling the system to magnetism to induce quantum anomalous Hall phase, or by subjecting the system to a strong perpendicular magnetic field inducing the quantum Hall effect. In the latter case, the field is applied perpendicular to the TI surface rather than parallel as in the case of threading flux through the wire. Therefore, the surface state is gapped, rotational symmetry is explicitly broken, and the discussion does not benefit from considering the cylindrical geometry, but rather is easier to carry out for a rectangular wire.

A strong field applied perpendicular to a TI rectangular wire breaks the spectrum of the surface states into Landau levels with unique characteristics stemming both from the Dirac-like behaviour of the particles and from the fact that the surface has no boundary [26]. While the top and bottom surfaces are gapped, the side surfaces, which are parallel to the direction of the external field, remain gapless. In the absence of termination of the surface state, chiral quantum Hall edge states, analogous to those resulting from the presence of a confining potential in purely two-dimensional systems, exist on the sides surfaces; see Fig. 4.4.

A single Dirac fermion in a flat infinite space realizes the half-integer quantum Hall effect with $\nu = n + 1/2$ [41]. In the case of a compact surface with uniform doping, the top and bottom surfaces have the magnetic field pointing in the opposite direction compared with the surface normal, and are therefore in the opposite quantum Hall state: if the upper surface is in state ν, the lower is in state $\nu' = -\nu$ [26]. This results in quantum Hall plateaus with $\Delta\nu = \nu - \nu' = 2n + 1$ and an associated two terminal conductance of $\sigma_{xx} = (2n+1)e^2/h$. The side surfaces therefore host an odd number of chiral edge states. The Hall conductivity σ_{xy} depends on the detailed configuration of current and voltage leads [49, 50]. In principle, the doping of the top and bottom surfaces can be tuned separately, realizing all integer states $\nu = n + n' + 1$, including $\nu = 0$ [51–53].

The Landau level at the charge neutrality point is special in that it is a combination of electron-like and a hole-like Landau levels, while all higher or lower Landau levels are purely electron-like or hole-like. At charge neutrality, counter-propagating modes are therefore found close in both energy and space, allowing scalar disorder to couple them such that a $\nu = 0$ quantum Hall plateau is obtained in a narrow window of energies [54]. The width of this window increases with increasing disorder strength. This $\nu = 0$ state is distinct from the one induced by doping the top and bottom surfaces differently.

The chiral edge states can be probed in various ways; here we mention three. First, in the lowest Landau-level regime, only a single chiral edge state moves on each side surface, and they move in opposite direction on each surface. The direction in which the chiral mode moves depends on the doping. In a p–n junction, therefore, on each surface one obtains counter-propagating modes that meet in the transition region between the p and n halves of the junction. Since they cannot disappear, they instead travel along the p–n junction interface to the other side of the wire, where they can propagate away and into the lead. The obtained conductance depends on the overlap of the spin of the chiral states and the phase they pick up while crossing the junction; this latter phase can be controlled by a flux along the wire, realizing a Mach–Zehnder interferometer [55].

Second, the higher Landau levels have a characteristic non-monotonic dependence on the longitudinal momentum: the degenerate Landau levels are spilt as they turn into edge states and one of them dips in energy below the energy of the Landau level. This non-monotonic dispersion has a surprising effect on thermal transport. Namely, when applying the right temperature difference between two leads, one can obtain a particle current flowing from the cold reservoir to the hot, counter to intuition [56].

Finally, the single chiral mode limit is useful in probing topological superconductivity, since in this case the two-terminal conductance becomes a direct probe of the presence of Majorana modes [48]. Superconducting proximity effect and transport is the subject of the next section.

4.3 Topological Insulator Nanowires and Superconductivity

Proximity-induced superconductivity in materials with spin–orbit coupling is one of the promising schemes to engineer superconducting states with non-trivial topological properties [57, 58]. One of the first proposals for such an engineered phase is a topological insulator put in proximity with an s-wave superconductor [10, 59]. Such a construction is predicted to yield a one-dimensional topological superconductor at the edge of a quantum spin Hall sample, or a two-dimensional topological superconductor at the surface of a three-dimensional topological insulator.

Other prominent examples are one-dimensional nanowires made from materials such and InAs under the application of an external magnetic field, or magnetic chains, which under proximity effect can form an effective p-wave superconductor in one dimension [60, 61]. Recently, such systems have shown signatures consistent with the appearance of zero energy modes at their ends, a central characterizing feature of topological superconductors in one dimension [62–64]. This was shown both through transport as well as in Scanning Tunnelling Microscopy.

In this section, we review how TI in three dimensions formed into nanowires represents a novel and tuneable version of a quasi-one-dimensional topological superconductor, using the elements described in previous section. To this end, we begin by recalling the essential requirements for a normal one-dimensional system to become a topological superconductor under proximity effect.

4.3.1 Topological Superconducting Phases in One Dimension

Almost two decades ago, Kitaev put forward simple criteria for the emergence of topological superconductivity in one dimension, and formulated a topological invariant, which can be calculated from the lattice model representing it, which determines the fate of the superconducting phase [9]. Essentially, the main criteria require that the underlying normal system has an odd number of Fermi points in the right half of the Brillouin zone, as well as a finite gap when superconductivity is introduced. In the light of the discussion above, TI nanowires become immediate suspects for becoming topological superconductors in one dimension when pierced with one half of a flux quantum.

The topological invariant characterizing the one-dimensional lattice system, known as the Majorana number, is most generally defined as

$$\mathcal{M} = \text{sgn}\left[\text{Pf}\tilde{B}(k=0)\right] \text{sgn}\left[\text{Pf}\tilde{B}(k=\pi)\right] \quad (4.5)$$

Here, Pf stands for Pfaffian, and $\tilde{B}$ is the Hamiltonian matrix of a (quasi-) one-dimensional system expressed in a Majorana basis. In the limit of small pairing potential Δ, this expression reduces to a much simpler one:

$$\mathcal{M} = (-1)^\nu \quad (4.6)$$

where ν is an integer counting the number of Fermi points. A non-trivial phase is labelled by $\mathcal{M} = -1$ and is expected to have zero modes when surfaces are introduced.

The first to consider this topological invariant in the context of TI nanowires were Cook and Franz [27], predicting that a cylindrical nanowire combined with superconductivity is expected to have a non-trivial Majorana number when the flux through the wire is close to π.

The Hamiltonian of a cylindrical wire in the presence of a pairing potential induced by proximity to an s-wave superconductor is given by

$$H^{(n_v)} = \left[-i\sigma_x \partial_x + \sigma_y(-i\partial_s + \eta\,\tau_z)2\pi/P - \mu\right]\tau_z + \Delta_0\theta(-x)e^{-i\tau_z n_v s}\tau_x, \quad (4.7)$$

where Δ_0 is the superconducting pair potential induced by an adjacent bulk s-wave superconductor, assumed to be a constant. The phase of the order parameter represents an important degree of freedom and is allowed to wind around the wire. This winding has been explicitly singled out here in the exponential factor $e^{-i\tau_z n_v s}$, where n_v denotes the number of vortices with a core that is co-aligned with the wire's axis.

The importance of phase winding in cylindrical wires was stressed in [48] and can be easily argued by considering the band structure of the wire as the spectrum of (4.7) with and without a vortex, i.e. the difference between $n_v = 0$ and $n_v = 1$. The full characterization of the phase diagram of the wire is obtained by considering, in parallel, the topological invariant, and the energy gap in the spectrum. In order to obtain a topological superconductor, two conditions must be met simultaneously: the spectrum must have a finite gap, and the Majorana number must be equal to -1. As is evident from Fig. 4.5, in the absence of a vortex, the spectrum is gapless, although a calculation of the Majorana number will yield a non-trivial value.

To understand the role of the vortex, we remind ourselves that an s-wave pairing potential couples fermion states to form cooper pairs of zero total angular momentum and spin at the Fermi energy. In order for this pairing to become effective, energy bands of particles and holes in the BdG spectrum must cross at the Fermi energy with the appropriate quantum numbers. Considering the normal state band structure discussed in Sects. 4.2.1 and 4.2.2, we note that the energy bands crossing at the fermi energy have a mismatch of angular momentum, hence s-wave pairing cannot open a gap at the chemical potential. However, if a winding of the order parameter is introduced, it can act to compensate for the mismatch of angular momentum, enabling the opening of a finite gap. Note, however, that this sharp statement strongly relies on the rotational symmetry of the problem, and may soften when this symmetry is structurally violated.

4.3.2 Boundaries and Interferences: Zero Modes

Once the emergence of topological superconductivity is well established, two questions immediately arise. The first concerns the fate of boundary states, and the second concerns signatures of them on experimentally observable quantities. It is well known that topological superconductors support Majorana edge or boundary modes, and their presence at the surface of topological insulators is no exception.

The two-dimensional topological superconductor formed at the surface of a proximitized three-dimensional topological insulator is a well-studied phase [2, 3, 10]. It crucially differs from the prototypical two-dimensional topological superconductor in two dimensions, the p-wave superconductor, by the fact that it respects time-

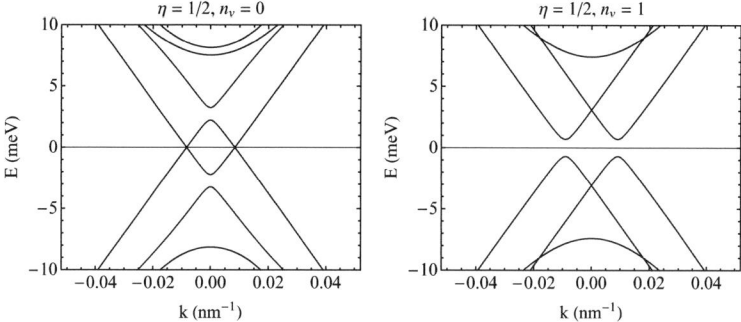

Fig. 4.5 Spectrum of the BdG Hamiltonian equation (4.7) describing the surface state of a cylindrical TI nanowires with proximity-induced superconductivity. The left panel is for $n_v = 0$, and the spectrum is gapless. The right panel represents the same spectrum, for $n_v = 1$, and is clearly gapped

reversal symmetry. It also differs from it by the fact that it has no natural boundary— the surface cannot terminate. In order to gain access to zero modes and boundary states, it is therefore necessary to interface a region with a topologically non-trivial superconducting gap with another region that has a different topological gap in order to trap gapless boundary states, or else create a topological defect such as a vortex or a Josephson junction.

The two mechanisms that can gap out the surface states of topological insulators are breaking charge conservation and time-reversal symmetry. Hence, it is expected that when two regions gapped by these are interfaced, gapless modes should arise [10, 65, 66]. Breaking time-reversal symmetry can be obtained in two ways: either by subjecting the system to a magnetic field that couples to the orbital motion of the particles, as discussed extensively in Sect. 4.2.4, or by magnetically doping the material and introducing an energy gap via Zeeman coupling to the particle's spin. Both mechanisms have been considered in the context of wires as ways to trap Majorana bound states.

Reference [27] considered a magnetic domain interfaced with superconductivity at the surface of a TI nanowire, following the realization of Fu and Kane that such an interface will host a chiral Majorana mode whose direction of propagation depends on the sign of the magnetization. In the case of a nanowire, this mode will form a compact loop around the wire, and therefore will have a discrete energy spectrum, tuneable via the boundary conditions. When half of a flux quantum threads the wire, this Majorana mode might have zero angular momentum and therefore a zero mode in its spectrum, which is protected since its counterpart is spatially separated from it and resides at the other end of the superconducting domain. Its wave function is exponentially localized in both regions over a length scale that is set by the two gaps: $\ell_\Delta = \Delta_0/v_F$, $\ell_m = |m|/v_F$, where $|m|$ is the amplitude of the magnetization.

As reviewed in Sect. 4.2.4, a magnetic field perpendicular to the surface of the TI introduces Dirac-like Landau levels, with the lowest one contributing a Hall conduc-

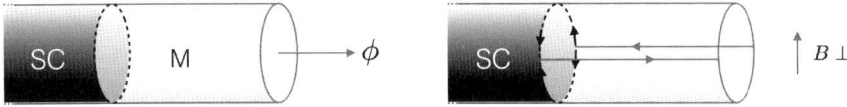

Fig. 4.6 An interface at the surface of a TI nanowire. On the left: boundary between superconductivity and magnetism traps a Majorana zero mode in the presence of flux. On the right: boundary between a gapped quantum Hall phase and a SC phase

Fig. 4.7 Unfolded geometry of the TI nanowire with a normal-superconducting interface in a perpendicular field

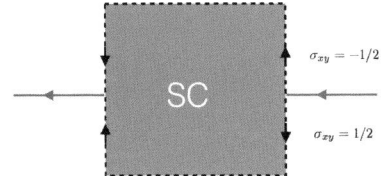

tance of $\sigma_{xy} = e^2/2h$, which one can understand as stemming from the chiral modes living on the side surfaces. Each fermionic chiral mode is predicted to be broken into two Majorana fermion chiral modes in the presence of superconductivity, since quite generically, chiral fermionic mode can always be trivially written as a superposition of two chiral co-propagating one-dimensional Majorana modes. In the absence of superconductivity, these are constrained to move together, a constraint that can only be removed in the presence of a pair potential.

The nanowire geometry in principle should allow to spatially separate these modes [48]. Consider a wire with normal-superconducting interface at some point x_0 along the wire. When the wire is subjected to a strong magnetic field (which we assume is fully screened in the superconducting region), chiral modes will flow on the side surfaces of the normal region to and from the normal-superconducting interface, as depicted in Fig. 4.6. At the interface, the chiral fermion modes are broken into two Majorana modes that flow in the top and bottom surfaces of the wire, along the interface. These modes break apart on one side surface and recombine on the other, forming a Majorana interferometer; see Fig. 4.7. The relative phase for the two chiral Majorana modes that encircle the wire can be tuned by introducing a vortex along the wire's axis.

4.3.3 *Transport Signatures of Topological Superconductivity*

The layout of the TI nanowire as described at the end of the previous section corresponds to a Majorana two path interferometer [48], where the two arms pick up a different phase that is directly correlated with the phase winding of the superconducting order parameter in the azimuthal direction (namely, around the wire), and is in one-to-one correspondence with the Majorana interferometer designed by Fu and Kane as well as Akhmerov, Nillson and Beenakker [65, 66]. However, there

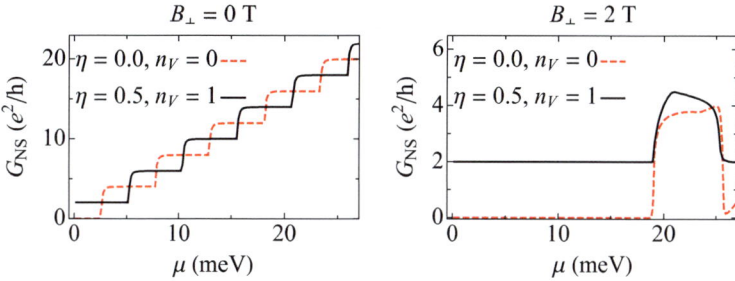

Fig. 4.8 The two terminal conductance across a normal-superconducting interface with (right) and without (left) a perpendicular magnetic field, without and without a vortex along the wire. A conductance plateau at $2e^2/h$ can clearly be seen when a vortex is present ($n_v = 1$) in both cases, but the plateaus is wider with the perpendicular field present. Figure taken from [48]

is a crucial difference between the these previous proposals and the one based on the nanowire geometry: realizing earlier proposals necessitates the use of magnetic domains of opposite magnetization on a single surface, as well as a magnetic field to generate phase winding for the two arms by threading vortices within a superconducting region enclosed by the interferometer's arms. The coexistence of these ingredients presents a great experimental challenge. Interfacing magnetism with a topological insulator surface state as means to create chiral modes has so far proven difficult, while small superconducting islands with a tuneable number of vortices are yet to be attempted.

The nanowire geometry seems to be a natural setting for the realization of the interferometer, one that bypasses some of the difficulties mentioned above. The small circumference of the nanowire allows the compactification of the chiral Majorana modes, while the flux through the wire enables the twisting of the boundary conditions for that mode through the formation of the vortex along the wire. It is therefore natural to look for the signatures of such a mode in plain two terminal conductance along the wire, measured across the interface between the SC and the normal state wire [29]. Indeed, theoretical predictions have been made for the two terminal conductance which predict either perfect Andreev reflection or perfect normal reflection [48]. The conductance of the wire, presented in Fig. 4.8, shows a clear hallmark of the interface Majorana mode: a flat $2e^2/h$ conductance plateau appears in the presence of a vortex along the wires core, which can be enhanced by the application of an external magnetic field. We stress that the perpendicular field is not a prerequisite for such a plateau to appear, but its presence enhances the chemical potential range in which the system is in the single mode regime, as demonstrated in Fig. 4.4, and required for topological superconductivity to appear.

Additional signatures of topological superconductivity in nanowires are predicted to appear in transport across Josephson junctions [57, 58, 67]. There, at a phase difference of π across the junction, a non-chiral Majorana mode is expected to be trapped within the junction provided, again, that a phase winding of π to the order parameter is properly introduced by a co-axial flux. The appearance of Majorana modes within Josephson junctions is expected to result in a 4π periodic current phase relation, provided no stray quasi-particles induce parity switches of the low lying state in the junction. Such a universal prediction is not unique to TI wires. The current phase relation should also have a distinct skewed shape in the absence of parity conservation, but perhaps the most striking signature is expected to appear in $I_c R_N$, namely the product of the critical current and the normal state resistance: $I_c R_N$ is expected to peak sharply when half of a flux quantum induced a vortex through the wire's core [67].

4.4 Technical Details: Transfer Matrix Technique

Assume we have a Dirac Hamiltonian of the form (in this section, we set $v_F = 1$)

$$H = \Gamma_x p_x + \Gamma_y p_y + \hat{V}(x, y), \tag{4.8}$$

and want to calculate the two terminal scattering matrix S. The $2N \times 2N$ matrices Γ_x and Γ_y depend on the problem at hand; for the single Dirac fermion in (4.1) we have $\Gamma_x = \sigma_x$ and $\Gamma_y = \sigma_y$, but for weak topological insulators they are certain 4×4 matrices. The potential term $\hat{V}$ is a matrix like the Γ's, constrained only by symmetries.

The scattering matrix relates incoming scattering states to outgoing scattering states [68]

$$\begin{pmatrix} \psi_L^{\text{out}} \\ \psi_R^{\text{out}} \end{pmatrix} = \begin{pmatrix} r & t' \\ t & r' \end{pmatrix} \begin{pmatrix} \psi_L^{\text{in}} \\ \psi_R^{\text{in}} \end{pmatrix}. \tag{4.9}$$

The transfer matrix M relates the wave function at two points [68]

$$\psi(x) = M(x, x')\psi(x'). \tag{4.10}$$

In the basis of scattering states, $M = M(L, 0)$ takes the form

$$\begin{pmatrix} \psi_R^{\text{in}} \\ \psi_R^{\text{out}} \end{pmatrix} = M \begin{pmatrix} \psi_L^{\text{in}} \\ \psi_L^{\text{out}} \end{pmatrix}, \tag{4.11}$$

and from the definition of the scattering matrix (4.9), M takes the form

$$M = \begin{pmatrix} t^{\dagger-1} & r't'^{-1} \\ -t'^{-1}r & t'^{-1} \end{pmatrix}. \tag{4.12}$$

Therefore, from the transfer matrix we obtain the scattering matrix; the two terminal conductance is given by the Landauer formula

$$G = \mathrm{tr}\, t^\dagger t. \tag{4.13}$$

The general Dirac equation (4.8) is transformed into the basis of scattering states by the unitary transformation U satisfying $U^\dagger \Gamma_x U = \Sigma_z$, where $\Sigma_z = \sigma_z \otimes \mathbb{1}_N$ is a diagonal matrix with the first N entries on the diagonal equal to 1 and the last N entries equal to -1. The Hamiltonian takes the form

$$H = \Sigma_z p_x + \tilde{\Gamma}_y p_y + \hat{\tilde{V}}(x, y). \tag{4.14}$$

where $\tilde{\Gamma}_y = U^\dagger \Gamma_y U$ and $\hat{\tilde{V}} = U^\dagger \hat{V} U$.

Using this form of the Hamiltonian to integrate the Dirac equation $H\psi = E\psi$, the transfer matrix takes the form

$$M = \mathcal{T}_x \exp\left\{ i\Sigma_z \int_0^L \left[E - \tilde{\Gamma}_y p_y - \hat{\tilde{V}}(x, y) \right] dx \right\}, \tag{4.15}$$

where $\mathcal{T}_x$ is the position ordering operator needed if the terms in the integral do not commute at different x, which happens for example if $\hat{V}$ depends on x. Using the transitive property $M(x_1, x_3) = M(x_1, x_2)M(x_2, x_3)$ and separating the integral into N_x equally spaced intervals, one can approximate the above expression by dropping the position ordering, which is valid if $L/N_x \ll \xi$ the correlation length of the potential $\hat{V}$, obtaining

$$M = \prod_{i=1}^{N_x} \exp\left\{ i\Sigma_z \int_{x_i}^{x_{i+1}} \left[E - \tilde{\Gamma}_y p_y - \hat{\tilde{V}}(x, y) \right] dx \right\} = \prod_{i=1}^{N_x} M_i, \tag{4.16}$$

with $x_1 = 0$ and $x_{N_x+1} = L$. This equation can be solved numerically in the basis of p_y momentum eigenstates in which case, for a fixed value of x, $\hat{\tilde{V}}$ is the $N_y \times N_y$ matrix

$$\hat{\tilde{V}}_{nn'}(x) = \int_0^P \frac{dy}{P} e^{i(q_n - q_{n'})y} \hat{\tilde{V}}(\mathbf{r}). \tag{4.17}$$

N_y is the total number of momentum modes included in the calculation and q_n are the eigenvalues of p_y. In principle, since we are in the continuum, N_y is infinite; in practice, however, the transmission of high transverse momentum modes is negligible and the matrix can be truncated at some cut-off momentum that is taken large enough that the conductance is independent of it. Similarly, N_x is increased until convergence is obtained.

The transfer matrix M has exponentially large and small eigenvalues, and the matrix product (4.16) is numerically unstable. The unitary scattering matrix in contrast has complex eigenvalues with unit amplitude. It is therefore useful to use the relation between transfer and scattering matrices to transform the transfer matrix M_i of the i-th interval into a scattering matrix S_i. The product of transfer matrices becomes a convolution of scattering matrices

$$S = \bigotimes_i S_i, \qquad (4.18)$$

with the convolution defined by

$$\begin{pmatrix} r_1 & t_1' \\ t_1 & r_1' \end{pmatrix} \otimes \begin{pmatrix} r_2 & t_2' \\ t_2 & r_2' \end{pmatrix} = \begin{pmatrix} r_1 + t_1' r_2 (1 - r_1' r_2)^{-1} t_1 & t_1'(1 - r_2 r_1')^{-1} t_2' \\ t_2 (1 - r_1' r_2)^{-1} t_1 & r_2' + t_2 r_1' (1 - r_2 r_1')^{-1} t_2' \end{pmatrix}. \qquad (4.19)$$

As a demonstration, consider the 2D Dirac Hamiltonian (4.1). Since it is translationally invariant, the position ordering can be dropped and the integral over x performed trivially since the integrand is independent of x, resulting in the transfer matrix

$$M = \exp\left[i\sigma_z \left(\mu + \sigma_y p_y\right) L\right], \qquad (4.20)$$

where we have taken $E = \mu$. From this expression, we obtain the scattering matrix. At the Dirac point $\mu = 0$, in particular,

$$M = U \begin{pmatrix} e^{p_y L} & 0 \\ 0 & e^{-p_y L} \end{pmatrix} U^\dagger, \qquad (4.21)$$

with $U^\dagger \sigma_x U = \sigma_z$. This finally gives

$$t = 1/\cosh(p_y L), \qquad (4.22)$$

consistent with [37].

4.5 Experimental Status and Outlook

Some of the theoretical aspects of the theory of the normal state transport in topological insulator nanowires presented here have already been tested, and some confirmed experimentally. Multiple groups have achieved Aharonov–Bohm interference in several nanowires of lengths ranging from hundred of nanometres to several microns, and circumference of approximately 200 nm long [30–36, 38]. Oscillations were observed in magnetic fields equivalent to up to 10 flux quanta threaded through the cross section of the wire, and signatures consistent with the emergence of a perfectly transmitted mode were also observed, showing up as an enhanced conductivity at half-integer flux quanta threaded through the wire.

In addition, 3DTI nanowires were also coupled to superconductivity. The Josephson effect was recently measured in $BiSbTeSe_2$ wires coupled to superconducting Niobium leads, displaying an anomalous behaviour indicating the formation of low energy Andreev bound states at the crossover from short junction to long junction behaviour [69]. The ability to resolve low energy modes is a promising step towards the realization and detection of Majorana bound states in such junctions. Nevertheless, it has been suggested that additional physics related to the Kondo effect might emerge in the presence of normal-superconducting interfaces [70], alluding to a different origin for the emergence of zero bias peaks in transport across such an interface. This certainly calls for additional exploration of transport in 3DTI nanowire-based heterostructure, both theoretically and experimentally.

Finally, the prospect of using 3DTI wires as a competitive platform for topological quantum computation is still being explored. A recent work has proposed an architecture made from coupled 3DTI nanowire-based Majorana box qubits, namely short segments of proximitized 3DTI wires connected by gapped 3DTI normal wire segments, as means to implement simple quantum operations on single qubits [71]. It will be both interesting as well as a challenge to bring such architectures to life both from the materials perspectives, as well as conceptually bridging the gap between these novel ideas and the limitations of the actual experimental system.

Acknowledgements We thank Fernando de Juan and Joel Moore for collaborations and Yong P. Chen for multiple discussion regarding the experimental systems. We would also like to thank Fernando de Juan for contributing Fig. 4.5 to this review. Work on this review was supported by the ERC Starting Grant No. 679722 and the Knut and Alice Wallenberg Foundation 2013-0093.

References

1. J.H. Bardarson, J.E. Moore, Quantum interference and Aharonov-Bohm oscillations in topological insulators. Reports Prog. Phys. **76**(5), 056501 (2013)
2. M.Z. Hasan, C.L. Kane, Colloquium: topological insulators. Rev. Mod. Phys. **82**(4), 3045–3067 (2010)
3. Xiao-Liang Qi and Shou Cheng Zhang, Topological insulators and superconductors. Rev. Mod. Phys. **83**(4), 1057–1110 (2011)
4. M. Zahid Hasan, J.E. Moore, Three-dimensional topological insulators. Annu. Rev. Condens. Matter Phys. **2**(1), 55–78 (2011)
5. C.-K. Chiu, J.C.Y. Teo, A.P. Schnyder, S. Ryu, Classification of topological quantum matter with symmetries. Rev. Mod. Phys. **88**(3), 035005 (2016)
6. C.L. Kane, E.J. Mele, Z2 topological order and the quantum spin hall effect. Phys. Rev. Lett. **95**(14), 146802 (2005)
7. C.L. Kane, E.J. Mele, Quantum spin hall effect in graphene. Phys. Rev. Lett. **95**(22), 226801 (2005)
8. B. Andrei Bernevig, T.L. Hughes, S. Cheng Zhang, Quantum spin hall effect and topological phase transition in HgTe quantum wells. Science **314**(5806), 1757–1761 (2006)
9. A. Yu Kitaev, Unpaired Majorana fermions in quantum wires. Phys.-Usp. **44**(10S), 131–136 (2001)
10. L. Fu, C. L. Kane, Superconducting proximity effect and Majorana fermions at the surface of a topological insulator. Phys. Rev. Lett. **100**(9), 096407 (2008)
11. J.H. Bardarson, J. Tworzydło, P.W. Brouwer, C.W.J. Beenakker, One-parameter scaling at the dirac point in graphene. Phys. Rev. Lett. **99**(10), 106801 (2007)
12. K. Nomura, M. Koshino, S. Ryu, Topological delocalization of two-dimensional massless dirac fermions. Phys. Rev. Lett. **99**(14), 146806 (2007)

13. P.M. Ostrovsky, I.V. Gornyi, A.D. Mirlin, Quantum criticality and minimal conductivity in graphene with long-range disorder. Phys. Rev. Lett. **98**(25), 256801 (2007)
14. S. Ryu, C. Mudry, H. Obuse, A. Furusaki, Z2 topological term, the global anomaly, and the two-dimensional symplectic symmetry class of anderson localization. Phys. Rev. Lett. **99**(11), 116601 (2007)
15. S. Hikami, A.I. Larkin, Y. Nagaoka, Spin-orbit interaction and magnetoresistance in the two dimensional random system. Prog. Theor. Phys. **63**(2), 707–710 (1980)
16. J.H. Bardarson, A proof of the Kramers degeneracy of transmission eigenvalues from antisymmetry of the scattering matrix. J. Phys. A: Math. Theor. **41**(40), 405203 (2008)
17. T. Ando, T. Nakanishi, R. Saito, Berry's phase and absence of back scattering in carbon nanotubes. J. Phys. Soc. Jpn. **67**(8), 2857–2862 (1998)
18. A.P. Schnyder, S. Ryu, A. Furusaki, A.W.W. Ludwig, Classification of topological insulators and superconductors in three spatial dimensions. Phys. Rev. B **78**(19), 195125 (2008)
19. S. Ryu, A.P. Schnyder, A. Furusaki, A.W.W. Ludwig, Topological insulators and superconductors: tenfold way and dimensional hierarchy. New J. Phys. **12**(6), 065010 (2010)
20. Z. Ringel, Y.E. Kraus, A. Stern, Strong side of weak topological insulators. Phys. Rev. B **86**(4), 045102 (2012)
21. R.S.K. Mong, J.H. Bardarson, J.E. Moore, Quantum transport and two-parameter scaling at the surface of a weak topological insulator. Phys. Rev. Lett. **108**(7), 076804 (2012)
22. J.H. Bardarson, P.W. Brouwer, J.E. Moore, Aharonov-Bohm oscillations in disordered topological insulator nanowires. Phys. Rev. Lett. **105**(15), 156803 (2010)
23. P.M. Ostrovsky, I.V. Gornyi, A.D. Mirlin, Interaction-induced criticality in Z2 topological insulators. Phys. Rev. Lett. **105**(3), 036803 (2010)
24. G. Rosenberg, H.M. Guo, M. Franz, Wormhole effect in a strong topological insulator. Phys. Rev. B **82**(4), 041104 (2010)
25. Y. Zhang, A. Vishwanath, Anomalous Aharonov-Bohm conductance oscillations from topological insulator surface states. Phys. Rev. Lett. **105**(20), 206601 (2010)
26. D.-H. Lee, Surface states of topological insulators: the dirac fermion in curved two-dimensional spaces. Phys. Rev. Lett. **103**(19), 196804 (2009)
27. A. Cook, M. Franz, Majorana fermions in a topological-insulator nanowire proximity-coupled to an s-wave superconductor. Phys. Rev. B **84**(20), 201105 (2011)
28. B. Béri, Dephasing-enabled triplet Andreev conductance. Phys. Rev. B **79**(24), 245315 (2009)
29. M. Wimmer, A.R. Akhmerov, J.P. Dahlhaus, C.W.J. Beenakker, Quantum point contact as a probe of a topological superconductor. New J. Phys. **13**(5), 053016 (2011)
30. H. Peng, K. Lai, D. Kong, S. Meister, Y. Chen, X.-L. Qi, S. Cheng Zhang, Z.-X. Shen, Y. Cui, Aharonov-Bohm interference in topological insulator nanoribbons. Nat. Mater. **9**(3), 225–229 (2010)
31. F. Xiu, L. He, Y. Wang, L. Cheng, L.-T. Chang, M. Lang, G. Huang, X. Kou, Y. Zhou, X. Jiang, Z. Chen, J. Zou, A. Shailos, K.L. Wang, Manipulating surface states in topological insulator nanoribbons. Nat. Nanotech. **6**(4), 216–221 (2011)
32. J. Dufouleur, L. Veyrat, A. Teichgräber, S. Neuhaus, C. Nowka, S. Hampel, J. Cayssol, J. Schumann, B. Eichler, O.G. Schmidt, B. Büchner, R. Giraud, Quasiballistic transport of dirac fermions in a Bi 2 Se 3 nanowire. Phys. Rev. Lett. **110**(18), 186806 (2013)
33. S.S. Hong, Y. Zhang, J.J. Cha, X.L. Qi, Y. Cui, One-dimensional helical transport in topological insulator nanowire interferometers **14**(5), 2815–2821 (2014)
34. S. Cho, B. Dellabetta, R. Zhong, J. Schneeloch, T. Liu, G. Gu, M.J. Gilbert, N. Mason, Aharonov–Bohm oscillations in a quasi-ballistic three-dimensional topological insulator nanowire. Nat. Comms. **6**, 7634 (2015)
35. L.A. Jauregui, M.T. Pettes, L.P. Rokhinson, L. Shi, Y.P. Chen, Magnetic field-induced helical mode and topological transitions in a topological insulator nanoribbon. Nat. Nanotech. **11**(4), 345–351 (2016)
36. B. Hamdou, J. Gooth, A. Dorn, E. Pippel, K. Nielsch, Surface state dominated transport in topological insulator Bi2Te3 nanowires. Appl. Phys. Lett. **103**(19), 193107 (2013)

37. J. Tworzydło, B. Trauzettel, M. Titov, Λ. Rycerz, C.W.J. Beenakker, Sub-poissonian shot noise in graphene. Phys. Rev. Lett. **96**(24), 246802 (2006)
38. J. Dufouleur, L. Veyrat, B. Dassonneville, E. Xypakis, J.H. Bardarson, C. Nowka, S. Hampel, J. Schumann, B. Eichler, O.G. Schmidt, B. Büchner, R. Giraud, Weakly-coupled quasi-1D helical modes in disordered 3D topological insulator quantum wires. Sci. Rep. **7**, 45276 (2017)
39. Y. Ran, A. Vishwanath, D.-H. Lee, Spin-charge separated solitons in a topological band insulator. Phys. Rev. Lett. **101**(8), 086801 (2008)
40. Y. Zhang, Y. Ran, A. Vishwanath, Topological insulators in three dimensions from spontaneous symmetry breaking. Phys. Rev. B **79**(24), 245331 (2009)
41. L. Brey, H.A. Fertig, Electronic states of wires and slabs of topological insulators: quantum Hall effects and edge transport. Phys. Rev. B **89**(8), 085305 (2014)
42. J. Ziegler, R. Kozlovsky, C. Gorini, M.H. Liu (Łł), S. Weishäupl, H. Maier, R. Fischer, D.A. Kozlov, Z.D. Kvon, N. Mikhailov, S.A. Dvoretsky, K. Richter, D. Weiss, Probing spin helical surface states in topological HgTe nanowires. Phys. Rev. B **97**(3), 035157 (2018)
43. T. Ando, H. Suzuura, Presence of perfectly conducting channel in metallic carbon nanotubes. J. Phys. Soc. Jpn. **71**(11), 2753–2760 (2002)
44. H.B. Nielsen, M. Ninomiya, Absence of neutrinos on a lattice. Nucl. Phys. B **185**(1), 20–40 (1981)
45. H.B. Nielsen, M. Ninomiya, Absence of neutrinos on a lattice. Nucl. Phys. B **193**(1), 173–194 (1981)
46. E. Xypakis, J.-W. Rhim, J.H. Bardarson, R. Ilan, Perfect transmission in rippled topological insulator nanowires (2017)
47. I.C. Fulga, F. Hassler, A.R. Akhmerov, C.W.J. Beenakker, Scattering formula for the topological quantum number of a disordered multimode wire. Phys. Rev. B **83**(15), 155429 (2011)
48. F. De Juan, R. Ilan, J.H. Bardarson, Robust transport signatures of topological superconductivity in topological insulator nanowires. Phys. Rev. Lett. **113**(10), 107003 (2014)
49. M. Sitte, A. Rosch, E. Altman, L. Fritz, Topological insulators in magnetic fields: quantum hall effect and edge channels with a nonquantized θ term. Phys. Rev. Lett. **108**(12), 126807 (2012)
50. E.J. König, P.M. Ostrovsky, I.V. Protopopov, I.V. Gornyi, I.S. Burmistrov, A.D. Mirlin, Half-integer quantum hall effect of disordered dirac fermions at a topological insulator surface. Phys. Rev. B **90**(16), 165435 (2014)
51. C. Brüne, C.X. Liu, E.G. Novik, E.M. Hankiewicz, H. Buhmann, Y.L. Chen, X.L. Qi, Z.X. Shen, S.C. Zhang, L.W. Molenkamp, Quantum hall effect from the topological surface states of strained bulk HgTe. Phys. Rev. Lett. **106**(12), 126803 (2011)
52. Y. Feng, X. Feng, O. Yunbo, J. Wang, C. Liu, L. Zhang, D. Zhao, G. Jiang, S. Cheng Zhang, K. He, X. Ma, Q.-K. Xue, Y. Wang, Observation of the zero hall plateau in a quantum anomalous hall insulator. Phys. Rev. Lett. **115**(12), 126801 (2015)
53. T. Morimoto, A. Furusaki, N. Nagaosa, Topological magnetoelectric effects in thin films of topological insulators. Phys. Rev. B **92**(8), 085113 (2015)
54. E. Xypakis, J.H. Bardarson, Conductance fluctuations and disorder induced $\nu=0$ quantum Hall plateau in topological insulator nanowires. Phys. Rev. B **95**(3), 035415 (2017)
55. R. Ilan, F. De Juan, J.E. Moore, Spin-based Mach-Zehnder interferometry in topological insulator pn junctions. Phys. Rev. Lett. **115**(9), 096802 (2015)
56. S.I. Erlingsson, A. Manolescu, G. Alexandru Nemnes, J.H. Bardarson, D. Sanchez, Reversal of thermoelectric current in tubular nanowires. Phys. Rev. Lett. **119**(3), 036804 (2017)
57. C.W.J. Beenakker, Search for Majorana fermions in superconductors. Annu. Rev. Condens. Matter Phys. **4**(1), 113–136 (2013)
58. J. Alicea, New directions in the pursuit of Majorana fermions in solid state systems. Reports Prog. Phys. **75**(7), 076501 (2012)
59. L. Fu, C.L. Kane, Josephson current and noise at a superconductor/quantum-spin-Hall-insulator/superconductor junction. Phys. Rev. B **79**(16), 161408 (2009)
60. R.M. Lutchyn, J.D. Sau, S. Das Sarma, Majorana fermions and a topological phase transition in semiconductor-superconductor heterostructures. Phys. Rev. Lett. **105**(7), 077001 (2010)

61. Y. Oreg, G. Refael, F. von Oppen, Helical liquids and Majorana bound states in quantum wires. Phys. Rev. Lett. **105**(17), 177002 (2010)
62. V. Mourik, K. Zuo, S.M. Frolov, S.R. Plissard, E.P.A.M. Bakkers, L.P. Kouwenhoven, Signatures of Majorana fermions in hybrid superconductor-semiconductor nanowire devices. Science **336**(6084), 1003–1007 (2012)
63. A. Das, Y. Ronen, Y. Most, Y. Oreg, M. Heiblum, H. Shtrikman, Zero-bias peaks and splitting in an Al–InAs nanowire topological superconductor as a signature of Majorana fermions. Nat. Phys. **8**(12), 887–895 (2012)
64. S. Nadj-Perge, I.K. Drozdov, J. Li, H. Chen, S. Jeon, J. Seo, A.H. MacDonald, B. Andrei Bernevig, A. Yazdani, Observation of Majorana fermions in ferromagnetic atomic chains on a superconductor. Science **346**(6209), 602–607 (2014)
65. L. Fu, C.L. Kane, Probing neutral majorana fermion edge modes with charge transport. Phys. Rev. Lett. **102**(21), 216403 (2009)
66. A.R. Akhmerov, J. Nilsson, C.W.J. Beenakker, Electrically detected interferometry of Majorana fermions in a topological insulator. Phys. Rev. Lett. **102**(21), 216404 (2009)
67. R. Ilan, J.H. Bardarson, H.-S. Sim, J.E. Moore, Detecting perfect transmission in Josephson junctions on the surface of three dimensional topological insulators. New J. Phys. **16**(5), 053007 (2014)
68. C.W.J. Beenakker, Random-matrix theory of quantum transport. Rev. Mod. Phys. **69**(3), 731–808 (1997)
69. M. Kayyalha, M. Kargarian, A. Kazakov, I. Miotkowski, V.M. Galitski, V.M. Yakovenko, L.P. Rokhinson, Y.P. Chen, Anomalous low-temperature enhancement of supercurrent in topological-insulator nanoribbon Josephson junctions: evidence for low-energy Andreev bound states (2017)
70. S. Cho, R. Zhong, J.A. Schneeloch, G. Genda, N. Mason, Kondo-like zero-bias conductance anomaly in a three-dimensional topological insulator nanowire. Sci. Rep. **6**(1), 21767 (2016)
71. J. Manousakis, A. Altland, D. Bagrets, R. Egger, Y. Ando, Majorana qubits in a topological insulator nanoribbon architecture. Phys. Rev. B **95**(16), 165424 (2017)

Chapter 5
Microwave Studies of the Fractional Josephson Effect in HgTe-Based Josephson Junctions

E. Bocquillon, J. Wiedenmann, R. S. Deacon, T. M. Klapwijk, H. Buhmann and L. W. Molenkamp

Abstract The rise of topological phases of matter is strongly connected to their potential to host Majorana bound states, a powerful ingredient in the search for a robust, topologically protected, quantum information processing. In order to produce such states, a method of choice is to induce superconductivity in topological insulators. The engineering of the interplay between superconductivity and the electronic properties of a topological insulator is a challenging task, and it is consequently very important to understand the physics of simple superconducting devices such as Josephson junctions, in which new topological properties are expected to emerge. In this chapter, we review recent experiments investigating topological superconductivity in topological insulators, using microwave excitation and detection techniques. More precisely, we have fabricated and studied topological Josephson junctions made of HgTe weak links in contact with Al or Nb contacts. In such devices, we have observed two signatures of the fractional Josephson effect, which is expected to emerge from topologically protected gapless Andreev bound states. We first recall the theoretical background on topological Josephson junctions, then move to the experimental observations. Then, we assess the topological origin of the observed features and conclude with an outlook toward more advanced microwave spectroscopy experiments, currently under development.

E. Bocquillon (✉)
Laboratoire Pierre Aigrain, École Normale Supérieure, Supérieure-PSL Research University, CNRS, Sorbonne Université, Université Paris Diderot-Sorbonne Paris Cité, 24 rue Lhomond, 75231 Paris Cedex 05, France
e-mail: erwann.bocquillon@lpa.ens.fr

J. Wiedenmann · H. Buhmann · L. W. Molenkamp
Physikalisches Institut (EP3), Institute for Topological Insulators,
University of Würzburg, Am Hubland, 97074 Würzburg, Germany

R. S. Deacon
Advanced Device Laboratory, Center for Emergent Matter Science, RIKEN,
2-1 Hirosawa, Wako-shi, Saitama 351-0198, Japan

T. M. Klapwijk
Kavli Institute of Nanoscience, Faculty of Applied Sciences,
Delft University of Technology, Lorentzweg 1, 2628 CJ Delft, The Netherlands

© Springer Nature Switzerland AG 2018
D. Bercioux et al. (eds.), *Topological Matter*, Springer Series in Solid-State Sciences 190, https://doi.org/10.1007/978-3-319-76388-0_5

5.1 Gapless Andreev Bound States in Topological Josephson Junctions

> In this first section, we recall the basic ingredients of induced superconductivity in topological insulators. The broken spin rotation symmetry in these systems results in the formation of a peculiar phase with a p wave symmetry. We briefly introduce an important consequence, namely the formation of zero-energy Majorana states. We then focus on topological Josephson junctions, which have been predicted to exhibit the *fractional Josephson effect*, first identified by Fu and Kane [1, 2] as a signature of topological superconductivity.

5.1.1 p Wave Superconductivity in 2D and 3D Topological Insulators

Proximity effect At the interface between a superconductor (S) of gap Δ and a normal (i.e. non superconducting) material (denoted N), the conversion of normal current into supercurrent (carried by Cooper pairs) and vice versa is mediated by Andreev reflections. When an electron incident from the N side with energy $\varepsilon < \Delta$ reaches the interface, a Cooper pair can be injected into the superconductor without breaking charge or energy conservation when combined with the retroreflection of a hole with energy $-\varepsilon$. This mechanism is called Andreev reflection and is a key notion that governs the physics of two electronic states: 'superconducting' and 'normal' interacting by exchange of electrons at the interface. This quantum process is, at the nanoscale of a Josephson junction, not localized at the interface. Its extension is given by the so-called coherence length $\xi = \frac{\hbar v_F}{\Delta}$ (v_F the Fermi velocity in the normal region) which measures, for a system without elastic scattering, how far correlations between paired electrons penetrate into the normal side. As a consequence, this length also naturally yields a *proximity effect* [3], i.e. the typical distance over which superconductivity can be induced in a normal conductor by a superconductor located nearby.

Induced p-wave superconductivity When a nearby conventional superconductor induces superconductivity in a topological phase, the symmetries and properties of the induced superconductivity are deeply influenced by the peculiar transport properties in this phase. In a vast majority of experimentally relevant cases, superconductivity is induced by a conventional superconductor (Al, Pb, Nb, NbTiN), in which superconductivity arises from s-wave-paired electrons of opposite spins. In contrast, spin rotation symmetry is broken in 2D and 3D topological insulators, since electrons have to abide the so-called spin-momentum locking: Electrons with opposite directions have opposite spins (in fact total angular momentum). Thus, topological phases give rise to induced "spinless" superconducting systems, since only one fermionic species (rather than two) is present and forms Cooper pairs. In other words, with spin rotation symmetry being broken in the topological phases, it must also be in the induced superconducting states and a so-called p-wave superconducting state with odd parity emerges [4–6].

Majorana bound states Such a p-wave superconductivity has several consequences, one of them being the existence of zero-energy modes known as Majorana bound states. In s-wave superconductors, the Bogoliubov quasiparticle operators read $\hat{\gamma}_s = u\hat{c}_\uparrow^\dagger + v\hat{c}_\downarrow$ (where $\hat{c}_\sigma$ are electron annihilation operators of spin $\sigma = \uparrow, \downarrow$), for which $\hat{\gamma}_s \neq \hat{\gamma}_s^\dagger$. In contrast, thanks to the lifted spin degeneracy, a p-wave superconducting state allows for excitations such that $\hat{\gamma}_p = u\hat{c}^\dagger + v\hat{c}$. The famous condition for Majorana excitations $\hat{\gamma}_p = \hat{\gamma}_p^\dagger$ can thus be fulfilled for $u = v$. While forbidden in conventional s wave superconductors, such states do exist in p-wave superconductors and constitute the realization of Majorana fermions emerging in a condensed matter system. They naturally lie at zero energy due to electron–hole symmetry and localize at system boundaries and topological defects (such as vortices). One can show [4] that they support non-abelian statistics, and as such hold promise for exotic fundamental physics, and application to topologically protected quantum computation.

Given the properties of Majorana states, it is natural to investigate interfaces between a topological phase in a normal state and a superconductor, for example, by tunnel spectroscopy. This route has led to intriguing observations of zero-bias anomalies in nanowires with strong spin–orbit coupling, in which similar physics should arise when a topological phase transition occurs under applied magnetic field along the axis of the nanowire [7, 8]. In the remainder of the chapter, we focus on a different approach to topological superconductivity, namely the study of Josephson junctions in topological insulators. The material system is here Hg(Cd)Te, which is commonly used for infrared detection. We in particular address how precursors of Majorana states alter the Josephson effect, and signal topological superconductivity.

5.1.2 Gapless Andreev Bound States in 2D and 3D Topological Insulators

The Josephson effect generically manifests itself as the occurrence of a phase-difference-dependent non-dissipative supercurrent in a weak link between two superconductors. The nature of the weak link influences the properties of the supercurrent, which can thus serve as a probe of superconductivity. In mesoscopic systems, in general short compared to the phase correlation length, the supercurrent and its properties can be obtained by solving a scattering problem [9], with a weak link represented as a scattering matrix, and at each ends, boundary conditions set by Andreev reflections (together with normal reflection at the interface, because dissimilar materials have different electronic properties [10]). A number of resonant states thus form, called Andreev bound states, and their energies $\varepsilon_n(\phi)$ depend on the superconducting phase

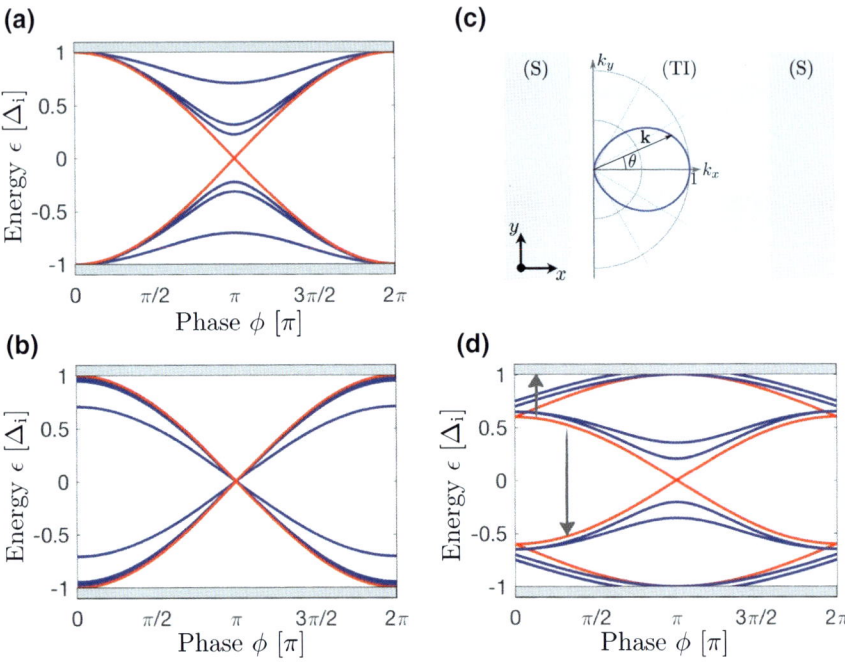

Fig. 5.1 Andreev bound states in 2D and 3D topological insulators—**a** conventional Andreev bound states for different transmission coefficients D: $D = 1$ in red and $D = 0.95, 0.9, 0.5$ in blue. **b** Gapless Andreev bound states in p-wave superconductors for the same transmissions. **c** Polar plot of the transmission D_θ as a function of angle θ and scheme of a Josephson junction. **d** More realistic picture of an Andreev spectrum with one topological mode (red) and two non-topological modes (blue). Two possible relaxation mechanisms (ionization to the continuum, energy relaxation) are depicted as gray arrows

difference ϕ between the two conductors. Major differences occur between the case of s- and p-wave superconductors [11]. We limit this discussion to the short junction limit, for which the length L of the junction is much shorter than the coherence length ξ.

Conventional Andreev bound states In a conventional Josephson system between two s-wave superconducting reservoirs, Andreev bound states can generically be written as

$$\varepsilon_s(\phi) = \pm \Delta_i \sqrt{1 - D \sin^2 \frac{\phi}{2}} \qquad (5.1)$$

where D is the transmission of the weak link in the normal state, and Δ_i the proximity induced gap. In Fig. 5.1a, $\varepsilon_s(\phi)$ is represented and shows that it is a 2π-periodic function of ϕ. An energy gap $2\Delta_i\sqrt{1-D}$ is opened at $\phi = \pi$ for any $D \neq 1$. As seen below, the limit $D \to 1$ is singular: The spectrum becomes $\varepsilon_s(\phi) = \pm \Delta_i \left|\sin\frac{\phi}{2}\right|$, and the Andreev doublet is gapless, with $\varepsilon_s(\pi) = 0$. This regime is approached in superconducting atomic point contacts, which exhibit high transparencies $D > 0.99$ [12, 13] and do not suffer from the use of dissimilar materials. Besides, in conventional systems, each Andreev doublet is in fact doubly degenerate as both spin species are active.

Gapless Andreev bound states in 2D topological insulators The Andreev bound states which form in topological weak links exhibit some remarkable differences as compared to the previous case, as shown in Fig. 5.1b. Generically, in a 1D geometry, Andreev bound states forming between p wave superconductors read [11, 14]:

$$\varepsilon_p(\phi) = \pm \Delta_i \sqrt{D} \sin \frac{\phi}{2} \qquad (5.2)$$

In the s wave case, the transmission D determines the avoided crossing at $\phi = \pi$. In contrast, in the p wave case, an imperfect transmission $D < 1$ opens a gap at $\phi = 0$ between the continuum and Andreev states, with an amplitude $\Delta_i(1 - \sqrt{D})$.

Such bound states indeed describe the solutions of a weak link fabricated from a 2D topological insulator [2]. There, spin-polarized 1D edge channels are responsible for the electrical transport. In such a system, back-scattering is forbidden, as long as time-reversal symmetry is preserved. Then, one finds $D = 1$, and the previous equation results in a unique 4π-periodic Andreev doublet with:

$$\varepsilon_{2D}(\phi) = \pm \Delta_i \sin \frac{\phi}{2} \qquad (5.3)$$

The degeneracy at $\phi = 0$ is then a manifestation of time-reversal symmetry. Conversely, when time-reversal symmetry is broken, the transmission is reduced ($D < 1$). The broken spin rotation symmetry results here in a lifting of the spin degeneracy: Gapless Andreev bound states are not spin degenerate as opposed to their conventional counterparts. The two states of the doublet correspond to opposite fermion parities. The level crossing at $\phi = \pi$ is a manifestation of topology and is as such protected, and the gapless Andreev doublet (sometimes also called Majorana-Andreev bound states) can in fact be seen as the hybridization of two Majorana end states (see Sect. 5.1.1) bound at the two S–TI interfaces.

Superconducting Klein tunneling in 3D topological insulators The Andreev energy spectrum in the 2D geometry of surface states in a 3D topological insulator is slightly richer. For a bar of width W, normal transport occurs through $N = W/\lambda_F$ modes, which results in N Andreev doublets in a Josephson junction (see Fig. 5.1c). For a wide junction, these doublets are indexed by the transverse momentum k_y or, equivalently, by the angle θ such that $\cos \theta = \sqrt{1 - \frac{k_y^2}{k_F^2}}$, and typically read as [15]:

$$\varepsilon_{3D}(\phi) = \pm \Delta_i \sqrt{1 - D_\theta \sin^2 \frac{\phi}{2}} \qquad (5.4)$$

where D_θ is a θ-dependent transmission. This generalized 2D transmissivity reflects the topological and Dirac nature of the charge carriers [15, 16] and can be written as:

$$D_\theta = \frac{\cos^2 \theta}{1 - \frac{\sin^2 \theta}{1+Z^2}} \quad (5.5)$$

where Z is a parameter characterizing the scattering (described here as a potential barrier). This system intrinsically hosts a single 4π-periodic mode together with many 2π-periodic ones which may simultaneously manifest in the Josephson response of the device. Indeed, a single topological Andreev doublet occurs at transverse momentum $k_y = 0$, $\theta = 0$ and is immune to back-scattering (thus has perfect transmission) as $D_{\theta=0} = 1$ regardless of Z. On the contrary, a large number $\approx N$ of non-topological oblique modes ($k_y \neq 0$, $\theta \neq 0$) have lower transmissions $D_{\theta \neq 0} < 1$. In that sense, the topological protection of the zero mode constitutes a superconducting analogue to Klein tunneling. Though the Andreev bound states are not protected from scattering for $\theta = 0$, they still feature the spin-momentum locking and may be called helical just as the topological surface states.

Beyond the short junction limit The preceding results are all obtained in the limit $L \ll \xi$, which for a single transport channel results in a unique Andreev doublet. Outside this regime, the situation is more complex, as more levels play a role in transport, with a typical level spacing of $\Delta \xi / L$. A schematic picture of a possible spectrum is presented in Fig. 5.1d. In experiments, the exact spectrum is not known and depends on parameters such as the length of the junction and the Fermi energy. Nevertheless, most features remain valid. In particular, there is in both 2D and 3D TIs a unique Andreev doublet with a protected level crossing at $\varepsilon = 0$ and $\phi = \pi$. It consequently exhibits 4π-periodicity. We refer the reader to [2, 17] for a more complete discussion.

5.1.3 Fractional Josephson Effect

5.1.3.1 Conventional and Fractional Josephson Effect

Conventional Josephson effect An Andreev bound state of energy $\varepsilon(\phi)$ carries a supercurrent, the amplitude of which is proportional to $\frac{\partial \varepsilon}{\partial \phi}$. The so-called current–phase relation expresses the relation between the supercurrent I_s and the superconducting phase difference ϕ between the two (undisturbed) macroscopic quantum phases of the superconductors on each side. It may be complicated when multiple ABS contribute, but its simplest expansion is:

$$I_s(\phi) = I_c \sin \phi + \text{higher harmonics} \quad (5.6)$$

with I_c the critical current of the junction, assumed to be a constant. The higher harmonics can in some cases represent an important contribution (e.g., high transmissions $D \to 1$). The main point here is however that it remains 2π-periodic in ϕ. When combined with the second Josephson equation $\frac{d\phi}{dt} = \frac{2eV}{\hbar}$, it is clear that a constant voltage V gives rise to an oscillating current $I_s(t) = I_c \sin(2\pi f_J t)$, with the conventional Josephson frequency $f_J = \frac{2eV}{h}$, which is currently the basis for the voltage standard.

Fractional Josephson effect The presence of topologically protected Andreev bound states with 4π-periodicity is expected to manifest itself as a *fractional Josephson effect* [11], by modifying the equations describing the junctions. Indeed, the current–phase relation now fundamentally reads:

$$I_{2D/3D}(\phi) = I_{4\pi} \sin \frac{\phi}{2} + I_{2\pi} \sin \phi + \text{higher harmonics} \qquad (5.7)$$

where $I_{4\pi}$ and $I_{2\pi}$ are two constants encoding the amplitude of the 4π- and 2π-periodic supercurrents. The Josephson supercurrent then oscillates with frequency $f_J/2 = \frac{eV}{h}$, hence the name *fractional Josephson effect* [11].

The fractional Josephson effect should have two clear signatures. First, under constant DC bias, the oscillating Josephson current should result in an observable dipolar Josephson emission at $f_J/2$, typically in the GHz range as $\frac{e}{h} \simeq 0.25\,\text{GHz}\,\mu\text{V}^{-1}$ which can be measured and analyzed using rf techniques (see Sect. 5.3.1). Secondly, when phase locking occurs between the internal junction dynamics and an external microwave excitation at frequency f, Shapiro steps [18] appear at discrete voltages given by $V_n = nhf/2e$, where n is an integer step index. In the presence of a sizable 4π-periodic supercurrent, an unconventional sequence of even steps (with missing odd steps) is expected, reflecting the doubled periodicity of the Andreev bound states [2, 19, 20] (see Sect. 5.3.2).

5.1.3.2 Obstacles to the Observation of the Fractional Josephson Effect

The above description must be carefully balanced out, as various phenomena can alter this simple picture.

Relaxation and thermodynamic limit The previous signatures of the fractional Josephson effect are based on the hypothesis that the occupation number of the gapless Andreev levels remain constant, so that $I_{4\pi}$ is unchanged over the full duration of the experiment. Due to quasiparticle poisoning or ionization to the continuum (depicted as gray arrows in Fig. 5.1d), the occupation of the 4π-periodic fluctuates, which in turn affects the periodicity of the Josephson effect.

For a time-independent phase ϕ, the occupation numbers reach the thermodynamical limit, and only the lower branches of Andreev bound states at $\varepsilon \leq 0$ are populated. The current is then 2π-periodic. It can indeed be shown that at equilibrium $I(\phi) = e\frac{\partial \varepsilon}{\partial \phi}(1 - 2f(\varepsilon)) \propto \sin \frac{\phi}{2} \tanh\left(\frac{\Delta_i}{2kT} \cos \frac{\phi}{2}\right)$, where f is the Fermi–Dirac distribution function. This 2π-periodic function is in fact identical to the expressions obtained for ballistic conventional Josephson junctions ($D \to 1$), and does not highlight the topological character of the induced superconductivity [21, 22].

Experiments relying on out-of-equilibrium dynamics are thus useful to provide evidence for the existence of gapless 4π-periodic Andreev bound states on timescales shorter than the equilibration time. Only on such short timescales can one in principle observe doubled Shapiro steps, or the anomalous Josephson emission at half the Josephson frequency $f_J/2$. We focus in the next sections on experiments focusing on dynamics in the GHz range. We also refer the reader to several works on the effects of relaxation mechanisms on the signatures of topological superconductivity [2, 23–26].

Coupling to the continuum It is also important to notice that time-reversal symmetry should in principle impose a Kramers degeneracy point at $\phi = 0, 2\pi, \ldots$. There, the gapless Andreev bound states are either connected to other Andreev states or to the bulk continuum [2, 27]. Bulk quasiparticles are then produced as ϕ is adiabatically advanced. This naturally leads to enhanced relaxation at these points, and suppresses the dissipationless and 4π-periodic character of the supercurrent, thus restoring a 2π-periodicity.

Surprisingly, these degeneracies are modified when electron–electron interactions are taken into account in a many-body picture. Then, the many-body Andreev spectrum is reorganized and gives rise to an effective 8π-periodic supercurrent when combined with electron interaction [28–32]. In that case, instead of the 4π-periodic fractional Josephson effect, one may expect to observe an 8π-periodic Josephson effect, with Shapiro steps only visible with index $n \equiv 0 \mod 4$, or emission at a quarter of the Josephson frequency $f_J/4$.

Landau–Zener transitions Another important *caveat* is the non-zero probability of Landau–Zener transitions (LZT) between Andreev bound states near an avoided level crossing. When the voltage V or equivalently the frequency f are sufficiently high, LZT can mimic an effective 4π-periodic Josephson effect, while the spectrum of Andreev states remains gapped, with only a small avoided crossing at $\phi = \pi$ [25, 33, 34]. Such LZTs have previously been observed in single Cooper pair transistors [35] and can in principle be distinguished from an intrinsic fractional Josephson effect by a strong voltage dependence of the emission or Shapiro step features.

In Sect. 5.4, special attention will be given to assessing the topological origin of the observed fractional Josephson effect and the role of the aforementioned mechanisms.

5.2 HgTe-Based Josephson Junctions and Experimental Techniques

Here, we first briefly introduce the reader to the geometry, fabrication, and basic properties of the devices. In particular, we show how we estimate parameters such as the induced gap Δ_i and the coherence length ξ. Then, we describe the setups operated to measure the response to ac excitations (Shapiro steps) or capture Josephson emission. Given their high mobilities and low intrinsic electron densities, we argue that HgTe-based 2D and 3D topological insulators thus appear as ideal base material to fabricate topological Josephson junctions and observe the manifestations of topological superconductivity. Finally, we conclude this section with a presentation of the experimental setups which allow for a simple, fast, and reliable measurement of the devices.

5.2.1 Fabrication of HgTe-Based Josephson Junctions

The junctions are fabricated from epitaxially grown layers of HgTe on a CdTe substrate for which the mobility and carrier density are evaluated from a Hall bar produced separately prior to the fabrication of Josephson junctions. Both 2D and 3D topological insulators can serve as weak links in Josephson junctions. However, these early devices made of 3D topological insulators suffer from lower mobilities caused by the absence of protective capping layer (CdHgTe), and from the absence of gate electrode to tune the electron density [36–38]. We briefly review below the main characteristics of the different devices.

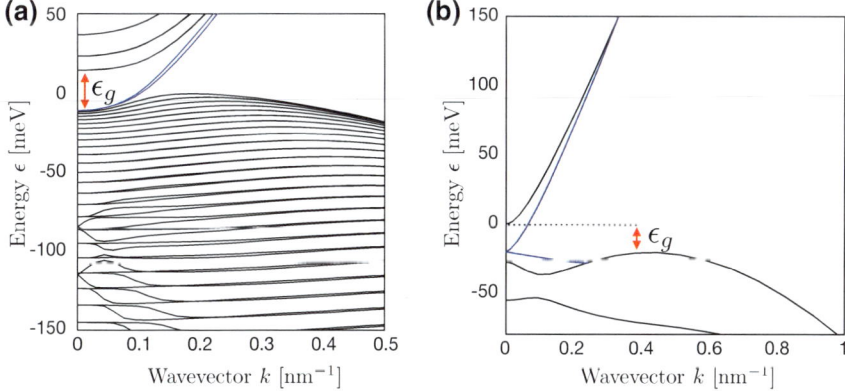

Fig. 5.2 Band structures of HgTe-based 2D and 3D topological insulators from $k.p$ simulations—**a** strained HgTe layer of 70 nm: Bulk 3D states are pictures in black, while 2D topological surfaces states appear in blue **b** HgTe quantum well of 7.5 nm: Bulk 2D states appear in black, while 1D topological edge states (not simulated) are indicated in blue

Strained HgTe as a 3D topological insulator The 3D topological insulators (TI) are obtained from coherently strained undoped HgTe layers of 60–90 nm thickness. The band inversion of HgTe enforces the existence of topological surface states, while strain opens a gap ($\epsilon_g \simeq 20$ meV) in the bulk of the material [39]. A typical band structure is shown in Fig. 5.2a for a strained 70-nm-thick HgTe layer. In previous works, we have proven the high quality of the topological states in this material [40, 41] and notably they entirely dominate electron transport up to very large electron densities [42, 43]. In the many devices tested, we find typical densities of $n_{3D} = 3 - 7 \times 10^{11}$ cm^{-2} and mobilities of $\mu_{3D} = 1 - 3 \times 10^4$ cm^2 V^{-1} s^{-1}. From these values, we calculate a mean free path of $l_{3D} \simeq 200$ nm.

HgTe quantum wells as a 2D topological insulator New lithography processes have enabled to fabricate Josephson junctions from thin quantum wells of HgTe, sandwiched between barrier layers of Hg$_{0.3}$Cd$_{0.7}$Te grown on a CdZnTe substrate [44]. As depicted in Fig. 5.2b, for thicknesses above a critical thickness $d_c \simeq 6.3$ nm, topological edge channels are expected between the conduction and valence band, separated by a small gap ϵ_g, with $\epsilon_g \simeq 10 - 30$ meV depending on growth parameters. The existence of topological edge states has been proven via transport measurements [45–47] and scanning-SQUID imaging [48]. The typical densities are $n_{2D} = 1 - 5 \times 10^{11}$ cm^{-2} with mobilities $\mu_{2D} = 3 - 5 \times 10^5$ cm^2 V^{-1} s^{-1}. As a consequence, 2D devices are expected to have a larger mean free path of $l_{2D} \simeq 2 \mu$m compared with the thick 3D layers.

Remarkably, it is possible to grow thinner quantum wells ($d < d_c$) that do not exhibit a band inversion and consequently do not host any topological edge channels. Outside of the gap region, such layers are extremely similar to thick quantum wells and exhibit the same typical densities and mobilities. They are as such ideal reference samples to benchmark the experimental techniques and observations in a trivial system. We will refer to such a reference sample in the rest of the chapter.

Geometry of the Josephson junctions The layout of the Josephson junctions is shown in Fig. 5.3 and is similar for both 2D TIs, 3D TIs and reference samples (apart from the absence of the gate and protective cap layer on the 3D sample). A rectangular mesa of HgTe is first defined. First the oxide and capping layers are etched, before superconducting contacts are deposited on the HgTe layer. Niobium has been sputtered on 3D TI samples, while Al is deposited (with a thin Ti adhesion layer) with standard evaporation techniques on 2D TI and reference samples. Upon the latter, a metallic gate electrode of Au is added between the Al contacts on an HfO$_2$ dielectric layer grown via atomic layer deposition (ALD) to control the electron density.

The superconducting stripes have a width of 1 μm. The HgTe mesa has a width $W = 2 - 4 \mu$m, determining the width of the weak link. In 3D topological insulators, it is advantageous to narrow down the mesa so as to reduce the number of bulk modes with $k_y \neq 0$, while in contrast a large mesa reduces the overlap of edge channels on opposite edges in 2D samples [49]. The length of the junctions has been varied between $L = 200$ nm and 600 nm.

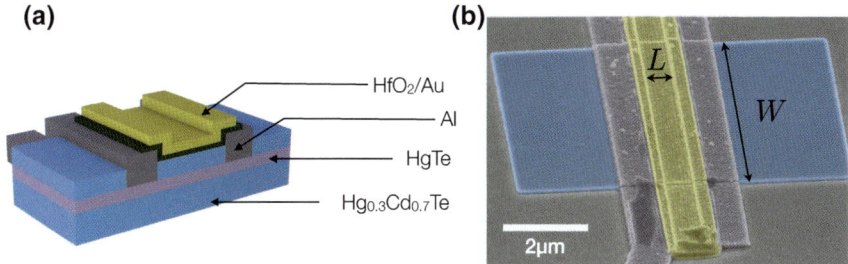

Fig. 5.3 Experimental realization of a HgTe-based topological Josephson junction—artist view (**a**) and colorized SEM picture (**b**) of a junction. The HgTe 2D topological insulator (in mauve) is sandwiched between two layers of $Hg_{0.3}Cd_{0.7}Te$ (in blue). The Al superconducting contacts are in dark purple, while the gate is in yellow and rests on a thin dielectric layer of HfO_2 (dark green). Devices realized on 3D TIs are similar, without the top gate and the $Hg_{0.3}Cd_{0.7}Te$ protective top layer

5.2.2 Basic Properties of HgTe-Based Josephson Junctions

Before moving to measurements specific to the fractional Josephson effect, the study of the current-voltage I–V curves of the junctions under DC bias provides some information on the microscopic parameters of the junctions that we review in this section.

I–V curve of Josephson junctions As mentioned earlier, the junctions based on 3D TIs do not have a gate, and their electron density is such that the number of transport modes lies typically between $N = 50$ and 200, depending on the sample width (with variations of about 30% for a given dimension). A typical I–V curve is presented in Fig. 5.4a and shows the expected behavior of a Josephson junction with a critical current of $I_c \simeq 5\,\mu A$. It exhibits hysteresis, as commonly reported [37]. We believe that the hysteresis is an intrinsic property of mesoscopic Josephson devices, which reflects the difference in the Josephson current amplitude in the static (DC) case compared to the dynamic (AC) case [50, 51].

In all devices, an excess current in the I–V curve is clearly visible, with an asymptote which does not go through the origin but is shifted toward higher currents. This excess current reflects the high probability of Andreev reflections in an energy window near the superconducting gap [10, 53] and underlines the high quality and reproducibility of our devices in line with previous observations [37, 38, 54].

An important parameter of the junction is the amplitude of the induced superconducting gap Δ_i. We have resorted to the study of the temperature dependence of the critical current and obtained estimates on the order of $\Delta_i = 100 - 350\,\mu eV < \Delta_{Nb}$, but with a large uncertainty given the lack of adequate theories [15]. The relevant coherence length for the quasi-ballistic weak link is then estimated $\xi = \sqrt{\frac{\hbar v_F l}{\pi \Delta_i}}$ in the range of 250–550 nm and is compatible with our observations of the decay of I_c with length. The 3D junctions are consequently in an intermediate regime $l \sim \xi \sim L$,

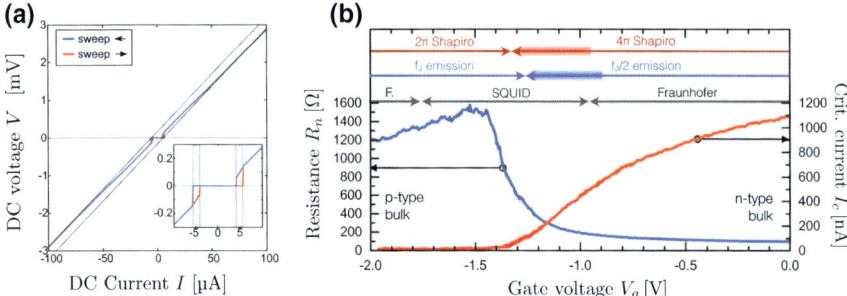

Fig. 5.4 DC characterization of HgTe-based Josephson junctions—**a** junction with a 3D topological insulator: I–V curve of a Josephson junction fabricated from a 3D TI, taken at base temperature $T \simeq 30$ mK. The asymptotes (gray solid lines) do not cross the origin, emphasizing the presence of an excess current. (Inset) Detailed view of the I–V curve, which exhibits hysteresis between the upward and downward sweep direction. **b** Junction with a 2D topological insulator: critical current I_c (red line) and normal state resistance R_n (blue line) as a function of gate voltage V_g. The red, gray, and blue arrows summarize the ranges where we observe the anomalous Josephson effect properties. The arrows are thicker where the emission at $f_J/2$ and the even sequences of Shapiro steps are the most visible. Panel a adapted from [52]

which is particularly hard to model, as the junctions reach neither the short (corresponding to $L \ll \xi$) nor the ballistic ($L \gg l$) limit.

Similarly, the study of I–V curves for 2D weak links yields the estimate $\Delta_i \simeq 80\,\mu\text{eV}$, compatible with the gap of the Al contacts ($\Delta_{Al} \simeq 170\,\mu\text{eV}$). Our junctions are consequently in an intermediate length regime $L \sim \xi$, given the estimated coherence length $\xi \simeq 600$ nm, but reach the ballistic limit $L \ll l$ owing to the large mean free path $l > 2\,\mu\text{m}$.

Mapping to the band structure in 2D topological insulator junctions The presence of a gate however enables to vary the electron density and to identify different transport regimes from the normal state resistance R_n and the critical current I_c. In agreement with the band structure presented in Fig. 5.2b, we distinguish three regimes. For gate voltages between $V_g = -1.1$ V and 0 V, R_n is low (below $300\,\Omega$) and I_c is large (above 200 nA). This signals the n-conducting regime, with a high mobility and high electron density in the plane of the junction. For gate voltages below $V_g = -1.7$ V, the normal state resistance tends to decrease slowly, indicating the p-conducting regime with a significantly lower mobility. The critical current I_c lies however below 50 nA. Between these two regimes, a peak in R_n (maximum around $1.5\,\text{k}\Omega$) and the quasi-suppression of I_c indicates the region where the QSH edge states should be most visible. The peak value of R_n is however much lower than the quantized value $h/2e^2 \simeq 12.9\,\text{k}\Omega$ and underlines the presence of residual bulk modes in the junction [55].

We here like to point out that the trivial narrow quantum wells used for reference samples exhibit a similar gate dependence for I_c and R_n. On the short length of the junctions, the gapped region (intermediate gate range) is not strongly insulating, maybe due to percolation transport due to disorder and residual charge puddles. In contrast, the gap region is observed to be strongly insulating on larger devices such as the Hall bars used for characterization of the layer properties.

The observations on 2D topological insulator junctions can be further validated by scrutinizing the response to a magnetic field perpendicular to the plane of the junction. Then, the superconducting phase difference ϕ becomes position-dependent [56], a property which helps revealing the spatial supercurrent distribution through modulations of the critical current I_c. When the electron density is high and the current flows uniformly in the 2D plane of the quantum well, the junction exhibits a conventional Fraunhofer pattern of the critical current versus magnetic field, which rapidly decays as the magnetic field increases. This indicates the n- and p-conduction regimes, respectively, for $V_g > -0.9$ V and $V_g < -1.8$ V. In contrast, the diffraction pattern is similar to that of a (DC) SQUID for V_g between -1.8 V and $V_g = -0.9$ V. It demonstrates that a large fraction of the supercurrent flows along the edges of the sample [55], as expected in the presence of QSH edge channels. We refer the reader to [57] for a detailed discussion.

5.2.3 Experimental Setups

Microwave setup for Josephson emission The simplest and most direct technique to measure Josephson radiation consists in measuring, after amplification, the emission spectra of the junctions with a spectrum analyzer. To this end, the junction is connected to a coaxial line and decoupled from the DC measurement line via a bias tee (see Fig. 5.5b). The rf signal is then amplified at both cryogenic and room temperature before being measured with a spectrum analyzer. The commercial rf components used in the readout line for our measurements limit the frequency range of detection to approximately 2–10 GHz. In spirit, this approach is similar to early measurements of Josephson emission using narrowband resonant cavities [59, 60], but with extended bandwidth thanks to microwave cryogenic amplifiers [61]. It contrasts with measurements using tunnel junctions as detectors [35, 62, 63]. This technique has the advantage of even wider bandwidths, but rely on the numerical deconvolution of modified I–V characteristics of Al tunnel junctions, the interpretation of which being difficult in some cases [64].

Microwave excitation for Shapiro steps The formation of Shapiro steps can be easily observed in the DC I–V characteristics when a Josephson junction is under rf excitation. The latter can be provided using either an open-ended coaxial cable (the end of which is placed a few millimeters from the sample), or by microbonding a lead to a microwave line, for example, via a directional coupler (see Fig. 5.5) to enable measurements of Josephson emission and Shapiro steps with a unique setup. In both geometries, frequencies in the range of 0–15 GHz are easily accessible, but the rf power supplied to the sample cannot easily be calibrated.

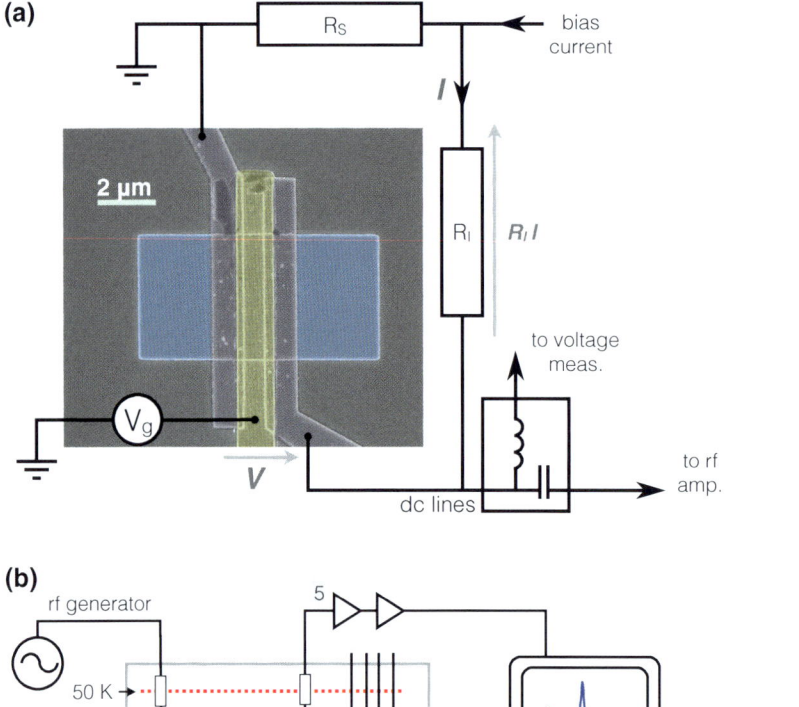

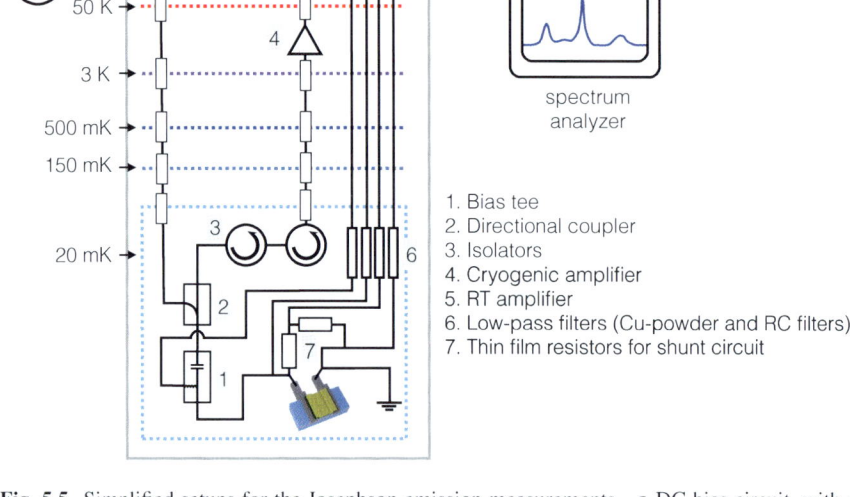

Fig. 5.5 Simplified setups for the Josephson emission measurements—**a** DC bias circuit, with a shunt resistance R_S to achieve a stable voltage bias and a resistance R_I to access the voltage V and current I. The rf signal is coupled to the amplification scheme via a bias tee. **b** Microwave amplification and detection setup, with cryogenic HEMT and room-temperature microwave amplifiers and a directional coupler to allow measurements of Shapiro steps and Josephson emission with a unique setup. Adapted from [58]

DC bias circuit An essential requirement for these measurements is to obtain a stable DC biasing of the junctions. We found that instabilities and hysteretic behavior occur at low voltages [57, 65] and therefore employ a small resistive shunt R_S (between 1 and 50 Ω) to enable a stable voltage bias. A small resistance R_I in series with the junction can be included to enable the measurement of the current I through the junction (Fig. 5.5a). With an adequately filtered fridge, we have been able to observe stable emission features down to about 1 GHz and Shapiro steps down to circa 500 MHz.

5.3 Experimental Observation of the Fractional Josephson Effect

> In this section, we review our observations of the fractional Josephson effect in the topological Josephson junctions based on HgTe, both in 2D and 3D topological insulators. We juxtapose the results obtained in the two systems to highlight their similarities, and compare them to the reference situation provided by a quantum well in the trivial regime.

5.3.1 Observation of Josephson Emission at $f_J/2$

5.3.1.1 Conventional Josephson Emission

We first focus on the investigation of Josephson emission. As a reference, we first discuss the case of the narrow HgTe quantum well in the trivial regime (Fig. 5.6, first line). In the first panel (Fig. 5.6a), the blue line indicates the I–V curve of the device. At zero bias on the junction, a background noise originating from black body radiation and parasitic stray noise from the environment is observed. It is subtracted from all measurements to isolate the contribution of the junction. When the junction is biased, a finite voltage V develops and the contribution of the junction appears, and it is plotted as a green line. The observed peak in the emission at $V \simeq 6\,\mu\text{V}$ corresponds to the matching of the detection frequency f_d with the Josephson frequency $f_J \simeq f_d = 3$ GHz. Sometimes, a second peak is observed at half this voltage, indicating a weak second harmonic at $2f_J$. The proportionality of f_J with V can be further verified by varying the detection frequency f_d, as shown in Fig. 5.6b. A single emission line is observed and fits perfectly with the theoretical prediction $f_J = \frac{2eV}{h}$. This constitutes the expected signature of the conventional Josephson effect, as already observed in the early days of Josephson physics [59, 60]. Besides, by varying the critical current I_c (with the gate voltage V_g), we verify that the amplitude A of the collected signal is proportional to $A \propto I_c$ with good agreement [66] and consequently reaches its minimal amplitude in the gap region.

5.3.1.2 Fractional Josephson Emission

In strong contrast, the junctions fabricated from 2D and 3D topological insulators reveal a strong emission peak at half the Josephson frequency $f_J/2$ (Fig. 5.6, second and third lines). This constitutes the most direct evidence of a 4π-periodic supercurrent flowing in these junctions [23].

Emission line at $f_J/2$ The observation of emission at half the Josephson frequency is illustrated for $f = 3$ GHz in panels Fig. 5.6c, e. As seen for the 3D TI, the emission at $f_J/2$ is sometimes concomitant with emission at f_J, depending on frequency or gate voltage. We detail below these aspects.

Besides, a recurring observation is that the linewidth of the emission line at $f_J/2$ is also larger (by up to a factor 10) than the conventional line at f_J. For instance in the quantum wells, both the topological and trivial devices exhibit a line at f_J with a typical width of $\delta V_{2\pi} \simeq 0.5 - 0.8\,\mu$V. The linewidth at $f_J/2$ exhibits values over a larger range, with widths in the range $\delta V_{4\pi} \simeq 0.5 - 8\,\mu$V.

Dependence on frequency We first discuss the data collected on the 3D topological insulator. In this device, we observe that the $f_J/2$ line is dominant for a large range of voltages (12–35 μV) or equivalently of frequencies (3–9 GHz). Outside that range, the conventional emission at f_J dominates. The data collected on the 2D topological insulator (Fig. 5.6b) is measured in the vicinity of the quantum spin Hall regime. In that case, the emission is clearly dominated by the $f_J/2$ line below $f = 5.5$ GHz, before the conventional line at f_J is recovered. We propose an interpretation of the influence of frequency in Sect. 5.4.

In both cases, the emission lines deviate from the expected emission lines and more complex structures with broadening, and multiple peaks are observed. We have identified resonant modes in the electromagnetic environment. In a dynamical Coulomb blockade situation, they possibly alter the emission spectrum in that range (see [58]) and are known to result in emission at $f_J/2$. Nevertheless, they cannot solely explain the fractional Josephson effect. Indeed, these are second-order processes in R_n/R_K (with $R_K = \frac{h}{e^2}$) and the amplitude of the $f_J/2$ line always remains of lesser amplitude than standard emission at f_J [67, 68].

Dependence on gate voltage (2D TI) As mentioned earlier, devices fabricated from 2D topological insulators enable to tune the electron density via the gate voltage V_g. We have clearly observed in these devices three regimes in the emitted power, which correlate with the expected band structure, as reported in Fig. 5.4b. When the gate voltage is above $V_g > -0.4$ V, we observe that emission occurs for $f_J/2$ at low frequency, but the conventional line is recovered and dominates above typically 5 GHz. These observations suggest transport in the conduction band of the quantum well, where gapless Andreev bound states have been seen to coexist with n-type conventional states, in agreement with previous observations and predictions [57, 69].

5 Microwave Studies of the Fractional ...

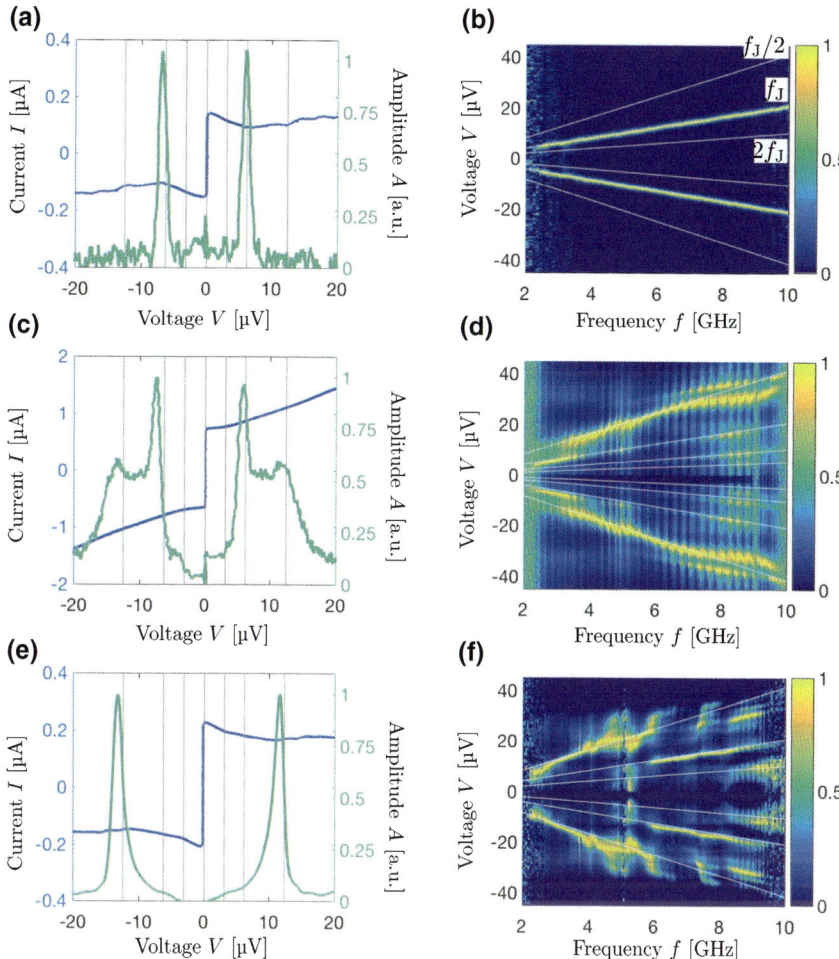

Fig. 5.6 Josephson emission in trivial and topological Josephson junctions—in the first column (**a**, **c**, **e**), an emission spectrum (amplitude A) taken at $f = 3$ GHz is plotted in green, alongside with the I–V curve of the device under consideration (depicted in blue). In the second column (**b**, **d**, **f**), the collected microwave amplitude A is presented as a colorplot, as a function of frequency f and voltage V. For better visibility, the data is normalized to its maximum for each frequency, and white guidelines indicate the f_J, $f_J/2$, and $2f_J$ lines (see panel **b**). The topologically trivial quantum well is shown in the first line (**a**, **b**), while the second (**c**, **d**) and third (**e**, **f**) show, respectively, the 3D topological weak links and the 2D ones (taken at $V_g = -0.55$ V). In panel **c**, the current I is actually the total bias current, sum of the current in the shunt resistor R_s and in the junction branches (Fig. 5.5a). The second resistor R_l has been here suppressed. This simplified circuit does not enable a proper measurement of the current in the junction only, but provides a correct readout of voltage V

However, in a narrower gate range $-0.8\,\text{V} < V_g < -0.6\,\text{V}$, one observes almost exclusively emission at half the Josephson frequency $f_J/2$ up to very high frequencies (circa 8–9 GHz). We attribute this observation to the quantum spin Hall regime, where edge states are the dominant transport channel. Finally for $V_g < -0.8\,\text{V}$, the Josephson radiation at $f_J/2$ is weakly visible, which suggests that the gapless Andreev modes more rapidly hybridize with bulk p-type conventional modes of the valence band. The overall gate voltage dependence is consistent with the expected band structure of a quantum spin Hall insulator, as presented in Fig. 5.2b, but a quantitative description of the features remains difficult due to the observed irregularities in the emission lines.

5.3.2 Observation of Even Sequences of Shapiro Steps]

5.3.2.1 Conventional Shapiro Response

We now turn to the second signature of the fractional Josephson effect, namely the observation of even sequences of Shapiro steps. Under microwave excitation at frequency f, the presence or absence of steps can be observed directly in the I–V curves, but are conveniently highlighted by binning the measurement data according to the voltage. For $V_n = nhf/2e$ with n integer, Shapiro steps then appear as peaks in the bin counts, and their amplitude then reflects the current range over which the DC voltage stays fixed. In Fig. 5.7, we present I–V curves for a given power of microwave irradiation (a, b, c), the histograms resulting from the binning as bar plots (d, e, f), and finally the same histograms in a colorplot function of voltage V and microwave power P_{RF}.

We first concentrate on the trivial Josephson junction as a reference situation (first column). Panels Fig. 5.7a, b illustrate for $f = 5.64\,\text{GHz}$ the observation of all steps (odd and even), regardless of gate voltage V_g, or excitation frequency f. The response is similar to that of other systems such as carbon nanotubes [70], graphene [71], Bi_2Se_3 [72] weak links, or the well-defined and meticulously analyzed case of atomic contacts [12]. The amplitude of steps (along the bin counts axis) and height (along the voltage axis) are both reduced with decreasing frequency, and a correct resolution of the steps is only possible down to circa 500 MHz. The evolution with power is shown in panel Fig. 5.7g). At zero microwave power ($P_{\text{RF}} = 0$), a single peak at $V = 0$ indicates the supercurrent. As P_{RF} increases, Shapiro steps appear, starting from low values of n, while the amplitude of the supercurrent ($n = 0$) decreases and eventually vanishes. For sufficiently high powers, an oscillatory pattern occurs in the amplitude of steps, as predicted for conventional Josephson junctions submitted to a voltage or current bias [56, 73].

We also point out that we occasionally observe (in trivial and topological weak links) the so-called subharmonic steps, i.e., steps for voltages $\frac{p}{q}\frac{hf}{2e}$ for $q = 2$ or 3, and p integer. These subharmonic steps are observed at high frequencies, in a regime where both conventional and topological weak links exhibit a conventional

5 Microwave Studies of the Fractional … 133

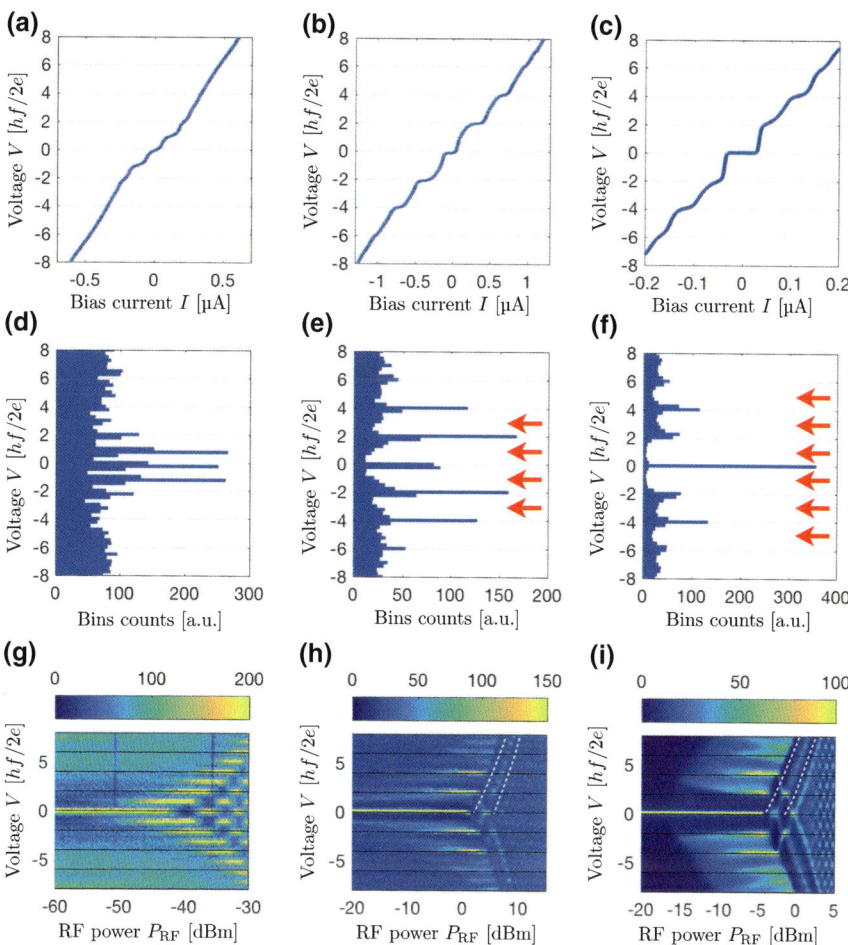

Fig. 5.7 Shapiro steps in trivial and topological Josephson junctions—in the first line (**a, b, c**), typical $I-V$ curves of the device under consideration are plotted. They exhibit Shapiro steps, the step index of which can be read from the normalized voltage V. In the second line (**d, e, f**), the histograms corresponding to the previous $I-V$ curves are shown as barplots and highlight the vanishing of odd steps (indicated by red arrows) in 2D and 3D topological weak links. Finally, the last line (**g, h, i**) presents colorplots obtained from the previous histograms, as a function of voltage V and microwave power P_{RF}. The first column (**a, d, g**) shows data from a reference non-topological device (for $f = 5.64$ GHz), the second column (**b, e, h**) from a 3D topological insulator ($f = 1$ GHz), and the last column (**c, f, i**) from a 2D topological insulator ($f = 1$ GHz)

Josephson effect (see next paragraphs). Such subharmonic steps are ubiquitous in Josephson junctions. They indicate a non-trivial phase-locking condition, namely $\phi(t + q/f) = \phi(t) + 2p\pi$. They are known to result from nonlinearities, stray capacitive coupling between the superconducting electrodes, or higher harmonics in the current–phase relations. The latter have been predicted [15] and detected [54] in our junctions.

5.3.2.2 Shapiro Response of Topological Josephson Junctions

As discussed in Sect. 5.1, for topological Josephson junctions, the presence of 4π-periodic supercurrents can in principle lead to the disappearance of odd steps, while even steps are preserved. First signs of the possible disappearance of the $n = 1$ step have been reported in etched InAs nanowires [74], driven by the predicted topological phase transition when a magnetic field is applied along the axis of the nanowire. Our observations made on the HgTe-based Josephson junctions [57, 65] have conclusively improved the data and exhibit the disappearance of several odd steps in devices made of 2D as well as 3D topological insulators.

Even sequence of Shapiro steps Data obtained on 3D topological insulators is presented in the second column of Fig. 5.7. The I–V clearly exhibits very strong $n = 0$ (supercurrent), 2, 4 steps, but the steps $n = 1$ and 3 are strongly suppressed. For higher voltages, the steps $n \geq 5$ are not resolved at such power. In the original work of Wiedenmann et al., only the $n = 1$ step was missing [65]. This new data thus shows a stronger 4π-periodic behavior and confirms that the disappearance of Shapiro steps is not related to hysteresis [75]. Similar features have been observed in devices fabricated from HgTe-based topological insulators exhibiting the QSH effect [57] and are summarized in the third column of Fig. 5.7. The linecut of panel c and the corresponding histogram (panel f) exhibits in particular the clear suppression of steps $n = 1, 3, 5$. At even lower frequencies, it has been possible to measure an even sequence of Shapiro steps up to $n = 10$.

Dark fringes in the oscillatory pattern The absence of odd steps is also remarkably clear on the colorplot of panel h of Fig. 5.7 where the steps $n = 1$ and 3 are suppressed, and $n = 5$ weakly visible (3D TI), and in panel i for which the steps $n = 1, 3, 5$ are suppressed (2D TI) for low microwave amplitude. Interestingly, the oscillatory pattern at higher microwave power is also modified: Darker fringes (highlighted with white dashes) occur from the suppression of the first and third maxima of the oscillations. They suggest the progressive transformation from a 2π- to a 4π-periodic pattern with a halved period of oscillations. Despite some unexplained deviations, the colorplot of the 2D TI confirms the absence of odd steps and exhibits even more pronounced dark fringes in the oscillatory pattern.

Dependence on frequency An important parameter that we emphasize now is the choice of the excitation frequency f. In analogy with the emission line at $f_J/2$, only visible at low frequencies, the even sequence of Shapiro steps is only observed when f is low. For high frequencies, the Shapiro response is conventional, and all step indices n are present. As f decreases, the odd steps progressively vanish, starting from low values of n. This is visible, for example, in Fig. 5.8. While odd steps are as visible as even ones at high frequencies ($f = 6.6$ GHz), they are progressively suppressed as f decreases. At the lowest accessible frequency ($f = 0.8\,\text{GHz}$), all odd steps up to $n = 9$ are absent. This remarkable effect of frequency is similar to the one observed in the Josephson emission features, for which the line at $f_J/2$ is mostly visible at low frequency. We analyze this behavior in Sect. 5.4.1.

5.4 Analysis: Assessing the Topological Origin of the Fractional Josephson Effect

The data summarized in this chapter exhibits two pieces of evidence of a strong 4π-periodic *fractional Josephson effect*, despite the obstacles to its observation listed in Sect. 5.1.3.2. We analyze in this section the possible (trivial or topological) origins of these features. First, we present an extended resistively shunted junction (RSJ) model that includes a 4π-periodic supercurrent. It enables a semiquantitative analysis of our experimental results and importantly yields an estimate of the amplitude of the 4π-periodic supercurrent, compatible with a topological origin. Then, we analyze more in depth the effects of time-reversal and parity symmetry breaking, and Landau–Zener transitions, following the discussion of Sect. 5.1.3.2.

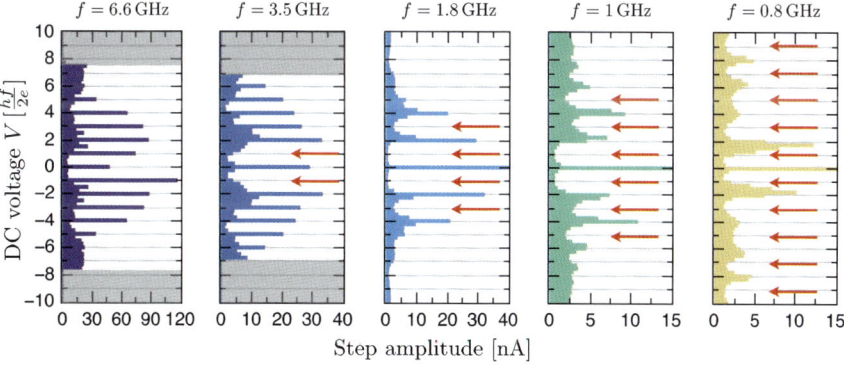

Fig. 5.8 Histograms of Shapiro steps—histograms of the voltage distribution obtained for different frequencies are shown. For a high frequency $f = 6.6\,\text{GHz}$, all steps are visible. For lower frequencies, steps $n = 1$ and 3 vanish ($f = 1.8\,\text{GHz}$), and up to $n = 9$ at $f = 0.8\,\text{GHz}$. Missing odd steps are highlighted by red arrows

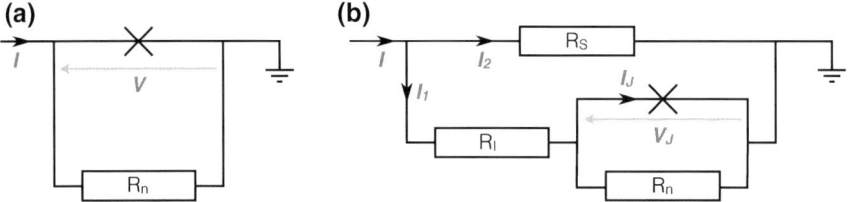

Fig. 5.9 Circuits for RSJ simulations—**a** standard RSJ representation, with the Josephson junction carrying a supercurrent $I_S(\varphi)$ in parallel with a shunt resistance R_n. **b** Modified RSJ circuit taking into account additional measurement setup with resistors R_I and R_S. Adapted from [58]

5.4.1 Modeling of a Topological Josephson Junction with 2π- and 4π-periodic Modes

It is possible to calculate, at a microscopic level, the Andreev spectrum of a topological Josephson junctions in the zero-bias equilibrium situation under various assumptions [2, 15, 17]. However, it is much more difficult to compute at the same elementary level the time-dependent response to a bias voltage or current that emerges from the nonlinear Josephson equations [76, 77], especially in the current-bias situation that is experimentally most relevant. To simulate the response of our devices, we turn to the RSJ model [73, 78, 79], in which we incorporate both 2π- and 4π-periodic supercurrents [33, 52, 57, 80].

Framework of the RSJ model and modeling The RSJ model and its variants are commonly used to define the time-averaged voltage measured when a Josephson junction is submitted to a dc current bias. The universally valid time evolution of the phase difference $d\phi/dt = 2eV/\hbar$ is combined with the current–phase relation $I_{2D/3D}(\phi)$ derived from microscopic models (Sect. 5.1.3). The junction is associated to a resistive shunt to capture the ohmic transport of electrons (Fig. 5.9a). Under current bias $I = I_{\text{dc}} + I_{\text{rf}} \sin(2\pi f t)$, one easily obtains a first-order ordinary differential equation:

$$\frac{\hbar}{2eR_n}\dot{\phi} + I_s(\phi) = I_{\text{dc}} + I_{\text{rf}} \sin(2\pi f t) \tag{5.8}$$

For different values of $I_{2\pi}$ and $I_{4\pi}$, we simulate the results of this equation using a simple Runge–Kutta algorithm (RK4). We obtain the I–V curve without ($I_{\text{rf}} = 0$) or with microwave excitation ($I_{\text{rf}} \neq 0$) to investigate Shapiro steps, or the Fourier spectrum of the voltage for the study of Josephson emission.

This model does not take into account all microscopic details, (e.g., R_n and I_s are assumed to be independent of voltage, which in reality may not be true). It nonetheless captures the key aspects of the dynamic Josephson current relevant to our observations. Besides, the more complicated bias circuit used to stabilize Josephson emission (Fig. 5.9b) can be readily implemented. This circuit is indeed described by 5.8, with the substitutions $R_n \to \tilde{R}_n$ such that $\frac{1}{\tilde{R}_n} = \frac{1}{R_n} + \frac{1}{R_S} + \frac{R_I}{R_S R_n}$, and $I_c \to \tilde{I}_c =$

$I_c\left(1 + \frac{R_L}{R_S}\right)$. Simulations performed in the standard RSJ model can be adapted to this new setup. The experimental data is then more naturally presented as a function of I_1 rather than I, which is obtained from $I_1 = \frac{I - \frac{V_J}{R_S}}{1 + \frac{R_L}{R_S}}$.

5.4.1.1 Simulations of the Response of a Topological Josephson Junction

We give in this section a summary of the major results of the simulations. A comprehensive presentation of the simulation can be found in [58, 65, 80].

Necessity of the 4π-periodic contribution This framework enables a complete analysis of our results. As expected, the simulations emphasize the absolute necessity of a 4π-periodic contribution to observe the vanishing of odd Shapiro steps or Josephson at half the Josephson frequency $f_J/2$. Indeed, the exact definition of the 2π-periodic supercurrent, namely the presence of higher harmonics ($\sin 2\phi$ terms for example), has marginal influence on the observed disappearance of the odd steps, but the presence of a 4π-periodic contribution is required to produce the signatures of the fractional Josephson effect.

This requirement contrasts with additional subharmonic Shapiro steps or Josephson emission at higher harmonics ($2f_J$) which naturally appear in the simulations. They result from either non-sinusoidal current–phase relations or also from the addition of a small capacitive coupling in the RSJ equations (see below) [81, 82].

Josephson emission spectrum and Shapiro steps The numerical solutions to the RSJ equations give access to the time-dependent quantity $\phi(t)$ and allow for the study of the Fourier spectrum of the voltage, which is (up to an unknown and frequency-dependent coupling factor) the quantity A plotted in Sect. 5.3.1. First setting the critical current I_c and the normal state resistance R_n to optimally fit the experimental I–V curve (see Fig. 5.10a), we adjust the ratio $I_{4\pi}/I_{2\pi}$ (keeping I_c constant) to obtain the best visual agreement between the experimental (Fig. 5.6f) and simulated (Fig. 5.10b) emission spectra.

In the present case, the simulated emission features reproduce semiquantitatively the observed ones. In particular, as in the measurements, the $f_J/2$ emission line dominates at low frequency, while the f_J takes over at higher frequencies. However, the crossover between the two regimes at $f \simeq 6$ GHz is however smoother than the measured one.

Similarly, we have been able to reproduce sequences of even Shapiro using the extended RSJ model. In particular, we observe the vanishing of all Shapiro steps only when a 4π-periodic supercurrent is present (see [65]). As the excitation frequency f decreases, we observe a transition from a conventional 2π-periodic Shapiro step pattern to a fractional 4π-periodic one. The crossover qualitatively describes our experimental observation, but in this model all odd steps vanish simultaneously, while our experiments exhibit a progressive disappearance starting from low values

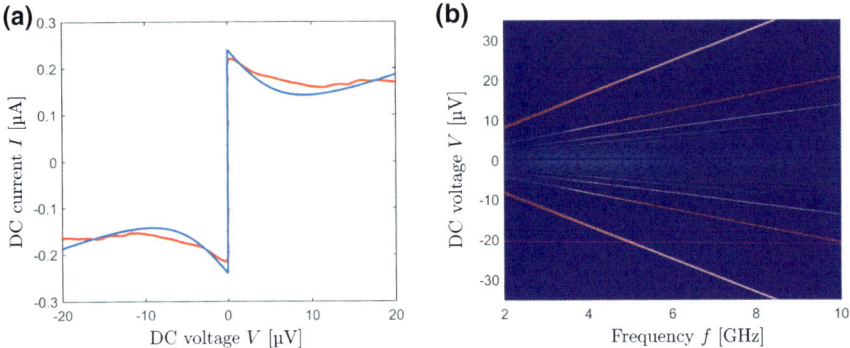

Fig. 5.10 Modeling Josephson Emission–**a** simulated I–V curve (blue) fitting measured data (red) The simulations are performed for $R_I = R_S = 24\,\Omega$ (nominal value of the resistors), $R_n = 130\,\Omega$, $I_{4\pi} = 100$ nA, and $I_c = 240$ nA. **b** Simulated Fourier transform of the voltage V in the junction, as function of frequency f and voltage V, with the same parameters as in **a**. Adapted from [58]

of the step index n (see Fig. 5.8). We discuss the peculiar effect of frequency in the next paragraph.

Role of frequency—Estimating the 4π-periodic supercurrent The puzzling dependence of the $f_J/2$ emission line or the sequence of Shapiro steps in fact signals the transition from 2π- to 4π-periodic dynamics and intrinsically results from the nonlinearities of the RSJ equation (5.8). It has first been elucidated and later clarified by Dominguez et al. [33, 80], and we refer the reader to these two publications for details. The transition is controlled by the 4π-periodic supercurrent $I_{4\pi}$ or equivalently the voltage $V_{4\pi} = R_n I_{4\pi}$ or the frequency scale $f_{4\pi} = \frac{2eR_n I_{4\pi}}{h}$. For an excitation frequency $f \ll f_{4\pi}$ (Shapiro steps) or a bias such that $V \ll V_{4\pi}$ (Josephson emission), the dynamics of the phase $\phi(t)$ is rather slow and very nonlinear: $V(t)$ is a very anharmonic function of t and becomes sensitive to the presence of the 4π-periodic component. There, signatures of the fractional Josephson effect are very prominent. On the opposite, for $f \gg f_{4\pi}$ or $V \gg V_{4\pi}$, $V(t)$ is sinusoidal and rather insensitive to the 4π-periodic contribution, so that the response of the device is comparable to conventional Josephson junctions.

The identification of the crossover frequency in both experiments consequently provides a criterion to estimate the amplitude $I_{4\pi}$ of the 4π-periodic supercurrent and compare it to theoretical expectations. We expect two modes to contribute in the 2D topological insulator (one edge mode on either edge), and one mode ($\theta = 0$) for the 3D topological insulator. A perfectly transmitted mode carries in the short junction limit a supercurrent of maximum amplitude $\frac{e\Delta_i}{\hbar}$. In all tested devices, we found amplitudes ranging from 50–300 nA. Knowing Δ_i, this corresponds to 1–5 modes. This is roughly compatible with theoretical predictions, though it slightly exceeds them. It is likely that the large uncertainty on Δ_i as well as the crude approximations of the RSJ model explain an improper estimate of $I_{4\pi}$. For example, we discuss below the influence of a capacitive coupling added to the RSJ equation (RCSJ model).

5.4.1.2 Beyond the RSJ Model

Our analysis has so far been based on an extended RSJ model taking into account a 4π-periodic contribution to the supercurrent. It provides a simple analysis and interpretation of our results, but several ingredients can be improved. The influence of a capacitive term in the RSJ equation can be important at microwave frequencies and can be easily accounted for in the resistively and capacitively shunted junction model (RCSJ). This has been investigated in a recent study [83], which demonstrates that $I_{4\pi}$ can be overestimated with the above criterion even for small capacitances. Though it is difficult to estimate quantitatively C, it may explain the discrepancies on $I_{4\pi}$ between theory and experiments. Besides, in that model, the odd Shapiro steps are observed to vanish one by one (starting from low values of the step index n) rather than altogether in the RSJ model. This appears to be in better agreement with our data. Based on our estimates [37, 57], the geometrical contribution to the capacitance between the two superconducting electrodes is however quite small. It has nevertheless been recently pointed out that Andreev bound states with high transparencies contribute to an intrinsic capacitance in mesoscopic devices and may, for example, explain the observed hysteresis in the DC I–V curves [51].

At this point, we would like to point out that it is easy to expand arbitrarily the nonlinear differential equation with additional terms. Instead, we believe that the most rigorous but also most challenging approach is to construct a full microscopic understanding of the dynamics of the Andreev levels, including relaxation processes, and hence the dynamics of the currents and voltages.

5.4.2 Time-Reversal and Parity Symmetry Breaking, and Landau–Zener Transitions

In this section, we review the influence of several mechanisms which are expected to obscure the fractional Josephson effect (Sect. 5.1.3.2) but cannot be easily taken into account in the preceding RSJ model.

5.4.2.1 Time-Reversal Symmetry Breaking

We have first pointed out that time-reversal symmetry in our devices imposes a degeneracy of Andreev bound states at $\phi \equiv 0 \bmod 2\pi$, which tends to restore the 2π-periodicity due to enhanced parity relaxation, or on the opposite create 8π-periodicity under the effect of electron–electron interactions. We have not detected any signal indicating an 8π-periodic Josephson effect (such as Josephson emission at $f_J/4$), but in contrast clearly observe a 4π-periodic fractional Josephson effect.

In our view, current models of topological Josephson junctions are partially inadequate and overlook important microscopic details. Though in our experiments time-reversal symmetry is not explicitly broken by a magnetic field or magnetic impurities, other mechanisms implicitly break time-reversal symmetry [29, 84–88], thus decoupling the Andreev spectrum from the continuum. Such mechanisms are, for example, in line with the observed weak stability of conductance quantization in the QSH regime [45, 46]. Descriptions based on hard superconducting gaps also overlook the complexity of the density of states in induced systems [89, 90], in which the behavior of Andreev bound states remains poorly understood.

More experiments are required to characterize the induced superconducting state, with techniques such as point contact Andreev spectroscopy, already successfully employed on 3D topological insulators [10, 52, 91].

5.4.2.2 Landau–Zener Transitions and Parity Relaxation Mechanisms

The role of Landau–Zener transitions and parity relaxation mechanisms must be emphasized as they are expected to strongly influence the response of Josephson junctions, in particular at low frequency. These two processes cannot, to our understanding, easily be disentangled. We subsequently evaluate in parallel both possibilities.

Low-frequency behavior First, non-adiabatic Landau–Zener transitions can enforce a fractional Josephson effect from (sufficiently) driven gapped states. As a result of the Josephson equation $\frac{d\phi}{dt} = \frac{2eV}{\hbar}$, the phase $\phi(t)$ will vary faster as the voltage (or current) bias increases. Second, parity relaxation mechanism defines a lifetime τ of the gapless Andreev bound states. When driven sufficiently rapidly, i.e., at characteristic frequencies f such that $f\tau \gg 1$, a topological Josephson junction can exhibit a 4π-periodic response. On long times, the system however thermalizes to a conventional 2π-periodic response. In both cases, the Josephson effect should be conventional at low bias/low frequency and turned into a fractional one above a crossover voltage V_c, as non-adiabatic processes are progressively activated.

Our observations show in 2D topological insulators that the two signatures of a fractional Josephson effect are observed down to the lowest observables frequencies ($\simeq 1$ GHz). Potential Landau–Zener transitions would thus be activated at a voltage $V_{LZ} \ll 4\,\mu$V. This sets an strict upper bound on a possible residual avoided crossing $\delta \ll \sqrt{\frac{V_{LZ}\Delta_i}{8\pi}} = 4\,\mu$eV [25, 92], equivalent to large transmissions $D \gg 0.995$. This strong constraint on the transmission thus tends to exclude Landau–Zener transitions as the origin for the 4π-periodic emission and suggests that the contributing bound states are indeed gapless.

Josephson junctions based on 3D topological insulators have mostly shown the same behavior (see [58]), with a strong $f_J/2$ emission line at low frequency. The device presented in this chapter (Fig. 5.6d) shows however a different behavior: A resurgence of the conventional f_J emission line is clearly visible for $f < 2\,\text{GHz}$, while the signal at $f_J/2$ vanishes. Assuming Landau–Zener transitions are responsible for this transition around $V_{LZ} \simeq 8\,\mu\text{V}$, we find a gap on the order of $\delta = 6\,\mu\text{V}$ and a transmission $D \simeq 0.995$ (with $\Delta_i = 100\,\mu\text{V}$).

Emission linewidths The finite lifetime of gapless Andreev bound states due to parity relaxation processes or Landau–Zener transitions can strongly affect the linewidth of the Josephson emission. For conventional Josephson radiation, the linewidth is in principle related to fluctuations in the pair or quasiparticle currents [93, 94] or can be dominated by the noise in the environment [66]. The linewidth at $f_J/2$ can additionally reflect the influence of parity relaxation mechanisms [20, 23].

A difference in the linewidths of the two emission lines is clearly visible in our data. In the quantum wells, the extracted width ($\delta V_{2\pi} \simeq 0.5\text{–}0.8\,\mu\text{V}$) of the f_J emission line can be converted to a coherence time $\tau_{2\pi} = \frac{h}{2e\delta V_{2\pi}} \simeq 3\text{–}4\,\text{ns}$. We find a shorter coherence time $\tau_{4\pi} \simeq 0.3\text{–}4\,\text{ns}$ for the $f_J/2$ emission line. The linewidth is found to increase when the gate voltage V_g is driven deeper in the conduction band. This may signal a decrease of lifetime when the 4π-periodic modes are coupled to an increasing number of 2π-periodic modes or to the continuum via ionization processes. We finally point out that the extracted lifetimes are consistent with the Shapiro steps being observable down to typically 0.5–1 GHz [57].

For the 3D topological device presented here, we find larger linewidths. The conventional f_J line with typically $\delta V_{2\pi} \simeq 2\text{–}3\,\mu\text{V}$ corresponding to $\tau_{2\pi} \simeq 0.7\text{–}1\,\text{ns}$, while the fractional line has a width of $\delta V_{4\pi} \simeq 4\text{–}7\,\mu\text{V}$, yielding $\tau_{4\pi} \simeq 0.25\text{–}0.5\,\text{ns}$. The difference of time scales between the 2D and 3D devices has been observed on several junctions but is not understood yet.

5.5 Summary, Conclusions, and Outlook

We end this chapter with a summary of our observations, our conclusions, and an outlook toward future experiments.

Summary and conclusions The existence of 4π-periodic supercurrents has been demonstrated from two sets of observations in HgTe-based Josephson junctions in both a 2D and 3D topological insulator regime. First, these junctions show even sequences of Shapiro steps, with several missing odd steps (step indices $n = 1, 3, 5$ missing). Secondly, they also exhibit strong Josephson emission at half the Josephson frequency $f_J/2$ [58]. In contrast, reference devices made of graphene or trivial HgTe quantum wells do not show any of the above features, but a conventional Josephson response. Besides, we have presented a model for the junctions based on an extended RSJ model. It provides a simple yet efficient and plausible explanation of the transition from a 4π- to a 2π-periodic response with frequency or bias. This model yields the amplitude of the 4π-periodic supercurrent, found to be compatible with theoretical predictions, though this estimate is subject to a large uncertainty. Our analysis of the emission features also sets very strong bounds on the possible influence of Landau–Zener activation and provides additional information on the lifetime of the Andreev bound states. Finally, in junctions based on 2D topological insulators, the two signatures of the fractional Josephson effect are found to be concomitant (with respect to gate voltage). As presented in Fig. 5.4b, they are more clearly visible when the current flow is mostly along the edges of the sample, and the bulk bands are depleted. The 4π-periodic contribution is also detected in the whole n-conduction band. This suggests that the 4π-periodic edge modes exist in parallel with bulk modes on the n-side. This interpretation is consistent with previous observations and predictions for HgTe [48, 55, 69].

All in all, our observations thus strongly favor the presence of gapless Andreev bound states in our topological Josephson junctions, as initially predicted by Fu and Kane [1, 2]. These devices, built from the well-characterized HgTe topological insulators, thus appear as first steps toward the development of a reliable platform for the future realization of Majorana end states and possibly scalable Majorana qubits.

Future objectives The exact microscopic properties of the induced superconducting state, the presence of purely ballistic modes regardless of topology, or the role of time-reversal and parity symmetry breaking remains partially unclear. While our studies provide clear evidence of a supercurrent with 4π-periodicity, a direct spectroscopy of such gapless Andreev bound states (ABS) is still missing. It is highly desirable as it would confirm the topological origin of the 4π-periodicity and offer direct proof of the existence of gapless Majorana-Andreev bound states, as well as allow to verify the robustness of the topological protection.

This calls for a microscopic description of the induced superconductivity and of the dynamics of Josephson transport, including relaxation processes. Recent works have tackled this challenging program in topological systems [76, 95], but the description of the induced superconductivity remains in many cases rudimental.

From an experimental point of view, some recent works have for example focused on point contact Andreev spectroscopy [10, 52, 91] to probe the proximity effect in topological insulators in S–TI junctions. Future experiments will consequently focus on collecting more direct information on the Andreev spectrum forming in Josephson junctions, beyond their manifestation in the Josephson effect.

A first method consists in the study of the current–phase relation, which can be measured in asymmetric SQUIDs [96, 97]. By tuning the electron density via a gate, the goal is to identify the contribution of the topological modes. For example, the linear susceptibility $\frac{\partial I}{\partial \Phi}$ (with Φ the magnetic flux) at high frequency is a very sensitive probe able to reveal the topologically protected level crossing at $\phi = \pi$ [98]. Furthermore, the investigation of the switching statistics [29, 96] around the critical current provides a means to prove that both states of a topological Andreev doublet have different parities.

The Andreev spectrum of a Josephson junction can also be obtained by means of tunneling spectroscopy, as already demonstrated in carbon nanotubes [99] or graphene [100]. A reliable tunnel barrier could, for example, be obtained from hexagonal boron nitride flakes. The junction is then controlled via a phase bias mode by including the junction in a SQUID geometry, such that the phase difference ϕ across the junction is directly set by the magnetic flux through the ring.

A third method relies on microwave spectroscopy techniques: The absorption or emission of microwaves when at resonance with a transition in the Andreev spectrum is monitored [13, 101, 102]. It can be performed in a SQUID geometry to bias the phase, with the SQUID inductively coupled to a microwave transmission line. Passing a microwave signal through the waveguide yields absorption lines of the Andreev spectrum of the junction. Additionally, the emission lines of the junction can also be measured. Topological Majorana bound states (MBS) then show various characteristic features [29]. First, since parity is invariable under photon absorption or emission, the transition between both states of the topological doublet should be strongly suppressed and only transitions involving the continuum should be visible. Second, the dispersion of the absorption/emission lines (as function of the magnetic flux) reflects the special 4π-periodicity of the bound states. A natural follow-up to these experiments is then the exploration of topological transmons [103] as a step toward braiding of Majorana qubits.

Acknowledgements We warmly thank the editors for their work and the opportunity to share our results. This work is supported by the German Research Foundation (Leibniz Program, SFB1170 Tocotronics) and the Elitenetzwerk Bayern program Topologische Isolatoren, and the EU ERC-AG Program (project 4-TOPS). EB acknowledges support from the Alexander von Humboldt foundation. TMK acknowledges support from RSF Grant Non 17-72-30036 of the Russian Federation and Advanced Research Grant of the EC No. 339306 (METIQUM). RSD gratefully acknowledges support from "Grants-in-Aid for scientific research" (No. 16H02204), from the Japan Society for the Promotion of Science.

References

1. L. Fu, C. Kane, Superconducting proximity effect and majorana fermions at the surface of a topological insulator. Phys. Rev. Lett. **100**(9), 096407 (2008)
2. L. Fu, C. Kane, Josephson current and noise at a superconductor/quantum-spin-Hall-insulator/superconductor junction. Phys. Rev. B **79**(16), 161408 (2009)
3. T.M. Klapwijk, Proximity effect from an Andreev perspective. J. Supercond. **17**(5), 593–611 (2004)
4. J. Alicea, New directions in the pursuit of Majorana fermions in solid state systems. Rep. Prog. Phys. **75**(7), 076501 (2012)
5. C.W.J. Beenakker, Search for Majorana fermions in superconductors. Ann. Rev. Condens. Matter Phys. **4**(1), 113–136 (2013)
6. R. Aguado, Majorana quasiparticles in condensed matter. Nuevo Cimento **40**, 523–593 (2017)
7. V. Mourik, K. Zuo, S.M. Frolov, S.R. Plissard, E.P.A.M. Bakkers, L.P. Kouwenhoven, Signatures of Majorana fermions in hybrid superconductor-semiconductor nanowire devices. Science **336**(6084), 1003–1007 (2012)
8. S.M. Albrecht, A.P. Higginbotham, M. Madsen, F. Kuemmeth, T.S. Jespersen, J. Nygård, P. Krogstrup, C.M. Marcus, Exponential protection of zero modes in Majorana Islands. Nature **531**(7593), 206–209 (2016)
9. C.W.J. Beenakker, Three "Universal" mesoscopic Josephson effects, in *Transport Phenomena in Mesoscopic Systems* (1992), pp. 235–253

10. G.E. Blonder, M. Tinkham, T.M. Klapwijk, Transition from metallic to tunneling regimes in superconducting microconstrictions: excess current, charge imbalance, and supercurrent conversion. Phys. Rev. B **25**(7), 4515–4532 (1982)
11. H.-J. Kwon, K. Sengupta, V.M. Yakovenko, Fractional ac Josephson effect in p- and d-wave superconductors. Eur. Phys. J. B Condens. Matter **37**, 349–361 (2003)
12. M. Chauvin, P. vom Stein, H. Pothier, P. Joyez, M.E. Huber, D. Esteve, C. Urbina, Superconducting atomic contacts under microwave irradiation. Phys. Rev. Lett. **97**, 067006 (2006)
13. L. Bretheau, Ç.Ö. Girit, H. Pothier, D. Esteve, C. Urbina, Exciting Andreev pairs in a superconducting atomic contact. Nature **499**(7458), 312–5 (2013)
14. H.-J. Kwon, V.M. Yakovenko, K. Sengupta, Fractional ac Josephson effect in unconventional superconductors. Low Temp. Phys. **30**(7), 613 (2004)
15. G. Tkachov, E.M. Hankiewicz, Helical Andreev bound states and superconducting Klein tunneling in topological insulator Josephson junctions. Phys. Rev. B **88**(7), 075401 (2013)
16. G. Tkachov, E.M. Hankiewicz, Spin-helical transport in normal and superconducting topological insulators. Phys. Status Solidi (b) **250**(2), 215–232 (2013)
17. C.T. Olund, E. Zhao, Current-phase relation for Josephson effect through helical metal. Phys. Rev. B **86**(21), 214515 (2012)
18. S. Shapiro, Josephson currents in superconducting tunneling: the effect of microwaves and other observations. Phys. Rev. Lett. **11**(2), 80–82 (1963)
19. M. Houzet, J.S. Meyer, D.M. Badiane, L.I. Glazman, Dynamics of Majorana states in a topological Josephson junction. Phys. Rev. Lett. **111**(4), 046401 (2013)
20. D.M. Badiane, L.I. Glazman, M. Houzet, J.S. Meyer, Ac Josephson effect in topological Josephson junctions. C. R. Phys. **14**(9–10), 840–856 (2013)
21. I.O. Kulik, A.N. Omelyanchuk, Properties of superconducting microbridges in the pure limit. Sov. J. Low Temp. Phys. **3**, 459 (1977)
22. C.W.J. Beenakker, H. van Houten, Josephson current through a superconducting quantum point contact shorter than the coherence length. Phys. Rev. Lett. **66**(23), 3056–3059 (1991)
23. D.M. Badiane, M. Houzet, J.S. Meyer, Nonequilibrium Josephson effect through helical edge states. Phys. Rev. Lett. **107**(17), 177002 (2011)
24. P. San-Jose, E. Prada, R. Aguado, ac Josephson effect in finite-length Nanowire junctions with Majorana modes. Phys. Rev. Lett. **108**(25), 257001 (2012)
25. D.I. Pikulin, Y.V. Nazarov, Phenomenology and dynamics of a Majorana Josephson junction. Phys. Rev. B **86**(14), 140504 (2012)
26. Shu-ping Lee, Karen Michaeli, Jason Alicea, Amir Yacoby, Revealing topological superconductivity in extended quantum spin Hall Josephson junctions. Phys. Rev. Lett. **113**(19), 197001 (2014)
27. C.W.J. Beenakker, D.I. Pikulin, T. Hyart, H. Schomerus, J.P. Dahlhaus, Fermion-parity anomaly of the critical supercurrent in the quantum spin-Hall effect. Phys. Rev. Lett. **110**(1), 017003 (2013)
28. F. Zhang, C.L. Kane, Time-reversal-invariant Z_4 fractional Josephson effect. Phys. Rev. Lett. **113**, 036401 (2014)
29. Y. Peng, Y. Vinkler-Aviv, P.W. Brouwer, L.I. Glazman, Felix von Oppen, Parity anomaly and spin transmutation in quantum spin Hall Josephson junctions. Phys. Rev. Lett. **117**, 267001 (2016)
30. H.-Y. Hui, J.D. Sau, 8π-periodic dissipationless ac josephson effect on a quantum spin-hall edge via a quantum magnetic impurity. Phys. Rev. B **85**, 014505 (2016)
31. Y. Vinkler-Aviv, P.W. Brouwer, F. von Oppen, Z_4 parafermions in an interacting quantum spin Hall Josephson junction coupled to an impurity spin. Phys. Rev. B **96**(19), 195421 (2017)
32. C.J. Pedder, T. Meng, R.P. Tiwari, T.L. Schmidt, Missing Shapiro steps and the 8π-periodic Josephson effect in interacting helical electron systems. Phys. Rev. B **96**(16), 165429 (2017)
33. F. Domínguez, F. Hassler, G. Platero, Dynamical detection of Majorana fermions in current-biased nanowires. Phys. Rev. B **86**(14), 140503 (2012)
34. J.D. Sau, E. Berg, B.I. Halperin, On the possibility of the fractional ac Josephson effect in non-topological conventional superconductor-normal-superconductor junctions (2012). Preprint at http://arxiv.org/abs/1206.4596

35. P.-M. Billangeon, F. Pierre, H. Bouchiat, R. Deblock, ac Josephson effect and resonant cooper pair tunneling emission of a single cooper pair transistor. Phys. Rev. Lett. **98**, 216802 (2007)
36. L. Maier, J.B. Oostinga, D. Knott, C. Brüne, P. Virtanen, G. Tkachov, E.M. Hankiewicz, C. Gould, H. Buhmann, L.W. Molenkamp, Induced superconductivity in the three-dimensional topological insulator HgTe. Phys. Rev. Lett. **109**(18), 186806 (2012)
37. J.B. Oostinga, L. Maier, P. Schüffelgen, D. Knott, C. Ames, C. Brüne, G. Tkachov, H. Buhmann, L.W. Molenkamp, Josephson supercurrent through the topological surface states of strained bulk HgTe. Phys. Rev. X **3**(2), 021007 (2013)
38. L. Maier, E. Bocquillon, M. Grimm, J.B. Oostinga, C. Ames, C. Gould, C. Brüne, H. Buhmann, L.W. Molenkamp, Phase-sensitive SQUIDs based on the 3D topological insulator HgTe. Phys. Scr. **T164**(1), 014002 (2015)
39. L. Fu, C. Kane, Topological insulators with inversion symmetry. Phys. Rev. B **76**(4), 045302 (2007)
40. C. Brüne, C.X. Liu, E.G. Novik, E.M. Hankiewicz, H. Buhmann, Y.L. Chen, X.L. Qi, Z.X. Shen, S.C. Zhang, L.W. Molenkamp, Quantum Hall effect from the topological surface states of strained bulk HgTe. Phys. Rev. Lett. **106**(12), 126803 (2011)
41. C. Brüne, C. Thienel, M. Stuiber, J. Böttcher, H. Buhmann, E.G. Novik, C.-X. Liu, E.M. Hankiewicz, L.W. Molenkamp, Dirac-screening stabilized surface-state transport in a topological insulator. Phys. Rev. X **4**(14), 041045 (2014)
42. A. Inhofer, S. Tchoumakov, B.A. Assaf, G. Fève, J.M. Berroir, V. Jouffrey, D. Carpentier, M.O. Goerbig, B. Plaçais, K. Bendias, D.M. Mahler, E. Bocquillon, R. Schlereth, C. Brüne, H. Buhmann, L.W. Molenkamp, Observation of Volkov–Pankratov states in topological HgTe heterojunctions using high-frequency compressibility. Phys. Rev. B **96**(19), 195104 (2017)
43. S. Tchoumakov, V. Jouffrey, A. Inhofer, E. Bocquillon, B. Plaçais, D. Carpentier, M.O. Goerbig, Volkov–Pankratov states in topological heterojunctions. Phys. Rev. B **96**(20), 201302 (2017)
44. B.A. Bernevig, T.L. Hughes, S.-C. Zhang, Quantum spin Hall effect and topological phase transition in HgTe quantum wells. Science **314**(5806), 1757–61 (2006)
45. M. König, S. Wiedmann, C. Brüne, A. Roth, H. Buhmann, L.W. Molenkamp, X.-L. Qi, S.-C. Zhang, Quantum spin Hall insulator state in HgTe quantum wells. Science **318**(5851), 766 (2007)
46. A. Roth, C. Brüne, H. Buhmann, L.W. Molenkamp, J. Maciejko, X.-L. Qi, S.-C. Zhang, Nonlocal transport in the quantum spin Hall state. Science **325**(5938), 294 (2009)
47. C. Brüne, A. Roth, H. Buhmann, E.M. Hankiewicz, L.W. Molenkamp, J. Maciejko, X.-L. Qi, S.-C. Zhang, Spin polarization of the quantum spin Hall edge states. Nat. Phys. **8**(6), 485–490 (2012)
48. K.C. Nowack, E.M. Spanton, M. Baenninger, M. König, J.R. Kirtley, B. Kalisky, C. Ames, P. Leubner, C. Brüne, H. Buhmann, L.W. Molenkamp, D. Goldhaber-Gordon, K.A. Moler, Imaging currents in HgTe quantum wells in the quantum spin Hall regime. Nat. Mater. **12**(9), 787–791 (2013)
49. B. Zhou, H.-Z. Lu, R.-L. Chu, S.-Q. Shen, Q. Niu, Finite size effects on helical edge states in a quantum spin-Hall system. Phys. Rev. Lett. **101**(24), 246807 (2008)
50. A.V. Galaktionov, A.D. Zaikin, Fluctuations of the Josephson current and electron-electron interactions in superconducting weak links. Phys. Rev. B **82**, 184520 (2010)
51. D.S. Antonenko, M.A. Skvortsov, Quantum decay of the supercurrent and intrinsic capacitance of Josephson junctions beyond the tunnel limit. Phys. Rev. B **92**(21), 214513 (2015)
52. J. Wiedenmann, E. Liebhaber, J. Kübert, E. Bocquillon, P. Burset, C. Ames, H. Buhmann, T.M. Klapwijk, L.W. Molenkamp, Transport spectroscopy of induced superconductivity in the three-dimensional topological insulator HgTe. Phys. Rev. B **96**(16), 165302 (2017)
53. T.M. Klapwijk, G.E. Blonder, Explanation of subharmonic energy gap structure in superconducting contacts. Phys. B+C **109–110**, 1657–1664 (1982)
54. L. Maier, C.A. Watson, J.R. Kirtley, C. Gould, G. Tkachov, E.M. Hankiewicz, C. Brüne, H. Buhmann, L.W. Molenkamp, K.A. Moler, Nonsinusoidal current-phase relationship in Josephson junctions from the 3D topological insulator HgTe. Phys. Rev. Lett. **114**(6) (2015)

55. S. Hart, H. Ren, T. Wagner, P. Leubner, M. Mühlbauer, C. Brüne, H. Buhmann, L.W. Molenkamp, A. Yacoby, Induced superconductivity in the quantum spin Hall edge. Nat. Phys. **10**(9), 638–643 (2014)
56. M. Tinkham, *Introduction to Superconductivity* (Dover Publications, New York, 2004)
57. E. Bocquillon, R.S. Deacon, J. Wiedenmann, P. Leubner, T.M. Klapwijk, C. Brüne, K. Ishibashi, H. Buhmann, L.W. Molenkamp, Gapless Andreev bound states in the quantum spin Hall insulator HgTe. Nat. Nanotechnol. **12**, 137–143 (2017)
58. R.S. Deacon, J. Wiedenmann, E. Bocquillon, F. Domínguez, T.M. Klapwijk, P. Leubner, C. Brüne, E.M. Hankiewicz, S. Tarucha, K. Ishibashi, H. Buhmann, L.W. Molenkamp, Josephson radiation from gapless Andreev bound states in HgTe-based topological junctions. Phys. Rev. X **7**(2), 021011 (2017)
59. I.K. Yanson, V.M. Svistunov, I.M. Dmitrenko, Experimental observation of the tunnel effect for Cooper pairs with the emission of photons. Sov. Phys. JETP **21**, 650 (1965)
60. N.F. Pedersen, O.H. Soerensen, J. Mygind, P.E. Lindelof, M.T. Levinsen, T.D. Clark, Direct detection of the Josephson radiation emitted from superconducting thin-film microbridges. Appl. Phys. Lett. **28**, 562–564 (1976)
61. R.J. Schoelkopf, J. Zmuidzinas, T.G. Phillips, H.G. LeDuc, J.A. Stern, Measurements of noise in Josephson-effect mixers. IEEE Trans. Microw. Theory Tech. **43**(4), 977–983 (1995)
62. R. Deblock, E. Onac, L. Gurevich, L.P. Kouwenhoven, Detection of quantum noise from an electrically driven two-level system. Science **301**, 203 (2003)
63. E. Onac, F. Balestro, B. Trauzettel, C.F.J. Lodewijk, L.P. Kouwenhoven, Shot-noise detection in a carbon nanotube quantum dot. Phys. Rev. Lett. **96**, 026803 (2006)
64. D. Laroche, D. Bouman, D.J. van Woerkom, A. Proutski, C. Murthy, D.I. Pikulin, C. Nayak, R.J.J. van Gulik, J. Nygård, P. Krogstrup, L.P. Kouwenhoven, A. Geresdi, Observation of the 4π-periodic Josephson effect in InAs nanowires (2017). Preprint available at https://arxiv.org/abs/1712.08459
65. J. Wiedenmann, E. Bocquillon, R.S. Deacon, S. Hartinger, O. Herrmann, T.M. Klapwijk, L. Maier, C. Ames, C. Brüne, C. Gould, A. Oiwa, K. Ishibashi, S. Tarucha, H. Buhmann, L.W. Molenkamp, 4π-periodic Josephson supercurrent in HgTe-based topological Josephson junctions. Nat. Commun. **7**, 10303 (2016)
66. K.K. Likharev, *Dynamics of Josephson Junctions and Circuits* (Gordon and Breach Science Publishers, Philadelphia, 1986)
67. T. Holst, D. Esteve, C. Urbina, M.H. Devoret, Effect of a transmission line resonator on a small capacitance tunnel junction. Phys. Rev. Lett. **73**, 3455–3458 (1994)
68. M. Hofheinz, F. Portier, Q. Baudouin, P. Joyez, D. Vion, P. Bertet, P. Roche, D. Esteve, Bright side of the coulomb blockade. Phys. Rev. Lett. **106**, 217005 (2011)
69. X. Dai, T. Hughes, X.-L. Qi, Z. Fang, S.-C. Zhang, Helical edge and surface states in HgTe quantum wells and bulk insulators. Phys. Rev. B **77**, 125319 (2008)
70. J.-P. Cleuziou, W. Wernsdorfer, S. Andergassen, S. Florens, V. Bouchiat, Th Ondarçuhu, M. Monthioux, Gate-tuned high frequency response of carbon nanotube Josephson junctions. Phys. Rev. Lett. **99**(11), 117001 (2007)
71. H.B. Heersche, P. Jarillo-Herrero, J.B. Oostinga, L.M.K. Vandersypen, A.F. Morpurgo, Bipolar supercurrent in graphene. Nature **446**(7131), 56–9 (2007)
72. L. Galletti, S. Charpentier, M. Iavarone, P. Lucignano, D. Massarotti, R. Arpaia, Y. Suzuki, K. Kadowaki, T. Bauch, A. Tagliacozzo, F. Tafuri, F. Lombardi, Influence of topological edge states on the properties of Bi_2Se_3/Al hybrid Josephson devices. Phys. Rev. B **89**(13), 134512 (2014)
73. P. Russer, Influence of microwave radiation on current-voltage characteristic of superconducting weak links. J. Appl. Phys. **43**(4), 2008 (1972)
74. L.P. Rokhinson, X. Liu, J.K. Furdyna, The fractional ac Josephson effect in a semiconductor/superconductor nanowire as a signature of Majorana particles. Nat. Phys. **8**(11), 795–799 (2012)
75. A. De Cecco, K. Le Calvez, B. Sacépé, C.B. Winkelmann, H. Courtois, Interplay between electron overheating and ac Josephson effect. Phys. Rev. B **93**(18), 180505 (2016)

76. D. Sun, J. Liu, Quench dynamics of Josephson current in a topological Josephson junction. Phys. Rev. B **97**, 035311 (2018)
77. Y.-H. Li, J. Song, J. Liu, H. Jiang, Q.-F. Sun, X.C. Xie, Doubled Shapiro steps in a topological Josephson junction. Phys. Rev. B **97**(4), 045423 (2018)
78. W.C. Stewart, Current-voltage characteristics of Josephson junctions. Appl. Phys. Lett. **12**, 277 (1968)
79. D.E. McCumber, Effect of ac impedance on dc voltage-current characteristics of superconductor weak-link junctions. J. Appl. Phys. **39**(7), 3113 (1968)
80. F. Domínguez, O. Kashuba, E. Bocquillon, J. Wiedenmann, R.S. Deacon, T.M. Klapwijk, G. Platero, L.W. Molenkamp, B. Trauzettel, E.M. Hankiewicz, Josephson junction dynamics in the presence of 2π- and 4π-periodic supercurrents. Phys. Rev. B **95**, 195430 (2017)
81. M.J. Renne, D. Polder, Some analytical results for the resistively shunted Josephson junction. Rev. de Phys. Appl. **9**(1), 25–28 (1974)
82. A. Valizadeh, M.R. Kolahchi, J.P. Straley, On the origin of fractional Shapiro steps in systems of Josephson junctions with few degrees of freedom. J. Nonlinear Math. Phys. **15**(sup3), 407–416 (2008)
83. J. Picò-Cortés, F. Domínguez, G. Platero, Signatures of a 4π-periodic supercurrent in the voltage response of capacitively shunted topological Josephson junctions. Phys. Rev. B **96**(12), 125438 (2017)
84. A. Ström, H. Johannesson, G.I. Japaridze, Edge dynamics in a quantum spin Hall state: effects from Rashba spin-orbit interaction. Phys. Rev. Lett. **104**, 256804 (2010)
85. F. Crépin, J.C. Budich, F. Dolcini, P. Recher, B. Trauzettel, Renormalization group approach for the scattering off a single Rashba impurity in a helical liquid. Phys. Rev. B **86**, 121106 (2012)
86. J.I. Väyrynen, M. Goldstein, Y. Gefen, L.I. Glazman, Resistance of helical edges formed in a semiconductor heterostructure. Phys. Rev. B **90**, 115309 (2014)
87. S. Essert, V. Krueckl, K. Richter, Two-dimensional topological insulator edge state backscattering by dephasing. Phys. Rev. B **92**, 205306 (2015)
88. J. Wang, Y. Meir, Y. Gefen, Spontaneous breakdown of topological protection in two dimensions. Phys. Rev. Lett. **118**, 046801 (2017)
89. N.B. Kopnin, A.S. Melnikov, Proximity-induced superconductivity in two-dimensional electronic systems. Phys. Rev. B **84**(6), 064524 (2011)
90. N.B. Kopnin, A.S. Mel'nikov, I.A. Sadovskyy, V.M. Vinokur, Weak links in proximity-superconducting two-dimensional electron systems. Phys. Rev. B **89**(8), 081402 (2014)
91. M.P. Stehno, N.W. Hendrickx, M. Snelder, T. Scholten, Y.K. Huang, M.S. Golden, A. Brinkman, Conduction spectroscopy of a proximity induced superconducting topological insulator. Semicond. Sci. Technol. **32**(9), 094001 (2017)
92. P. Virtanen, P. Recher, Microwave spectroscopy of Josephson junctions in topological superconductors. Phys. Rev. B **88**, 144507 (2013)
93. M.J. Stephen, Theory of a Josephson oscillator. Phys. Rev. Lett. **21**, 1629–1632 (1968)
94. A.J. Dahm, A. Denenstein, D.N. Langenberg, W.H. Parker, D. Rogovin, D.J. Scalapino, Linewidth of the radiation emitted by a Josephson junction. Phys. Rev. Lett. **22**, 1416–1420 (1969)
95. J.-J. Feng, Z. Huang, Z. Wang, Q. Niu, Landau–Zener effect induced hysteresis in topological Josephson junctions (2018). Preprint available at https://arxiv.org/abs/1801.05099
96. M. Zgirski, L. Bretheau, Q. Le Masne, H. Pothier, D. Esteve, C. Urbina, Evidence for long-lived quasiparticles trapped in superconducting point contacts. Phys. Rev. Lett. **106**(25), 257003 (2011)
97. R. Delagrange, R. Weil, A. Kasumov, M. Ferrier, H. Bouchiat, R. Deblock, 0-π quantum transition in a carbon nanotube Josephson junction: universal phase dependence and orbital degeneracy. Phys. Rev. B **93**(19), 195437 (2016)
98. A. Murani, A. Chepelianskii, S. Guéron, H. Bouchiat, Andreev spectrum with high spin-orbit interactions: revealing spin splitting and topologically protected crossings. Phys. Rev. B **96**(16), 165415 (2017)

99. J.-D. Pillet, C.H.L. Quay, P. Morfin, C. Bena, A. Levy, Yeyati, P. Joyez, Andreev bound states in supercurrent-carrying carbon nanotubes revealed. Nat. Phys. **6**(12), 965–969 (2010)
100. L. Bretheau, J.I-J Wang, R. Pisoni, K. Watanabe, T. Taniguchi, P. Jarillo-Herrero, Tunnelling spectroscopy of Andreev states in graphene. Nat. Phys. **13**(8), 756–760 (2017)
101. O. Astafiev, A.M. Zagoskin, A.A. Abdumalikov, Y.A. Pashkin, T. Yamamoto, K. Inomata, J.S. Nakamura, Y. Tsai. Resonance fluorescence of a single artificial atom. Science **327**(5967), 840–843 (2010)
102. D.J. Van Woerkom, A. Proutski, B. Van Heck, D. Bouman, J.I. Väyrynen, L.I. Glazman, P. Krogstrup, J. Nygård, L.P. Kouwenhoven, A. Geresdi, Microwave spectroscopy of spinful Andreev bound states in ballistic semiconductor Josephson junctions. Nat. Phys. **13**(9), 876–881 (2017)
103. B. van Heck, T. Hyart, C.W.J. Beenakker, Minimal circuit for a flux-controlled Majorana qubit in a quantum spin-Hall insulator. Phys. Scr. **T164**, 8 (2014)

Chapter 6
Common and Not-So-Common High-Energy Theory Methods for Condensed Matter Physics

Adolfo G. Grushin

Abstract This chapter is a collection of techniques, warnings, facts and ideas that are sometimes regarded as theoretical curiosities in high-energy physics but have important consequences in condensed matter physics. In particular, we describe theories that have the property of having finite but undetermined radiative corrections that also happen to describe topological semimetallic phases in condensed matter. In the process, we describe typical methods in high-energy physics that illustrate the working principles to describe a given phase of matter and its response to external fields.

6.1 Introduction: What This Chapter Is and What It Is Not

Imagine you are (good) theoretical high-energy physicist and you come up with a fantastic theory: the F-theory. As a good theorist you know that any theory that aspires to describe the universe has to comply with those symmetries that are verified up to experimental precision, e.g. Lorentz symmetry. This constraint comes with slightly less freedom to devise new testable theories, but also with a typically overlooked positive side that we will dive into: those same constraints save theories from apparently fatal ambiguities.

If, alternatively, you are a (good) theoretical condensed matter physicist, you have the freedom to come up with theories that violate fundamental symmetries of nature so long as you justify such effective scenario in a sufficiently realistic context. This freedom comes with a price; those ambiguities that high-energy theorists disposed of, can emerge when calculating observables, which however, should be well defined

---

A. G. Grushin (✉)
Université Grenoble Alpes, CNRS, Institut Néel, 38000 Grenoble, France
e-mail: adolfo.grushin@neel.cnrs.fr

A. G. Grushin
Department of Physics, University of California Berkeley, Berkeley, CA, USA

© Springer Nature Switzerland AG 2018
D. Bercioux et al. (eds.), *Topological Matter*, Springer Series in Solid-State Sciences 190, https://doi.org/10.1007/978-3-319-76388-0_6

objects. Their direct experimental relevance forces us to address them, and in doing so, sometimes we can explore an exotic land in between high-energy and condensed matter physics.

This chapter is a hopefully coherent and motivated compilation of different theoretical facts that deal with and, in the best case scenario, fix those ambiguities. Due to their historical context, they are not typically treated in field theory books despite that they keep being useful in the study of condensed matter, and very particularly topological phases. This chapter is motivated and tailored to the study of current research in topological semimetals of different kinds, a focus that serves to emphasize that keeping in mind these examples can prepare the reader for (a small part of) the unknown.

Finally, a disclaimer. Due to the short nature of this Chapter, it mostly uses physically motivated plausibility arguments rather than formal arguments or proofs. Along the way I will try to guide the interested reader towards the relevant formal literature as specifically as possible, but avoiding severe computations in favour of physical intuition. More generally, the reader is referred to the numerous reviews for details of Weyl semimetal physics and anomalies (e.g. [1, 2] and references therein) as well as other chapters of this volume as a backup of what is discussed here.

6.2 Lorentz Breaking Field Theories

In this section, we will define a simple field theory that we will use to exemplify some of the methods we will discuss. This theory is simple but it can be used to understand a large fraction of the Weyl semimetal literature [3, 4]. Moreover, it has many interesting features and can be promoted, with intuitive generalizations, to describe other topological phases such as nodal semimetals.

6.2.1 One Useful Field Theory: Lorentz Breaking QED

Consider the following 4×4 Hamiltonian in 3D momentum space spanned by the vector $\mathbf{k} \in \mathcal{R}^3$

$$\mathcal{H}_0^{\mathbf{k}} = \begin{pmatrix} b_0 + \boldsymbol{\sigma} \cdot (\mathbf{k} - \mathbf{b}) & m \\ m & -b_0 - \boldsymbol{\sigma} \cdot (\mathbf{k} + \mathbf{b}) \end{pmatrix}. \quad (6.1)$$

Here $\boldsymbol{\sigma}$ is the vector of Pauli matrices for a spin-1/2 degree of freedom and $b_\mu = (b_0, \mathbf{b})$ is a constant four-vector. The matrices in this representation will be termed Γ to distinguish them from the Dirac matrices below and serve to define a more compact representation of the above that reads

$$\mathcal{H}_0^{\mathbf{k}} = \mathbf{k} \cdot \boldsymbol{\Gamma} + \Gamma_5 b_0 - \mathbf{b} \cdot \boldsymbol{\Gamma}^b + m\Gamma_0, \quad (6.2)$$

where $\Gamma = \sigma \otimes \tau_3$, $\Gamma_5 = \sigma_0 \otimes \tau_3$, $\Gamma^b = \sigma \otimes \tau_0$ and $\Gamma_0 = \sigma_0 \otimes \tau_1$, with σ_0 and τ_0 being 2×2 identity matrices. The Hamiltonian density (6.1) acts on a four-component spinor that, for future convenience, we can write in terms of two component spinors $\Psi^\dagger = (\Psi_R^\dagger, \Psi_L^\dagger)$. In high-energy physics, it is more common to use the action

$$S = \int d^4 k \bar{\Psi} (\slashed{k} - m + \gamma_5 \slashed{b}) \Psi, \tag{6.3}$$

where we have used the custom high-energy notation $\bar{\Psi} = \Psi^\dagger \gamma_0$[1] and Feynman's slashed notation $\slashed{k} = \gamma^\mu k_\mu$ with $k_\mu = (\omega, \mathbf{k})$ and $\mu = 0, 1, 2, 3$. This notation is not strictly necessary, but it will help us connect with the high-energy literature. Deducing (6.3) from (6.1) is straightforward if we have the Lagrangian density $\mathscr{L}$ since $S = \int d^4 k \mathscr{L}$. We can use the standard relation between Lagrangian density and Hamiltonian density $\mathscr{L}(t, \mathbf{k}) = \pi \dot{\Psi}_\mathbf{k} - \Psi_\mathbf{k}^\dagger \mathscr{H}_\mathbf{k}^0 \Psi_\mathbf{k}$. Remembering that the generalized momentum in this case is $\pi = i\Psi_\mathbf{k}^\dagger$ and going to the frequency domain, we can write $\mathscr{L}(\omega, \mathbf{k}) = \Psi_\mathbf{k}^\dagger \gamma_0 (\gamma_0 \omega - \gamma_0 \mathscr{H}_0^\mathbf{k}) \Psi_\mathbf{k}$ using the matrix multiplying m in (6.1) ($\gamma_0 = \sigma_0 \otimes \tau_1$) which satisfies $\gamma_0^2 = 1$. You can check that the Dirac matrices with our choice (6.1) are given by

$$\gamma^0 = \begin{pmatrix} 0 & \sigma_0 \\ \sigma_0 & 0 \end{pmatrix}, \ \gamma^j = \begin{pmatrix} 0 & \sigma^j \\ -\sigma^j & 0 \end{pmatrix}, \ \gamma^5 = \begin{pmatrix} -\sigma_0 & 0 \\ 0 & \sigma_0 \end{pmatrix}, \tag{6.4}$$

or, alternatively, $\gamma = i\sigma \otimes \tau_2$, $\gamma_0 = \Gamma_0$ and $\gamma_5 = \Gamma_5$. Equations (6.1) and (6.3) are the central objects of this chapter and contain the same information. In what follows we will use them interchangeably.

6.2.1.1 Spectrum and Symmetries

Lets break down the properties of this Hamiltonian by choosing some easy limits. The most familiar should be the one where $b_\mu = 0$ but $m \neq 0$ (see Fig. 6.1 upper left panel). This is the Dirac Hamiltonian where the spinor Ψ satisfies the Dirac equation

$$(\gamma_\mu k^\mu - m)\Psi = (E\gamma_0 - \mathbf{k} \cdot \boldsymbol{\gamma} - m)\Psi = 0. \tag{6.5}$$

Solving for E it is easy to find that the spectrum is gapped and has two degenerate bands $E_\pm = \pm\sqrt{|\mathbf{p}|^2 + m^2}$. The explicit form of its eigenstates can be found in any quantum field theory book (see [5] or, for a more condensed matter perspective, [6]). An important property of the Hamiltonian equation (6.1) is that it is time-reversal and inversion symmetric. These symmetries are represented by $\mathscr{T} = i\sigma_2 \otimes \tau_0$, and $\mathscr{I} = \gamma_0$, operators respectively, which can be explicitly be checked to commute with (6.1) with $b_\mu = 0$.

---

[1] This object is sometimes referred to as the Dirac adjoint. Its form is helpful to define objects that are Lorentz scalars such as $\bar{\Psi}\Psi$.

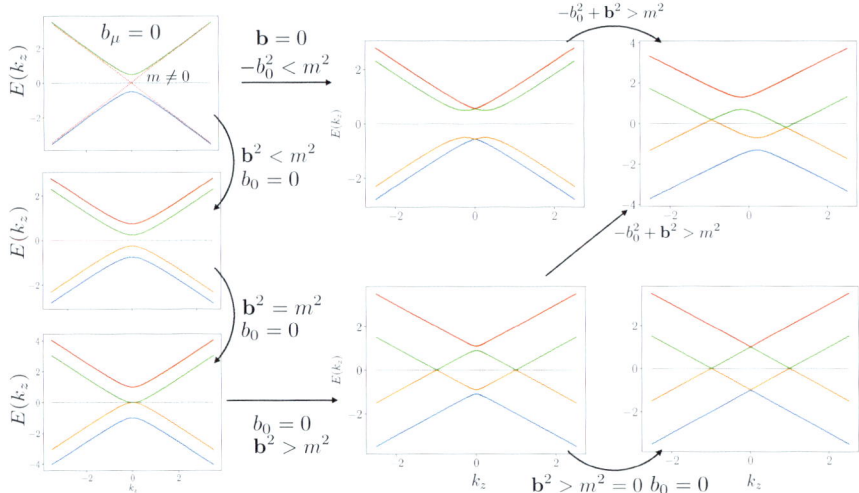

Fig. 6.1 Band strutture of Hamiltonian given in (6.2) for different values of the parameters. Whenever $-b^2 = -b_0^2 + \mathbf{b}^2 < m^2$, the spectrum is gapped (see main text), while it is in semimetallic phase otherwise

Turning off the mass ($m = 0$), exposes another very useful symmetry, known as chiral symmetry. In this limit, the Hamiltonian equation (6.1) decomposes in two 2×2 blocks

$$\mathcal{H}_{\mathbf{k}} = \pm \boldsymbol{\sigma} \cdot \mathbf{k}. \tag{6.6}$$

This is the Hamiltonian for two Weyl fermions Ψ_R and Ψ_L. The spinors Ψ_R and Ψ_L are known as right and left chiralities, and are eigenstates of γ_5 with eigenvalues ± 1. The eigenvalue of γ_5 is referred to as chirality, and it is a good quantum number for Weyl spinors; γ_5 commutes with the Hamiltonian and thus it is a symmetry. The chiral symmetry can be expressed as the invariance of a Hamiltonian against the continuous $U(1)$ transformation

$$\begin{aligned} \Psi &\to e^{\frac{i}{2}\theta \gamma_5} \Psi, \\ \bar{\Psi} &\to \bar{\Psi} e^{\frac{i}{2}\theta \gamma_5}, \end{aligned} \tag{6.7}$$

which is a symmetry of (6.3) since

$$\{\gamma_5, \gamma_\mu\} = 0. \tag{6.8}$$

This symmetry becomes particularly explicit in the basis (6.4), which is thus known as the basis. Projecting a Dirac spinor into a Weyl spinor can be done by the projector $\mathcal{P}_{\pm} = \frac{1}{2}(1 \pm \gamma_5)$

$$\Psi_{R/L} = \frac{1}{2}(1 \pm \gamma_5)\Psi. \tag{6.9}$$

Physically, the meaning of chirality will become clearer when we couple the theory to an external electromagnetic field; for instance, each chirality propagates in opposite directions when subject to a magnetic field. For now, we can regard this symmetry as a the mathematical statement of invariance under (6.7).

Before moving forward, a small word of warning. The concepts of chirality and helicity are not equivalent and often confused. A state with definite helicity is a two spinor that is the eigenstate of the helicity operator

$$\frac{1}{2}\frac{\boldsymbol{\sigma} \cdot \mathbf{p}}{|\mathbf{p}|}. \tag{6.10}$$

This operator is not, in general, a Lorentz invariant object and changes between references frames. A massive particle with positive helicity in a given frame can be seen by another observable in a different frame with negative helicity. Only when $m = 0$, the helicity is independent of the reference frame. In this case, it is possible to show that the states with well-defined helicity have a well-defined chirality as well, and the two notions coincide (see Sect. 7.4 in [7]).

Although both the Dirac Hamiltonian, defined by (6.1) with $b_\mu = 0$, and the Weyl Hamiltonian equation (6.6) satisfy time-reversal and inversion symmetries, only the Weyl Hamiltonian posesses chiral symmetry. From (6.1) with $b_\mu = 0$ but $m \neq 0$ notice that the two chiralities are coupled, ceasing to be chiral and resulting in a gapped spectrum. Unlike time reversal or inversion symmetries, chiral symmetry is not a fundamental symmetry of any material but rather an emergent low energy symmetry. Therefore, one should expect that $m \neq 0$ in physical realizations of this Hamiltonian, and thus, we might conclude that a system described by low-energy Weyl fermions is a very fined tuned situation.

There are in fact two possibilities to protect the Weyl fermions from gapping out due to m. The first is a very physical option in condensed matter: if additional symmetries are imposed (e.g. point group symmetries), they endow the two chiralities with extra quantum numbers that we can use to impose that $m = 0$ by symmetry, and the two Weyl fermions remain decoupled. This special case is a Dirac semimetal, where the two Weyl fermions of (6.6) live at the same point in the Brillouin zone but remain decoupled. A material that falls into the Dirac semimetal class is Na_3Bi [8] and corresponds to the dashed red lines in the upper left panel of Fig. 6.1.

In this chapter, we will be interested in a second and richer possibility to stabilize Weyl fermions that does not require additional symmetries. The idea is that separating them in phase space (energy–momentum space) will effectively stabilize them, since a large momentum transfer would be needed to couple them, preventing a gap from opening. To implement this separation, we use b_μ. The spectrum now will depend generically on the relative size of $b_\mu b^\mu = b^2 = b_0^2 - \mathbf{b}^2$ with respect to m^2 (see Fig. 6.1).

To analyse each case, start from a massless Dirac Hamiltonian equation (6.6), i.e. $m = 0$, $b_\mu = 0$. In this case, $b^2 = m^2 = 0$ and the theory is gapless. If we add a small space-like $b_\mu = (0, \mathbf{b})$, we can diagonalize (6.1) to see that the masless Dirac cone splits into two Weyl nodes at zero energy that also cross at higher energies (see Fig. 6.1 lower right panel). For momenta close to each Weyl node, the spectrum is still given by (6.6) if we measure the momentum relative to the Weyl node. The Weyl node separation in this case is $\delta \mathbf{K} = 2\mathbf{b}$. Now add a small mass such that $-b^2 > m^2$. Such small mass hybridizes the Weyl nodes only at high-energies as shown in Fig. 6.1 lower central panel. The distance between Weyl nodes in momentum space now changes to

$$\delta \mathbf{K} = 2\mathbf{b}\sqrt{1 - \frac{m^2}{\mathbf{b}^2}}. \tag{6.11}$$

Note that, as long as $\mathbf{b}^2 > m^2$, the phase is gapless and the square root is real valued. If we keep increasing m, the nodes start to approach until they annihilate at $\mathbf{b}^2 = m^2$. When $\mathbf{b}^2 < m^2$, there is a gap between all four bands, reaching the massive Dirac limit when $\mathbf{b}^2 = 0$.

Adding a small b_0 does not change the basic picture (see Fig. 6.1 upper central and right panels). A finite b_0 will shift the Weyl nodes along the energy axis and the condition for gaplessness becomes $-b^2 = -b_0^2 + \mathbf{b}^2 > m^2$. If this is satisfied, the Weyl node separation in energy momentum space can be written compactly as

$$\delta K_\mu = 2 b_\mu \sqrt{1 - \frac{m^2}{|b^2|}}. \tag{6.12}$$

With this condition, note in particular that for a time-like $b^\mu = (b_0, 0)$, the spectrum is always gapped.

In order to connect with physical systems, it is important to note a few important symmetry properties of the Hamiltonian (6.1). First, the spatial part $\mathbf{b}$ breaks time reversal since it couples to the Hamiltonian as a Zeeman term $\mathbf{b} \cdot \boldsymbol{\sigma}$. One can check that explicitly by applying the operator that implements time-reversal symmetry defined above, $\mathcal{T} = i\sigma_2 \otimes \tau_0$. Therefore, the coupling of $\mathbf{b}$ can be physically regarded as a zero-field magnetization which is a finite expectation value of a field.[2]

Second, the time-like part b_0 breaks inversion (or parity), which one can check by applying the inversion operator $\mathcal{I} = \gamma_0$ to the b_0 term in (6.3). From (6.1), it is evident as well that it enters similar to a chirality-dependent energy offset. This parameter can arise from inversion breaking spin–orbit coupling (e.g. see [9]) but in general can have several physical origins to be traced back to microscopic inversion breaking perturbations. However, it is important to note that b_0 is not, technically a chemical potential: (6.1) is an equilibrium Hamiltonian and b_0 is a parameter of it (it

---

[2]This is a statement which is particularly evident in the Burkov–Balents model [9], one of the first models of Weyl semimetals.

is observable!), unlike the chemical potential, which is introduced as a gauge field (see [2, 10] for a discussion).

Finally and most importantly, a finite b_μ breaks Lorentz symmetry. Note that, since b_μ is a constant vector by assumption, it chooses a preferred direction in space–time and considering the above we have identified this vector as a background expectation value. Therefore, it is not allowed to transform as a Lorentz vector under Lorentz transformations. This specific type of Lorentz transformation, the one that leave background fields invariant while changing the coordinate frame, is referred to as particle-Lorentz transformation. It is meant to distinguish it from Lorentz frame transformations where the fields do change; for instance, a particle that experiences only magnetic field will be seen by an observer in another frame experiencing both a magnetic field and an electric field. Our theory is actually invariant under these global changes (see [11] for more discussion on this issue) but is not invariant under particle-Lorentz transformations.[3]

6.2.1.2 Coupling to Electromagnetism: QED

The above symmetry considerations, summarized in Table 6.1, combined with our previous analysis of the spectrum implies that in order to have a Weyl phase in this model, we need to satisfy two conditions: time-reversal symmetry must be broken through $\mathbf{b} \neq \mathbf{0}$ and b_μ must be space-like ($\mathbf{b}^2 > b_0^2$) with $-b^2 > m^2$. If only one of the two conditions is satisfied, the system with $m \neq 0$ will always be gapped. Therefore in the theory (6.1), a finite mass is not equivalent to being an insulator, unlike in the simple Dirac equation.

The conditions in which the Hamiltonian enters a Weyl semimetal phase will have consequences when we calculate the response of a Weyl semimetal to an external electromagnetic field. This will require introducing an external electromagnetic gauge field A_μ with the usual minimal (Peierls) substitution $k_\mu \to k_\mu - eA_\mu$ which results in

$$S[A] = \int d^4k \bar{\Psi}(\slashed{k} - m - e\slashed{A} + \gamma_5 \slashed{b})\Psi. \tag{6.13}$$

This form is very suggestive: it tells us that b_μ couples to a Dirac fermion similarly to an electromagnetic gauge field, but it distinguishes the two chiralities due to the presence of γ_5. Of course, this was already apparent in (6.1). For a high-energy theorist, it is very tempting to regard b_μ as the chiral or axial electromagnetic field A_5^μ used in high-energy literature [12]. However, there is an important difference: b_μ is itself an observable and it is a parameter in the Hamiltonian, rather than an external field. The first issue affects our gauge freedom to change b_μ, while the second has

---

[3] The difference between particle and global Lorentz transformations is simple when thinking about a particle in a box experiencing the action of gravity $\mathbf{g}$. The vector $\mathbf{g}$ sets a prefer direction, so performing a rotation, which is a transformation belonging to the Lorentz group, will leave the system invariant only if we rotate the box and the field. Rotating the box only (particle-Lorentz transformation) breaks Lorentz invariance due to the fixed direction of $\mathbf{g}$.

Table 6.1 Summary of the symmetry properties of the different terms in (6.1). $\mathcal{T}$, $\mathcal{I}$ and Λ denote time-reversal, inversion and (particle) Lorentz symmetry, respectively. When all parameters are nonzero, the Weyl node separation is set by all of them through (6.12)

	$\mathcal{T}$	$\mathcal{I}$	Λ	Physical meaning
m	Yes	Yes	Yes	Band gap when $b_\mu = 0$
$2\mathbf{b}$	No	Yes	No	Weyl node separation in momentum space when $m = 0$. Magnetization
$2b_0$	Yes	No	No	Weyl node separation in energy space when $m = 0$. Spin orbit coupling

consequences for out-of-equilibrium responses such as the chiral magnetic effect [2, 10, 13].

The beauty of Hamiltonian (6.1) and the corresponding action (6.13) is that with a few parameters, they capture the band structure and response of Dirac and Weyl semimetals, as well as a Dirac (trivial or topological) insulator. In the high-energy physics community, this theory is known as Lorentz breaking quantum electrodynamics and has been thoroughly studied in the context of theories beyond the standard model of particle physics [11, 14]. It was recognized early on that it can describe Weyl semimetals as well, establishing a connection between these seemingly different types of systems [3, 4, 15–17]. In the following, we will take advantage of the existing high-energy field theory knowledge to infer some properties of the Weyl semimetal phase, but before doing so, we will discuss some generalizations.

6.2.2 Generalizations of Lorentz Breaking Field Theories

There are many interesting ways to generalize the action (6.13), anticipating its connection to condensed matter. One quantity that has been missing, and is the first and simplest addition to the theory, is the Fermi velocity. In general, the Fermi velocity will be anisotropic and so one can include its effect as a diagonal matrix $M^\mu{}_\nu = \text{diag}(1, v_x, v_y, v_z)$, such that (6.13) is promoted to:

$$S[A] = \int d^4k \bar{\Psi} (\gamma_\mu M^\mu{}_\nu k^\nu - m - e\slashed{A} + \gamma_5 \slashed{b})\Psi. \tag{6.14}$$

This factor will slightly mess up the isotropy of our equations, but it is important in order to recover known lattice expressions [15]. Fortunately, it is not unusual that when calculating response functions, we can factor these out by rescaling the momenta, but this is not always true (i.e. when higher-order radiative corrections are involved). The chirality is simply the determinant of the matrix $M^\mu{}_\nu$, since it can be shown to control the sign of the dispersion relation [18].

Additionally, note that considering an even number of copies of (6.14) with opposite values of **b** can restore time-reversal symmetry. Recall that **b** enters like a magnetization, so if we superimpose two magnetizations with opposite directions we effectively restore time-reversal symmetry. To again avoid the different copies from gapping out, we can separate them in momentum space, thereby breaking inversion, but preserving time-reversal symmetry. This is in fact the case of most of the Weyl semimetals found so far, which break inversion but respect time-reversal symmetry by realizing $N > 2$ pairs of Weyl nodes. Since (6.14) can be regarded as the building block of time-reversal symmetric Weyl semimetals, we will not consider these cases here, although they can have richer phenomenology [1].

In fact, the matrix M^μ_ν is actually one of the many generalizations of QED that have been studied. Generally, one could aim to exhaust all matrices and write down all possible terms that break Lorentz invariance using the 16 matrices in the 4×4 subspace. As we have seen, there are five Dirac matrices in $3 + 1$ dimensions labelled γ^μ and γ_5. Explicitly, γ_0 is even under time reversal and inversion while $\boldsymbol{\gamma}$ are odd under both since they multiply the momentum **k**. The chiral matrix γ_5 is a product of all so its odd under inversion and time reversal. To span the full space, one includes the 10 matrices resulting from $\sigma^{\mu\nu} = \frac{i}{2}[\gamma^\mu, \gamma^\nu]$. Together with the identity, they span the full space of 4×4 matrices. With this information, we can construct a pretty general theory

$$S[A] = \int d^4k \bar{\Psi}(\tilde{\Gamma}_\mu k^\mu - \tilde{m})\Psi, \qquad (6.15)$$

where we have promoted $\gamma_\mu \to \tilde{\Gamma}^\mu$ such that

$$\tilde{\Gamma}_\mu = \gamma_\mu + \Gamma^{LV}_\mu + \Gamma^{CPTV}_\mu, \qquad (6.16)$$
$$\Gamma^\mu_{LV} = M^\mu_\nu \gamma^\nu + d^\mu_\nu \gamma_\mu \gamma_5, \qquad (6.17)$$
$$\Gamma^\mu_{CPTV} = e^\mu + f^\mu \gamma_5 + g^{\mu\nu\lambda}\sigma_{\nu\lambda}, \qquad (6.18)$$

and the mass term $m \to \tilde{m}$ such that

$$\tilde{m} = m + m_5 \gamma_5 + \gamma^\mu a_\mu + b_\mu \gamma^\mu \gamma_5 + H^{nm}\sigma_{nm}, \qquad (6.19)$$

The vector a_μ is not very interesting, since it can be absorbed into a redefinition of the fields ($\Psi \to e^{ia_\mu x^\mu}\Psi$). Many semimetals, including Weyl's and nodal lines, and their phase transitions to trivial phases can be captured only with the generalized mass term (6.19). For instance, a nice exercise is to compare (6.19) with the terms discussed by [9]. You will notice that some of the terms in $\tilde{m}$ lead to nodal line semimetals, materials which have a gapless 1D line node in three-dimensional momentum space. However, the Lorentz breaking generalization (6.15) does not include Type-II Weyl semimetals. These will be discussed briefly in Sect. 6.5.

6.3 Field Theories on The Lattice

Quantum field theories are always effective [19]. This means that they are valid below or above some energy scale, that is sometimes referred to as a cut-off, be it infrared or ultraviolet. In condensed matter, this observation is particularly important since there is always an underlying lattice that regularizes the theory after some cut-off scale. The existence of the lattice in condensed matter naturally links with the attempt of studying gauge theories in the lattice [20]. In this section, we discuss some types of lattice generalizations for field theories.

There are many types of lattice fermions, since there are many ways of reproducing the same low-energy physics from the continuum. Here, we will mention three different constructions that we will call "simple" lattice fermions, Wilson fermions and Ginsparg–Wilson (or GW) fermions. Out of the three, Wilson fermions have gained the most popularity in condensed matter, since they are the basis to understand many topological phases of matter. Other types of lattice fermions that we will not cover include staggered fermions or twisted mass fermions (see [20]).

6.3.1 "Simple" Lattice Fermions

These are based on the most naive way of regularizing a Dirac fermion on the lattice. They are based on the simple mapping

$$k_i \to \sin(k_i a),$$
$$m \to M, \tag{6.20}$$

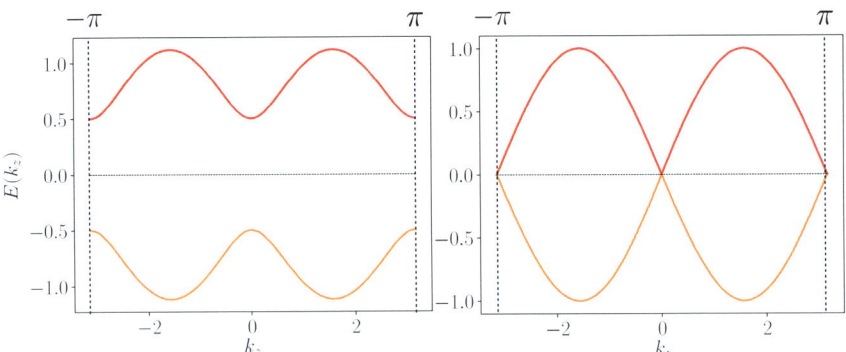

Fig. 6.2 Massive (left) and massless (right) "simple" lattice fermions defined by the mapping (6.20) applied to (6.1) with $b^\mu = 0$. Each band is doubly degenerate and the gap is set by M. When M is zero, there are 2^d gapless doublers

where a is a lattice constant. It is based on the intuition that close to $k_i = 0$ we will recover the Dirac Hamiltonian equation (6.1) with $b^\mu = 0$. Applying this to the Dirac Hamiltonian defined by (6.1) with $b^\mu = 0$ results in the spectrum shown in Fig. 6.2. Now try to set $M = 0$. If this mapping was to be a good description of the continuum quantum field theory at low energies, we would like to recover one single massless Dirac fermion, which we know is invariant under chiral symmetry (6.7) so long as $M = 0$. However, close to zero energy this simple substitution leads to many copies of Dirac fermions; 2^d fermions in d space dimensions (see Fig. 6.2). This doubling of solutions is known as the fermion doubling "problem". What this means is that applying simply (6.20) results in massless fermions that always come in pairs, since chiral symmetry is a symmetry in the lattice. In fact, even if we try to be smart and use the projector operator (6.9) to create one chiral fermion, applying to it the map (6.20) will always result in pairs of chiral fermions with opposite chiralities. As will be clear later on, this prevents any kind of anomaly to be present; each doubler will contribute with an opposite sign to the anomaly, since the theory on the lattice is anomaly free. This collection of facts is known as the Nielsen Ninomiya theorem [21, 22]. If can be stated as follows: if a theory is unitary, local and translational invariant, there is no way to avoid the fermion doubling problem unless we sacrifice chiral symmetry in the limit $M \to 0$.

6.3.2 Wilson Fermions

A solution to the doubling problem where chiral symmetry is sacrificed is offered by Wilson fermions. Wilson fermions break chiral symmetry by gapping out the doublers at the corners of the Brillouin zone with the mapping

$$k_i \to \sin(k_i),$$
$$m \to M - \sum_i \sin^2(k_i/2). \tag{6.21}$$

The last term makes sure that the gap is finite irrespective of M at high symmetry momentum points, except $\Gamma = (0, 0, 0)$ where the gap does vanish when $M = 0$. In condensed matter, the last mass term is sometimes written using the identity $\sin^2(k_i/2) = \frac{1}{2}(1 - \cos(k_i))$. Because of the last term in (6.21) the chiral transformation (6.7) is no longer a symmetry when $M \to 0$, but the theory is free of doublers.

Wilson fermions are a constant source of inspiration for constructing models of topological phases. A phenomenologically rich two-dimensional (2D) Wilson fermion is

$$\mathcal{H}_{CI} = \sin(k_x)\sigma_x + \sin(k_y)\sigma_y + \big(M - \cos(k_x) - \cos(k_y)\big)\sigma_z. \tag{6.22}$$

This is the simplest Chern insulator model, and its main property is that it breaks time-reversal symmetry and has a finite Hall effect. It is therefore one of the simplest topological phases (see other chapters in this volume). A 3D topological insulator is in fact a 3D Wilson fermion:

$$\mathcal{H}_{\text{TI}} = \sin(k_x)\Gamma_1 + \sin(k_y)\Gamma_2 + \sin(k_z)\Gamma_3 \tag{6.23}$$
$$+ \left(M - \cos(k_x) - \cos(k_y) - \cos(k_z)\right)\Gamma_0. \tag{6.24}$$

where we used the Γ_i and Γ_0 defined below (6.2). Of course, as we change dimensions, the discrete symmetry representations change, and different models respect different symmetries. The properties of these two models, their symmetries and relation to quantum field theories can be found, for example, in [23].

Now we are in place to construct a lattice generalization of the theory (6.1) using the Wilson fermion rules. Using (6.21), we can write our Lorenz breaking QED in the lattice as [13]

$$\mathcal{H}_{\text{WSM}} = \mathcal{H}_{\text{TI}} + b_i \Gamma_i^b + b_0 \Gamma_5, \tag{6.25}$$

where $\mathcal{H}_{\text{TI}}$ was defined by (6.23) and Γ_i^b and Γ_5 are defined under (6.2). This simple Hamiltonian has a very rich phase diagram including weak, strong, trivial insulators as well as Weyl semimetals with 1, 2 or 3 pairs of Weyl fermions [24]. It can thus very easily help to describe interfaces between topological insulating and semimetallic phases by promoting its parameters to be space dependent [24, 25].

6.3.3 Ginsparg–Wilson Fermions

Finally, a small note on a way of solving the fermion doubling problem less familiar in the context of condensed matter physics using a different kind of lattice fermions. These types of lattice fermions are known as Ginsparg–Wilson (GW) fermions, which preserve chiral symmetry up to lattice artefacts. The exact symmetry they possess is a generalization of the symmetry (6.7) that can be written as

$$\begin{aligned}\Psi &\to e^{\frac{i}{2}\theta\gamma_5(1-\frac{a}{2}D)}\Psi, \\ \bar{\Psi} &\to \bar{\Psi}e^{\frac{i}{2}\theta\gamma_5(1-\frac{a}{2}D)},\end{aligned} \tag{6.26}$$

where a is the lattice constant. They acquire this symmetry if we define the GW fermion as a type of non-local Dirac fermion

$$S = \sum_{x,y} \bar{\Psi}_x \left(D_{x,y} - m\delta_{x,y}\right)\Psi_y, \tag{6.27}$$

where $D_{x,y}$ is a non-local lattice operator that is required to satisfy the commutator relationship

$$\{\gamma_5, D\} = aD\gamma_5 D, \tag{6.28}$$

that recovers (6.8) when we take the limit of $a \to 0$. This construction is quite interesting since it allows to study fermions on a lattice with chiral symmetry. Many of the properties of massless Dirac fermions translate upon the replacement $\gamma_5 \to \gamma_5(1 - \frac{a}{2}D)$. The properties match those of the continuum theory, albeit differences of order $\mathcal{O}(a)$ should be expected. A lesson to take from this is that, sometimes, corrections of order $\mathcal{O}(a)$ can be crucial to understand the linear response of a certain phase. One specific form for the operator D which is local and free of doublers was found over a decade after the GW proposal [26], and is referred to as overlap fermion. The explicit form of D for overlap fermions will not be given here, but can be found easily in standard textbooks [20].

6.4 Quantum Field Theories Can be Finite But Undetermined

Jackiw, among others, noticed that some quantum field theories have radiative corrections that are superficially divergent, but are finite (see [27] for a review, which we will follow closely in this section). They are therefore regularization dependent, and thus ambiguous! One could ask: Why should we care? Anyway we could can argue that renormalizable and super-renormalizable field theories should be supplemented by a measurement and non-renormalizable field theories are already pathological (in a very definite sense!). Such measurements set a renormalization scale and give us boundary conditions to solve the flow equations for the coupling constant [5]. The difference here is that the constants do not necessarily flow but do need an experimental input. No big deal right?

But let's step back for a moment. Imagine that one of these field theories actually describes low-energy electrons in a material (or whichever degree of freedom for that matter). It seems we would have a problem; our low-energy field theory would not tell us what the values of some observables are, even if the theory is finite. It is tempting to say that, in condensed matter, the answer is simply that the lattice fixes the regularization rendering a finite result, which is certainly true. As it turns out, understanding the exact way this happens gives us plenty of useful information about the phase this theory describes. The kinds of field theories known so far that have this property are all tied to topological semimetallic phases of matter that exhibit quantum anomalies and thus the focus of the following sections.

6.4.1 A 1+1 D Example: The Schwinger Model

Let us work out a simple example first, massless QED in 1+1, or in other words, two counter-propagating one-dimensional chiral fermions. It is defined by the action

$$S[A] = \int d^2 k \bar{\Psi}(\slashed{k} - e\slashed{A})\Psi, \tag{6.29}$$

which also defines the propagator

$$G_k = \frac{i}{\slashed{k}}. \tag{6.30}$$

As per usual $\slashed{k} = k^\mu \gamma_\mu$ and the three necessary γ matrices can be taken to be the three Pauli matrices $\gamma_0 = \sigma_y$, $\gamma_1 = i\sigma_x$ and $\gamma_5 = \sigma_z$. This is a really simple theory of two chiral modes that disperse with energy $E_k = \pm k$. It as a linearization of a simple quadratic dispersion around the Fermi level as shown in Fig. 6.3 left panel. Imagine you want to find the response of this theory to an external electromagnetic field A_μ. You will have to calculate the expectation value of the current j^μ using perturbation theory in A_μ

$$\langle j^\mu(p) \rangle = \left\langle \frac{\delta S}{\delta A_\mu} \right\rangle = \Pi^{\mu\nu}(p) A_\mu + \ldots, \tag{6.31}$$

where $\Pi^{\mu\nu}$ is the polarization function. As described in the Appendix, the polarization function defines the effective action that governs linear response

$$S_{\text{eff}}[A] = \int d^4 p\, A_\mu(p) \Pi^{\mu\nu}(p) A_\nu(-p). \tag{6.32}$$

The polarization function is given by

$$i\Pi^{\mu\nu}(p) = e^2 \int \frac{d^2 k}{(2\pi)^2} \text{Tr}\left[\gamma^\mu G_k \gamma^\nu G_{k+p}\right] \tag{6.33}$$

$$= e^2 \int \frac{d^2 k}{(2\pi)^2} \text{Tr}\left[\gamma^\mu \frac{i}{\slashed{k}} \gamma^\nu \frac{i}{\slashed{k}+\slashed{p}}\right]. \tag{6.34}$$

and can be represented by the Feynman diagram in Fig. 6.3 right panel. In quantum field theory, the limits of integration are $\pm\infty$. This integral is, by power counting, logarithmically divergent, and thus we need to regularize it. Let us isolate the divergent part of the integral and evaluate the finite part. This can be done following any of the standard quantum field theory text books (e.g. [5], Chap. 7) only noting that in 1+1 dimensions, $\text{Tr}[1] = 2$ and that $\gamma_5 \gamma^\mu = \varepsilon^{\mu\nu} \gamma_\nu$. Formally, we can write the result as:

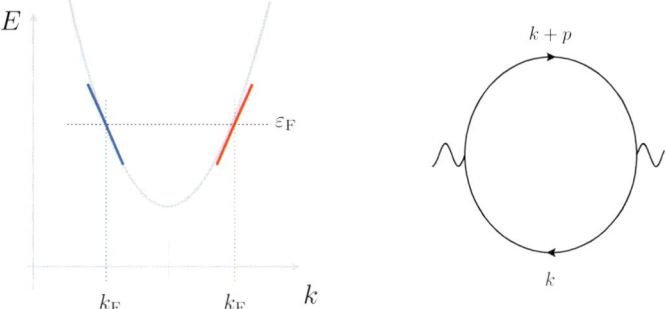

Fig. 6.3 The left panel shows how two chiral fermions described by the action of (6.29) arise from linearizing a quadratic band dispersion close to the Fermi level ε_F. The right panel shows the polarization bubble $\Pi^{\mu\nu}(p)$, where a solid line represents the Green's function

$$\Pi^{\mu\nu} = \Pi^{\mu\nu}_\infty + \Pi^{\mu\nu}_{\text{finite}}, \tag{6.35}$$

$$\Pi^{\mu\nu}_{\text{finite}} = \frac{1}{\pi}\left(\frac{1}{2}g^{\mu\nu} - \frac{p^\mu p^\nu}{p^2}\right), \tag{6.36}$$

$$\Pi^{\mu\nu}_\infty = \frac{a}{2\pi}g^{\mu\nu}. \tag{6.37}$$

To calculate the constant a, we could use for instance dimensional regularization, Pauli–Villars regularization or a high-energy cut-off. In doing so, we would realize that they all lead to a non-divergent result; a is a number, but this number depends on the regularization. What fixes the value of a is in fact the requirement that the theory should be gauge invariant. Gauge invariance is equivalent to charge conservation: the four-divergence of the current (6.31) should be zero. This implies that its four-divergence is zero, or in other words that $p_\mu \Pi^{\mu\nu} = 0$. Imposing this condition immediately sets $a = 1$ and we can breathe again![4]

However, one key point of this section is that for certain models, the requirement of gauge invariance is not enough to fix the undetermined coefficient. This can be illustrated by the chiral Schwinger model, defined as

$$S[A] = \int d^2k \bar{\Psi}\left(\slashed{k} - 2eP_R\slashed{A}\right)\Psi. \tag{6.38}$$

Remember that the projector $P_R = \frac{1}{2}(1 + \gamma_5)$ projects out the right chirality fermion. We can now ask what is the response of this model to an external field. A similar exercise as for the Schwinger model leads to [28]

---

[4] Dimensional regularization and Pauli–Villars in fact automatically give a transverse photon ($a = 1$) since they both preserve gauge invariance.

$$\Pi^{\mu\nu} = \Pi^{\mu\nu}_\infty + \Pi^{\mu\nu}_{\text{finite}}, \tag{6.39}$$

$$\Pi^{\mu\nu}_{\text{finite}} = -\frac{1}{\pi}(g^{\mu\alpha} + \varepsilon^{\mu\alpha})\frac{p_\alpha p_\beta}{p^2}(g^{\beta\nu} - \varepsilon^{\beta\nu}), \tag{6.40}$$

$$\Pi^{\mu\nu}_\infty = \frac{a}{\pi} g^{\mu\nu}. \tag{6.41}$$

Now notice that

$$p_\mu \Pi^{\mu\nu} = \frac{1}{\pi}\left(p^\nu(a-1) + p_\mu \varepsilon^{\mu\nu}\right); \tag{6.42}$$

the dimensionless constant a is not fixed! No value of a sets $p_\mu \Pi^{\mu\nu} = 0$. Note that if we add up to this result the left chirality ($\varepsilon^{\mu\nu} \to -\varepsilon^{\mu\nu}$), we recover the complete Schwinger model calculation and gauge invariance. A lesson we can already grasp is that theories with a single chiral fermion seems to have an inherent ambiguity to them. This statement is of course nothing but the statement that the conservation of chiral charge is anomalous and importantly, regularization dependent.

One could argue that the arbitrariness of a is not a problem. The lattice theory always has two fermions of opposite chirality so we will always recover gauge invariance in the lattice (i.e. the Schwinger model). Although this is in general true, note that (i) the two chiral fermions can in principle be probed independently (e.g. if they are realized at different edges of a sample) and (ii) even when chirality is restored, the answer can be intrinsically ambiguous, and not determined by the bulk theory, as we will see in the next subsection.

6.4.2 A 3+1 D Example: Lorentz Breaking QED

In Sect. 6.2, we introduced the following Lorentz breaking QED theory

$$S[A] = \int d^4k\, \bar{\Psi}(\gamma_\mu M^\mu{}_\nu k^\nu - m - e\slashed{A} + \gamma_5 \slashed{b})\Psi \tag{6.43}$$

which we argued describes a Weyl semimetal if $-b^2 > m^2$. Since experiments can probe the response of these materials to external perturbations, this section is devoted to calculating such response in linear order (for technical details, see [15]).

As in QED, the coupling to the external electromagnetic field is $j^\mu A_\mu$, where j_μ is the current operator, defined by the free fermionic action, $j_\mu = \frac{\delta S}{\delta A_\mu}$. Taking care of $M^\mu{}_\nu$ containing the Fermi velocities, we have

$$j^\mu = M^\mu{}_\alpha \bar{\psi}_\mathbf{k} \gamma^\alpha \psi_\mathbf{k}. \tag{6.44}$$

Using (6.31), we can define the polarization function in linear response $\Pi^{\mu\nu}(p,b)$ which is given by

$$\langle j^\mu \rangle = \langle M^\mu{}_\alpha M^\nu{}_\beta \bar{\psi}_{\mathbf{k}} \gamma^\alpha \psi_{\mathbf{k}} \bar{\psi}_{\mathbf{k}} \gamma^\beta \psi_{\mathbf{k}} \rangle A_\nu$$
$$= M^\mu{}_\alpha M^\nu{}_\beta \Pi^{\alpha\beta} A_\nu. \qquad (6.45)$$

The polarization function $\Pi^{\mu\nu}$ is the usual photon self-energy bubble diagram

$$\Pi^{\mu\nu}(p,b) = \frac{e^2}{v_x v_y v_z} \int \frac{dk^4}{(2\pi)^4} \mathrm{Tr}\left\{\gamma^\mu G(k,b) \gamma^\nu G(k+p',b)\right\}. \qquad (6.46)$$

that contains two Green's functions, this time defined by

$$G(k,b) = \frac{i}{\slashed{k} - m - \slashed{b}\gamma_5}. \qquad (6.47)$$

The velocity prefactor $1/v_x v_y v_z$ appears by rescaling the integrated loop momentum k. This rescaling defines $p'^\mu = M^\mu{}_\nu p^\nu$, the rescaled external four-momentum vector. Note that in this case, the polarization function depends not only on the momentum p_μ but also on b_μ so it can be expressed as $\Pi^{\mu\nu}(b,p)$. It can be separated into odd and even parts with respect to the interchange $\mu \leftrightarrow \nu$. Anticipating where the ambiguity in this theory lies, we will be interested in the odd part, which can be defined as

$$\Pi^{\mu\nu}_{\mathrm{odd}}(b,p) = \varepsilon^{\mu\nu\rho\sigma} p_\rho b_\sigma K(p,b,m), \qquad (6.48)$$

where $K(p,b,m)$ is a scalar function and $\varepsilon^{\mu\nu\rho\sigma}$ is the Levi-Civita fully antisymmetric tensor.[5] In a remarkably beautiful paper, Perez-Victoria showed how to calculate $\Pi^{\mu\nu}_{\mathrm{odd}}(b,p)$ to all orders in b [30]

$$\Pi^{\mu\nu}_{\mathrm{odd}} = \frac{e^2}{v_x v_y v_z} \varepsilon^{\mu\nu\rho\sigma} p'_\rho \begin{cases} C_\sigma & \text{if } -b^2 \leq m^2 \\ C_\sigma - \dfrac{b_\sigma}{2\pi^2}\sqrt{1 - \dfrac{m^2}{b^2}} & \text{if } -b^2 \geq m^2 \end{cases}, \qquad (6.49)$$

where C_σ is a finite but undetermined constant four-vector [30–32]. Introducing (6.49) into (6.45), one obtains the response of the Weyl semimetal to an external electromagnetic field

$$\langle j^\mu_{\mathrm{odd}} \rangle = \frac{e^2 M^\mu{}_\alpha M^\nu{}_\beta}{v_x v_y v_z} \varepsilon^{\alpha\beta\rho\sigma} p'_\rho A_\nu \begin{cases} C_\sigma & \text{if } -b^2 \leq m^2 \\ C_\sigma - \dfrac{b_\sigma}{2\pi^2}\sqrt{1 - \dfrac{m^2}{b^2}} & \text{if } -b^2 \geq m^2 \end{cases}, (6.50)$$

Again we find a finite but undetermined result, which looks quite complicated. There are several considerations that can help us digest this calculation, and come to terms with this ambiguity. First, lets see what we can learn from the terms that do

---

[5]The even part has been also calculated (see [29]), but the discussion on its physical implications lies outside of the topic of this short chapter.

not involve C. The interesting regime in this case occurs for a space-like b_μ such that $\mathbf{b}^2 > b_0^2$, which corresponds to a gapless theory (see Sect. 6.2). A simple limiting case is $b_0 = 0$. Then using that the gauge potential can be expressed as $A_i = E_i/\omega$ in terms of the electric field E_i, we can recognize that the spatial current is

$$\mathbf{j} \propto \sqrt{1 - \frac{m^2}{b^2}} \mathbf{b} \times \mathbf{E} = \delta \mathbf{K} \times \mathbf{E}, \tag{6.51}$$

where we recovered the Weyl node separation $\delta \mathbf{K}$ from (6.11). In other words, the part of the Hall conductivity that is independent of C (and thus not-ambiguous) is proportional to the Weyl node separation. This is good since therefore this calculation can recover a known result. But then, what is the role (and the correct value!) of C?

Of course, we should expect that a decent lattice theory has to fix C in some way. As we will now discuss, the answer is not unique, and this is quite physical. First, lets convince ourselves that we should not be surprised that C indeed can be arbitrary. Recall that $\Pi^{\mu\nu}$ determines the effective action $S_{\text{eff}}[A]$ through (6.32). Inserting the form of the odd part of the polarization function (6.48) into (6.32) we can write

$$S_{\text{eff}}[A]_{\text{odd}} = \int d^4 p \ A_\mu(p) \left[\varepsilon^{\mu\nu\rho\sigma} p_\rho b_\sigma K(p, b, m) \right] A_\nu(-p). \tag{6.52}$$

You may recognize this action as a Chern–Simons action, which in real space has the schematic form $\varepsilon b A \partial A$. One might recall that Chern–Simons terms can only occur in odd space time dimensions, which is not our case. It is the existence of a finite four-vector b_μ which allows us to write (6.52). This type of functional form is known as the Carroll–Field–Jackiw (CFJ) term [33]

$$\mathscr{L}_{\text{CFJ}} = c_\mu \varepsilon^{\mu\nu\rho\sigma} F_{\nu\rho} A_\sigma, \tag{6.53}$$

where c_μ is a constant. It is named after the three physicists that considered it as an extension of Maxwell's electrodynamics that broke Lorentz invariance. This addition to Maxwell's equations has very interesting consequences, including, but not limited to, a Faraday effect, birefringence [33] or even a repulsive Casimir effect [15, 34].

There are two important mutually related features of a Chern–Simons action: (i) it is not gauge invariant, and (ii) it describes a system with a Hall effect. The latter is of course consistent with (6.51). The former gives us a hint of why it is ambiguous. Imagine that by choosing a gauge invariant regulator, we impose the gauge invariance of the Lagrangian density. Then the whole CFJ term is zero, since a Chern–Simons Lagrangian density is gauge non-invariant. However, if we are less strict and choose that only the action should be gauge invariant, a term like the CFJ can survive. The reason is that in an infinite system, we hide the gauge non-invariant terms that live at the surface. The difference between imposing gauge invariance of the Lagrangian density or the full action is equivalent to ask whether we, through the regulator, impose gauge invariance at all momenta or only at $p^\mu = 0$ respectively, since (schematically) the zero momentum Lagrangian is the real space action $\mathcal{L}(q=0) = \int dx \mathcal{L} = S_{\text{eff}}$. Pauli–Villars or dimensional regularization[6] imposes gauge invariance at all momenta and thus prohibits the appearance of the CFJ term. Other, less strict regularizations, however, will allow this term to exist, since they will only impose gauge invariance at the level of the action.

Additionally, the sole fact that we are dealing with a Chern–Simons action points to the fact that any surface term can alter the value of C, and thus, the whole term is ambiguous even ignoring the above regularization ambiguity. This statement can in fact be proven using the path integral approach known as the Fujikawa formalism [32]. So then, how does a lattice fix C?

One can make use of the fact that if b_μ is time-like, the response is completely determined by C. We know from the band structure that this state is an insulator which may or may not have a Hall conductivity. Haldane calculated the general form of the Hall conductivity in 3D [35, 36]

$$\sigma_{H,ij}^{3D} = \varepsilon_{ijm} \frac{K_H^m}{2\pi} \frac{e^2}{h}, \qquad (6.54)$$

giving an explicit expression for $\mathbf{K_H}$

$$\mathbf{K_H} = \nu \mathbf{G} + \sum_i \int_{S_i} \frac{\mathbf{k}_F \cdot \mathcal{F}}{2\pi} + \sum_{i\alpha} \int_{\partial S_i} \frac{\mathbf{G}_{i\alpha} \mathcal{A}}{2\pi}. \qquad (6.55)$$

The first term is the contribution from occupied states in the Brillouin zone $\mathbf{G}$ a reciprocal lattice vector. The last two terms encode Fermi surface contributions. They involve S_i and ∂S_i, which parametrize the Fermi surface sheet i and its boundary respectively, and $\mathbf{k}_F$, which is the Fermi momentum. The functions $\mathcal{F}$ and $\mathcal{A}$ are the Berry curvature and Berry connection respectively.

Since a time-like b_μ results in an insulator, the last two Fermi surface terms necessarily vanish. For such a simple insulator, $\mathbf{G} = \nu \mathbf{G}_0$ [35] which can be interpreted as the conductivity of a stacking of 2D Hall insulators with "filling" ν on the lattice planes stacked by $\mathbf{G}_0$ [37]. By comparing with the time-like case of (6.50) (i.e. the upper row), this fixes $C_\mu = (0, \nu \mathbf{G}_0)$.[7] We have found a 3D quantum Hall insulator.

The insulating phase borders a Weyl semimetal phase that is described by a space-like b_μ. In the simplest case where $b_0 = 0$, the Fermi surfaces S_i are point-like and have no boundary, which excludes the last term in (6.55).[8] They also have $\mathbf{k}_F = 0$, so does this would mean that the second term in (6.55) is excluded and the Hall conductivity is again $\nu \mathbf{G}_0$. However, it was noted in [36] that $\mathbf{k}_F$ it is ambiguous under

---

[6]It should be noted that dimensional regularization results in complications arising due to the ambiguity of the definition of γ_5 in odd space–time dimensions.

[7]The $\mu = 0$ component can be understood to be fixed to zero by the fact that there is no chiral magnetic effect in equilibrium [13, 38].

[8]There are subtleties with this statement for finite systems due to possible non-trivial edge states, but we will not discuss them here.

the change $\mathbf{k}_F \to \mathbf{k}_F + \text{constant}$ when time reversal is broken. So even if $\mathbf{k}_F = 0$ the second term in (6.55) has a contribution from all insulator planes perpendicular to the Weyl node separation. This sets $\mathbf{K}_H = 2\delta\mathbf{K}e^2/h + \nu\mathbf{G}_0$. Thus, by comparing with the space-like case of (6.50) (i.e. the lower row) we can fix $C_\sigma = (0, \nu\mathbf{G}_0)$ consistent with our previous result. We note that there can be other equivalent ways to understand this fixing in finite systems, using the topological surface states known as Fermi arcs, that contribute to the last term in (6.55), but we will not discuss that here [36].

To summarize, the ambiguity in the low-energy theory tells us that the Weyl fermion separation should have been measured from a reciprocal lattice vector $b_i \to G_i - b_i$. It is nothing but the physical result that b_i is only defined modulo a lattice vector. This is equivalent to allowing a term in the action that looks like the CFJ term

$$S_{\text{eff}}[A]_G = \sum_i \int d^4 p \, A_\mu(p) \left[\varepsilon^{\mu\nu\rho\sigma} G_\sigma^i p_\rho\right] A_\nu(-p), \tag{6.56}$$

where $G_\mu = (0, \mathbf{G})$ where $\mathbf{G}$ are integrals of the Berry curvature of each disjoint group of occupied bulk bands below the Fermi level.

6.4.3 Connections to the Chiral Anomaly

In this section, we will connect the above with the chiral anomaly, which has been thoroughly discussed both in high-energy physics and in condensed matter (see [2] for a focused review). Without dwelling too much on the details, we will focus on its ambiguities and discuss briefly how they are fixed.

Consider again (6.46) and expand the Green's functions to lowest orders in b_μ. Using that $G(k, b) \sim G(k, 0) + iG(k, 0)\gamma^\mu \gamma_5 b_\mu G(k, 0)$, we have that the first nontrivial order is

$$\Pi^{\mu\nu}(p, b) \sim e^2 \int \frac{dk^4}{(2\pi)^4} \text{Tr}\{\gamma^\mu G(k) \gamma^\nu G(k+p) \gamma^\alpha \gamma_5 G(k+p)\} b_\alpha$$
$$+ \{\mu \leftrightarrow \nu, \, p \leftrightarrow -p\} \equiv \Gamma^{\mu\nu\alpha}(p, q = -p) b_\alpha. \tag{6.57}$$

We have identified the integrand as a triangle diagram, shown in Fig. 6.4 lower right panel, with a particular kinematics $\Gamma^{\mu\nu\alpha}(p, q = -p)$ with two vector vertices and one axial vertex (see for instance [39]). With this particular kinematic, one can also isolate a divergent part of $\Gamma^{\mu\nu\rho}(p, -p)$ that depends on the regulator.

$$\Gamma^{\mu\nu\rho}(p, -p)_{\text{undet}} \sim a\varepsilon^{\mu\nu\rho\sigma} p_\sigma, \tag{6.58}$$

where a is finite but undetermined. If b_μ is a constant, then we are safe since $p_\mu \Pi^{\mu\nu} = 0$ and charge is conserved. However, this conservation law is satisfied regardless of the value of a so gauge invariance in fact does not fix a in any way; this is the ambiguity analysed in the previous section.

This ambiguity is in fact inherited from the chiral anomaly which is itself determined by the triangle diagram $\Gamma^{\mu\nu\rho}(p,q)$. To understand this, lets first recall how the chiral anomaly works. We can start with two decoupled chiral fermions in $3+1$ dimensions, which from our knowledge of previous sections we can write as

$$S[A] = \int d^4k \bar{\Psi}(\slashed{k} - e\slashed{A})\Psi, \tag{6.59}$$

which is a generalization of the 1+1D action of (6.29). To see how the chiral anomaly emerges, we can follow the arguments developed in [40]. Choose $A_\mu = (A_0, \mathbf{A})$ such that $\mathbf{A} = B_z x \mathbf{e}_y$ sets a magnetic field of magnitude B_z along the $\mathbf{e}_z$ direction, where $\mathbf{e}_i$ is the unit vector in i-direction (with $i = x, y, z$). The Hamiltonian corresponding to (6.59) is simply the Weyl Hamiltonian equation (6.6) that describes two Weyl fermions of chirality $\chi = \pm$ coupled to the gauge field

$$\mathcal{H}_0^\chi = \chi v_F (\mathbf{k} - e\mathbf{A}) \cdot \boldsymbol{\sigma}, \tag{6.60}$$

where we can set $v_F = 1$ for simplicity. Defining the magnetic length $l_B = 1/\sqrt{eB_z}$ and the creation and annihilation operators

$$a_{k_y} = \frac{1}{\sqrt{2}}\left(\frac{x - k_y l_B^2}{l_B} + ik_x l_B\right), \tag{6.61a}$$

$$a_{k_y}^\dagger = \frac{1}{\sqrt{2}}\left(\frac{x - k_y l_B^2}{l_B} - ik_x l_B\right), \tag{6.61b}$$

which obey $[a_{k_y}, a_{k_y}^\dagger] = 1$, we can write the Hamiltonian in the $|k_y\rangle$ basis

$$\langle k_y | \mathcal{H}_0^\chi | k_y' \rangle = \delta_{k_y, k_y'} \chi v_F \begin{pmatrix} k_z & i\sqrt{2} a_{k_y}^\dagger / l_B \\ -i\sqrt{2} a_{k_y} / l_B & -k_z \end{pmatrix}. \tag{6.62}$$

This form of the Hamiltonian allows us to label the eigenvalues of $a_{k_y}^\dagger a_{k_y}$ by n, the Landau level quantum number. The spectrum of (6.62) comprises particle–hole symmetric bands with dispersion $E_{0,n>0}^\chi(k_z) = \pm\chi\sqrt{v_F^2 k_z^2 + 2n/l_B^2}$ and a chiral linearly dispersing lowest Landau level $E_{0,n=0}^\chi(k_z) = \chi v_F k_z$, as illustrated in Fig. 6.4. Notice that the chiral Landau level dispersion is exactly the dispersion relation of the 1+1D field theory (6.29). The important difference is that the bands are independent of the momentum eigenvalue k_y, and thus, they are extremely degenerate. Each Landau level, including the chiral ones, has degeneracy

$$N_{\text{LL}} = \frac{L_x L_y B_z}{2\pi/e}, \tag{6.63}$$

where L_i is the length of the system in i-direction. At high magnetic fields, the low-energy physics is determined by the gapless lowest Landau level only. Thus, a single Weyl fermion in a strong magnetic field in the $\mathbf{e}_z$ direction is described by a macroscopically degenerate set of right- or left-moving chiral electrons with a one-dimensional dispersion $E^\chi_{0,n=0}(k_z)$. Nearly without any calculation, we can read off the effect of an electric field $\mathbf{E} = E_z\mathbf{e}_z$, set for instance by the time-dependent gauge field $A_\mu = (0, 0, 0, E_z t)$, on the chiral Landau levels. Minimal substitution requires that $k_z \to k_z - eA_z = k_z - eE_z t$ and tells us that the states from two chiral branches $\pm k_z$ are created or destroyed at a rate $dk/dt = eE$ (see Fig. 6.4). If we count the charge imbalance between left and right taking into account the Landau level degeneracy, we arrive to

$$\partial_t(n_+ - n_-) = N_{LL}\frac{1}{2\pi}\frac{dk}{dt} = \frac{e^2}{4\pi^2\hbar}\mathbf{E}\cdot\mathbf{B}, \qquad (6.64)$$

where we have restored $\hbar$. This is in fact the anomalous conservation equation for the chiral current in the absence of currents, that is expressed in general as

$$\partial_\mu j_5^\mu = \frac{e^2}{4\pi^2\hbar}\mathbf{E}\cdot\mathbf{B}, \qquad (6.65)$$

where $j_5^\mu = j_L^\mu - j_R^\mu$. This result is nothing but the 3+1D generalization of the non-conservation of chiral charge (6.42). The total charge is conserved, but their difference is not, just as happened with (6.42); gauge invariance is recovered when summing over chiralities.

Even though this derivation in terms of Landau levels is physically transparent, we could have obtained this from a diagrammatic perspective which now use to connect to our previous results. Notice that (6.65) can be seen as arising from a Feynman diagram shown in Fig. 6.4, where two legs represent the gauge fields that will compose $\mathbf{E}$ and $\mathbf{B}$ and one represents the chiral current j_5^μ. The triangle amplitude, which determines the conservation of the currents, enters the vacuum expectation value of the chiral current to second order in the external field

$$j_5^\mu(l) = e^3 \int \frac{d^4q}{(2\pi)^4}\frac{d^4p}{(2\pi)^4}\Gamma^{\mu\nu\rho}(p,q)\delta(l-(p+q))A_\nu(p)A_\rho(q). \qquad (6.66)$$

A similar argument will allow us to write a contribution to j^μ in terms of $\Gamma^{\mu\nu\rho}$. Thus demanding that $\Gamma^{\mu\nu\rho}(p,q)$ is transverse in all of its indices is required to conserve both currents. This amounts to ask that its contraction with all momenta vanishes. However, owing to the existence of the ambiguous contribution, one can show that its contractions take the form [2, 41]

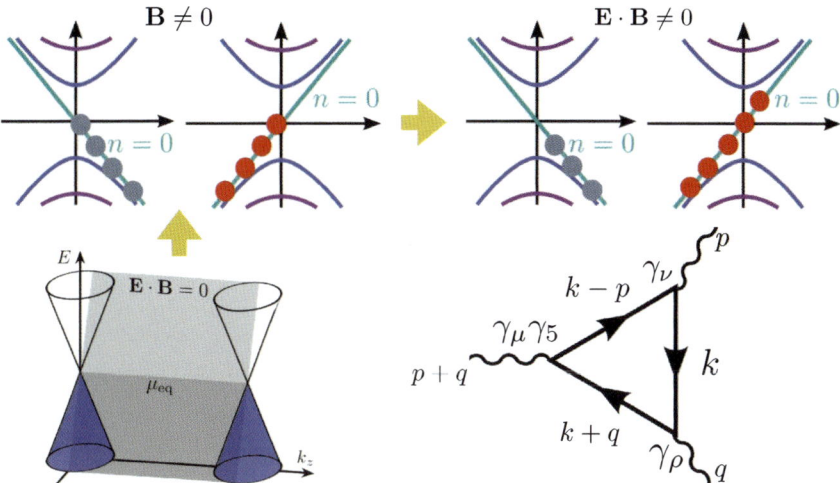

Fig. 6.4 The magnetic field breaks the two 3+1D Weyl fermions spectrum (bottom left panel) into Landau levels which include two chiral modes dispersing along the field (top left panel). Applying an electric field parallel to the magnetic field (upper right panel) turns left movers (blue circles) into right movers (red circles), defining the chiral anomaly. Diagrammatically, the chiral anomaly stems from a triangle diagram shown in the bottom right corner $\Gamma^{\mu\nu\rho}(p,q)$

$$(p_\mu + q_\mu)\Gamma^{\mu\nu\rho}(p,q) = \frac{(1+a)}{4\pi^2}\varepsilon^{\nu\rho\alpha\beta}p_\alpha q_\beta, \qquad (6.67)$$

$$p_\nu \Gamma^{\mu\nu\rho}(p,q) = \frac{(1-a)}{8\pi^2}\varepsilon^{\mu\rho\alpha\beta}p_\alpha q_\beta, \qquad (6.68)$$

where a again parametrizes the regularization-dependent terms. However, unlike in Lorentz breaking QED, the ambiguity can be fixed by demanding gauge invariance, which imposes $a = 1$. In contrast, for the particular kinematics that leads to (6.58), gauge invariance is always a symmetry, independent of a.

6.5 Beyond Weyl Fermions

One could ask whether the above considerations can help us to study more exotic emergent fermions, such as Type-II (or over-tilted) Weyl semimetals or three-, four-, six- and eightfold fermions. This section will not address this question fully, but will give two examples of what is possible.

The first example is Type-II Weyl fermions which have an over-tilted cone, such that the Fermi surface has a hole and an electron pocket that touch at a protected point [42]. Type-II Weyl fermions seem to break Lorentz invariance; however, they can be understood as Type-I fermions in space–times which have a non-Minkowski metric, which is defined by the tilt-vector [43, 44]. A simple Hamiltonian that realizes this state is [45]

$$\mathcal{H}_{\pm} = v_{\perp}(\pm k_x \sigma_x + k_y \sigma_y) + v_z(k_z - b_z)\sigma_z + w(k_z - b_z)\sigma_0, \quad (6.69)$$

where w is the tilt parameter. The last term induces a time-like component to the velocity matrix $M^{\mu}{}_{\nu}$ defined in (6.14), which can be reinterpreted as a background metric. To see this, compare (6.69) to a Weyl fermion in curved space–time

$$\mathcal{L} = \sigma^{\alpha} e^{\mu}_{\alpha} \partial_{\mu}, \quad (6.70)$$

where $\sigma_{\alpha} = (\sigma_0, \boldsymbol{\sigma})$ and the tetrads e^{α}_{μ} define the metric $g^{\mu\nu} = \eta^{ab} e^{\mu}_a e^{\nu}_b$. This comparison leads to the definition of the line element

$$ds^2 = g_{\mu\nu} dx^{\mu} dx^{\nu} = -dt^2 + \frac{1}{v_{\perp}^2}(dx^2 + dy^2) + \frac{1}{v_z^2}(dz - wdt)^2. \quad (6.71)$$

The tilt parameter w changes the untilted spectrum ($w = 0$) to a moving reference frame with speed w [45].

The second example is the collection of various other "Dirac-like" equations that describe particles beyond Weyl fermions in high energy physics. A particularly exotic one may describe the gravitino: it is the Rarita–Schwinger Lagrangian [46]

$$\mathcal{L} = \frac{1}{2} \bar{\psi}_{\mu} (\varepsilon^{\mu\rho\sigma\nu} \gamma_5 \gamma_{\rho} \partial_{\sigma} - i\sigma^{\mu\nu} m) \psi_{\nu}, \quad (6.72)$$

which describes fermions with spin-3/2. Recently, these types of fermions have been suggested to exist with fourfold degenerate crossings (see supplementary material in [47] and the spin-3/2 fermion described in [48]).

6.6 Conclusions

In this chapter, we discussed how Weyl semimetallic phases of matter and related systems are described by ambiguous field theories, which highlight interesting aspects of their responses to external fields. These ambiguities are connected to anomalies and are fixed by the lattice in interesting ways. The main take-home message is that combining high-energy theory literature with condensed matter phenomena can lead to interesting new insights on physical responses and a deeper understanding of both realms of physics. It is likely that many of these techniques serve as well to understand anomalies and ambiguities in related systems such as nodal line semimetals or multifold fermions [1] as well as higher-order responses which have recently shown interesting phenomenology [49].

Acknowledgements I would like to express gratitude to all collaborators and colleagues that over the years have shaped whatever you find useful in this chapter and are not to be blamed by any misstatement, especially Jens H. Bardarson, Jan Behrends, Alberto Cortijo, Yago Ferreiros, Felix Flicker, Roni Ilan, Fernando de Juan, Michael Kolodrubetz, Karl Landsteiner, Titus Neupert, Sthitadhi Roy, Jorn W. F. Venderbos and Maria A. H. Vozmediano. I am especially thankful to the organizers of the 2017 Topological Matter School in San Sebastian, Maia Vergniory, Reyes Calvo, Dario Bercioux and Jerome Cayssol, for which these lecture notes were conceived.

Appendix: Calculating the Effective Action

This is a standard quantum field theory method [5]. The effective action $S_{\text{eff}}[A]$ can be formally defined through the partition function

$$Z[A] = e^{i S_{\text{eff}}[A]} \equiv \int D[\Psi] e^{i S[A]}, \tag{6.73}$$

where we assume the action can be written as $S[A] = G_0 + J^\mu A_\mu$. In this chapter, we are interested in defining $S_{\text{eff}}[A]$ perturbatively in A. To do so, we write the partition function as

$$Z[A] = \det \left[G_0^{-1} - J_\mu A^\mu \right] \tag{6.74}$$
$$= \det \left[G_0^{-1} \right] \det \left[1 - G_0 J_\mu A^\mu \right]. \tag{6.75}$$

Using that $\det A = e^{i \operatorname{Tr} \ln A}$ and noting that the $\det \left[G_0^{-1} \right]$ will be an overall factor that will not contribute to the calculation of observables, we can define

$$S_{\text{eff}}[A] = \int \frac{d^d k}{(2\pi)^d} \sum_{n=0}^{\infty} \frac{-1}{n} \operatorname{Tr} \left[(G_0 J_\mu A^\mu)^n \right]. \tag{6.76}$$

The second-order term, responsible for linear response through (6.31), is

$$S_{\text{eff}}[A] = \int d^d k A_\mu(p) \Pi^{\mu\nu}(p) A_\mu(-p) \tag{6.77}$$

where [50]

$$\Pi^{\mu\nu}(p) = \int \frac{d^d k}{(2\pi)^d} \operatorname{Tr} \left[G_{k-p/2} J_k^\mu G_k J_{k+p/2}^\nu \right]. \tag{6.78}$$

References

1. N.P. Armitage, E.J. Mele, A. Vishwanath (2017). arXiv:1705.01111
2. K. Landsteiner, Acta Phys. Pol., B **47**, 2617 (2016). arXiv:1610.04413 [hep-th]

3. G.E. Volovik, JETP Lett. **70**, 1 (1999)
4. F.R. Klinkhamer, G.E. Volovik, Int. J. Mod. Phys. A **20**, 2795 (2005)
5. M.E. Peskin, D.V. Schroeder, *An Introduction to Quantum Field Theory* (Westview Press, 1995)
6. T.D. Stanescu, *Introduction to Topological Quantum Matter and Quantum Computation* (CRC Press, 2016)
7. A. Bettini, *Introduction to Elementary Particle Physics* (Cambridge University Press, Cambridge, 2008)
8. Z. Wang, Y. Sun, X.Q. Chen, C. Franchini, G. Xu, H. Weng, X. Dai, Z. Fang, Phys. Rev. B - Condens. Matter Mater. Phys. **85**, 195320 (2012). arXiv:1202.5636
9. A.A. Burkov, M.D. Hook, L. Balents, Phys. Rev. B **84**, 235126 (2011)
10. K. Landsteiner, Phys. Rev. B **89**, 075124 (2014)
11. D. Colladay, V.A. Kostelecký, Phys. Rev. D **55**, 6760 (1997)
12. R.A. Bertlmann, *Anomalies in Quantum Field Theory*, vol. 91 (Oxford University Press, Oxford, 2000)
13. M.M. Vazifeh, M. Franz, Phys. Rev. Lett. **111**, 027201 (2013)
14. D. Colladay, V.A. Kostelecký, Phys. Rev. D **58**, 116002 (1998)
15. A.G. Grushin, Phys. Rev. D **86**, 045001 (2012)
16. A.A. Zyuzin, S. Wu, A.A. Burkov, Phys. Rev. B **85**, 165110 (2012)
17. P. Goswami, S. Tewari, Phys. Rev. B **88**, 245107 (2013)
18. V. Aji, Phys. Rev. B **85**, 241101 (2012)
19. H. Georgi, Annu. Rev. Nucl. Part. Sci. **43**, 209 (1993)
20. H.J. Rothe, *Lattice Gauge Theories: An Introduction* (World Scientific Publishing Company, 2005)
21. H. Nielsen, M. Ninomiya, Nucl. Phys. B **185**, 20 (1981)
22. H. Nielsen, M. Ninomiya, Nucl. Phys. B **193**, 173 (1981)
23. X.-L. Qi, T.L. Hughes, S.-C. Zhang, Phys. Rev. B **78**, 195424 (2008)
24. A.G. Grushin, J.W.F. Venderbos, J.H. Bardarson, Phys. Rev. B **91**, 121109 (2015)
25. A.G. Grushin, J.W.F. Venderbos, A. Vishwanath, R. Ilan, Phys. Rev. X **6**, 041046 (2016)
26. R. Narayanan, H. Neuberger, Phys. Rev. Lett. **71**, 3251 (1993)
27. R. Jackiw, Int. J. Mod. Phys. B **14**, 2011 (2000)
28. R. Jackiw, R. Rajaraman, Phys. Rev. Lett. **54**, 1219 (1985)
29. B. Altschul, Phys. Rev. D **73**, 036005 (2006)
30. M. Pérez-Victoria, Phys. Rev. Lett. **83**, 2518 (1999)
31. R. Jackiw, V.A. Kostelecký, Phys. Rev. Lett. **82**, 3572 (1999)
32. J.-M. Chung, Phys. Rev. D **60**, 127901 (1999)
33. S.M. Carroll, G.B. Field, R. Jackiw, Phys. Rev. D **41**, 1231 (1990)
34. J.H. Wilson, A.A. Allocca, V. Galitski, Phys. Rev. B **91**, 235115 (2015)
35. F.D.M. Haldane, Phys. Rev. Lett. **93**, 206602 (2004)
36. F.D.M. Haldane (2014). arXiv:1401.0529 [cond-mat.str-el]
37. M. Kohmoto, B.I. Halperin, Y.-S. Wu, Phys. Rev. B **45**, 13488 (1992)
38. K. Landsteiner, Phys. Rev. B **89**, 075124 (2014b)
39. A. Zee, *Quantum Field Theory in a Nutshell: (In a Nutshell)*, 2nd edn. (Princeton University Press, Princeton, 2010)
40. H. Nielsen, M. Ninomiya, Phys. Lett. **130B**, 389 (1983)
41. M. Kaku, *Quantum Field Theory: A Modern Introduction* (Oxford University Press, Oxford, 1993)
42. A.A. Soluyanov, D. Gresch, Z. Wang, Q. Wu, M. Troyer, X. Dai, B.A. Bernevig, Nature **527**, 495 (2015)
43. G.E. Volovik, Sov. J. Exp. Theor. Phys. Lett. **104**, 645 (2016). arXiv:1610.00521 [cond-mat.other]
44. K. Zhang, G.E. Volovik (2016). arXiv:1604.00849 [cond-mat.other]
45. S. Guan, Z.-M. Yu, Y. Liu, G.-B. Liu, L. Dong, Y. Lu, Y. Yao, S.A. Yang, npj Quantum Mater. **2**, 23 (2017)
46. W. Rarita, J. Schwinger, Phys. Rev. **60**, 61 (1941)

47. B. Bradlyn, J. Cano, Z. Wang, M.G. Vergniory, C. Felser, R.J. Cava, B.A. Bernevig (2016). arXiv:1603.03093 [cond-mat.mes-hall]
48. P. Tang, Q. Zhou, S.-C. Zhang (2017). arXiv:1706.03817 [cond-mat.mtrl-sci]
49. F. de Juan, A.G. Grushin, T. Morimoto, J.E. Moore, Nat. Commun. **8**, 15995 (2017)
50. B.A. Bernevig, T.L. Hughes, *Topological Insulators and Topological Superconductors* (Princeton University Press, Princeton, 2013)

Chapter 7
Anomalies and Kinetic Theory

Alberto Cortijo

Abstract In this chapter, we will make an overview of the quantum anomalies, as quantities that are no longer conserved when passing from the classical to the quantum realm. We will focus on the chiral anomaly. The discussion will be made in terms of the semiclassical kinetic theory, where the classical Boltzmann transport equation is supplemented by the equations of motion that explicitly contain the Berry connection. In this regard, we will make explicit the connection between the chiral anomaly and the non-trivial topological structure of Weyl semimetals. We will make the discussion beyond the different relaxation time approaches that are commonly used in the literature. This approach introduces some mathematical complexities but also reveals some less known features of transport in Weyl semimetals. Finally, we will discuss other quantum anomalies that have been of interest recently in Condensed Matter Physics.

7.1 Introduction

One of the cornerstones in modern Physics is the idea of that for every symmetry a given system displays, there is an associated quantity that is *conserved*. Conservation means here that this quantity does not change with time upon the dynamical evolution of the system. In the classical realm, this statement is accurately demonstrated in the most celebrated Noether's theorem [1]. The standard examples usually quoted to exemplify this deep result are the conservation of linear momentum, if the system is invariant under translations, the conservation of angular momentum if the system is invariant under rotations, or the conservation of the electric charge if the system is gauge invariant.

Of course, it is highly desirable that these quantities are also conserved when going to the quantum world. Although the common sense tells us that this is obvious (no energy is popping out of nothing, and things do not spontaneously start to rotate

---

A. Cortijo (✉)
Materials Science Factory, Instituto de Ciencia de Materiales
de Madrid, CSIC, Cantoblanco, 28049 Madrid, Spain
e-mail: alberto.cortijo@csic.es

© Springer Nature Switzerland AG 2018
D. Bercioux et al. (eds.), *Topological Matter*, Springer Series in Solid-State
Sciences 190, https://doi.org/10.1007/978-3-319-76388-0_7

due to quantum fluctuations), it is not something trivial to prove, and in fact, it is not an universal statement. In this chapter, we will find and discuss an example of a quantity that is conserved at the classical level, but it is not in quantum mechanics. When this happens, we talk about *quantum anomalies*.

The reader can find a vast literature about quantum anomalies. Virtually any book dealing with Quantum Field Theory explains the notion of quantum anomalies. There is a lot of drama there: Quantum anomalies are not just funny oddities due to quantum mechanics. The models used to describe Nature are constructed following the rule that they must preserve some symmetries. If any of these symmetries are violated, we have the risk of having a theory not describing reality. Well, we will see that, although anomalies are a reality, in the best case, the spoiled symmetry is not a shot below the waterline of the theory, or if the symmetry is a fundamental one, the presence of an anomaly helps to develop the right theory when trying cancel it. Then we can make a virtue of need and think that quantum anomalies allow for interesting physical phenomena.

In the rest of this chapter, we will approach to the notion and consequences of quantum anomalies in the context of Condensed Matter Physics. As mentioned above, the notion of quantum anomaly can be found in every field theory book, but almost always oriented to Particle Physics, where these anomalies where originally discussed [2, 3], with very few notable exceptions [4]. This is essentially because until very recently, there were few examples where a quantum anomaly provided observable effects. The three well-known examples we had so far were the physics of superfluid ^{3}He [4], the Luttinger liquid [5], and the edge states of the Quantum Hall effect [6]. Nowadays, with the advent of gapless topological media, there has been a revival of the subject of quantum anomalies, and these systems offer new perspectives and possibilities to observe (and discuss) these *quantum oddities*.

Even in the case of dealing with gapless topological media, there are already several good reviews [7–9] on quantum anomalies, and my own knowledge of this subject has been obtained from them. So the reader interested in getting a better (and more profound) taste of anomalies is highly encouraged to study these references.

7.2 Chiral Anomaly in Weyl Semimetals

Let us first define gauge invariance in electromagnetism [10]. Let us consider a system collectively described by a matter field ψ_a, possessing a global symmetry consisting in that the system remains invariant under a change of phase: $\psi_a \to \psi'_a = e^{i\theta}\psi_a$. At the classical level, Noether's theorem tells us that, since this change of phase is a symmetry of the system, there will be an associated conserved current, $\partial_\mu J^\mu = 0$, or $\dot\rho - \nabla \cdot \mathbf{J} = 0$. The space integral of the zeroth component of J^μ, $Q = \int \rho d\mathbf{r}$, will be a conserved charge, that is, $\dot Q = 0$. The value of Q will not change under the dynamical evolution of the system. Now, if we go from the global symmetry to the local version of it, $\theta \to \theta(x)$, we will need to define a new field that keeps the local symmetry. If $\psi_a \to \psi'_a = e^{i\theta(x)}\psi_a$, we need to define a field A_μ that changes as $A_\mu \to A'_\mu = A_\mu - \partial_\mu \theta(x)$. In this way, the gauge symmetry is maintained in the local version.

Two important points to realize here are: First, the operator that couples to the gauge field A_μ is precisely the current J_μ obtained after applying the Noether theorem to the global symmetry. The second is that the symmetry is not spoiled due to the passage from a global symmetry to the local version (the variation of the action by local phases does not exactly give the same action as prior to the transformation, but the extra produced piece $\int d^4x \theta(x) \partial_\mu J^\mu$ vanish if $\partial_\mu J^\mu = 0$). Simply we have added a new dynamical field, the electromagnetic field, to restore the full symmetry. And so far everything is at the classical level, no quantum mechanics, yet. The moral of this story is if we want to analyze if any symmetry is no longer a symmetry after going to the quantum world, the starting point should be first global anomalies.

When we go to the quantum world, the philosophy is the same, but the mathematical statement is now that the quantity Q is an operator that commutes with the Hamiltonian, $[H, Q] = 0$.

Let us consider now a similar but different symmetry in classical massless fields, where the dynamics is governed by the Weyl Hamiltonian [9]:

$$H(\mathbf{k}) = s v \psi^+ \boldsymbol{\sigma} \cdot (\mathbf{k} - s\mathbf{b}) \psi, \qquad (7.1)$$

where $s = \pm 1$ and the vector $2\mathbf{b}$ is the vector separating the two Weyl nodes. Because Weyl semimetals are condensed matter realizations of fermions described by this Hamiltonian, the Nielsen–Ninomiya theorem ensures the presence of at least a couple if Weyl points with opposite chirality, denoted by the parameter s [11]. From (7.1), we can write the following Lagrangean, after using standard definitions of the γ−matrices, expanding around the $\mathbf{b}$, $\mathbf{k} \sim s\mathbf{b} + i\partial$, and constructing the adjoint spinor $\bar{\psi} = \psi^+ \gamma_0$ (we have set all the Fermi velocities to one):

$$\mathcal{L} = i \bar{\psi} \gamma^\mu \partial_\mu \psi. \qquad (7.2)$$

The use of the Weyl Lagrangean instead of the Hamiltonian is prompt to the application of Noether's theorem when the symmetries of the system are identified. Contrary to other symmetries shared with more conventional electronic systems, the Lagrangean (7.2) has a symmetry similar, but not equal, to the phase symmetry associated with electromagnetism. The Lagrangean (7.2) is invariant under the change $\psi \to \psi' = e^{i\theta \gamma_5} \psi$, and $\bar{\psi} \to \bar{\psi}' = \bar{\psi} e^{i\theta \gamma_5}$, since all the γ matrices anticommute with γ_5. The picture is as follows: Since now we have two Weyl nodes, we can choose to change the phase to be the same for both nodes (the $U(1)$ gauge invariance associated with electromagnetism) or to be opposite. This is why in the latter case the phase transformation goes with the γ_5 matrix, indicating that the phase is opposite for nodes with opposite chirality s. It is easy to see that in order for the chiral transformation to leave the action $S = \int d^4x \mathcal{L}$ invariant, the divergence of the corresponding current $J_5^\mu = \bar{\psi} \gamma_5 \gamma^\mu \psi$ must vanish: $\partial_\mu J_5^\mu = 0$. Although only valid for Weyl fermions, we can make use of the superficial similarity of the chiral symmetry with the electromagnetic phase symmetry. Let us define $J^\mu = J_L^\mu + J_R^\mu$, and $J_5^\mu = J_L^\mu - J_R^\mu$. The labels L and R stand for *left* and *right* (nothing special, just to remind that these currents have something to do with the notion of chirality). In this way, it is apparent that the current J^μ is associated with the total charge in the system, while J_5^μ represents the difference between the charges of different chirality. At the classical level, both currents, J_L^μ and J_R^μ, are separately conserved.

Now let us quote the final result we want to discuss. When promoting the classical Lagrangean (7.2) to the quantum realm, and in presence of electromagnetic fields, the chiral current is no longer conserved:

$$\partial_\mu J_5^\mu = \frac{e^2}{24\pi^2} \varepsilon^{\mu\nu\rho\sigma} F_{\mu\nu} F_{\rho\sigma}. \tag{7.3}$$

The physical content of the anomaly (7.24) strongly depends on the situation. In the context of High Energy Physics, the principal consequence of the anomaly (7.24) is to allow for the process $\pi^0 \to \gamma + \gamma$ [2, 3], that is, the decay of a neutral pion in two photons. In a condensed matter setting, we can notice that the product $\varepsilon^{\mu\nu\rho\sigma} F_{\mu\nu} F_{\rho\sigma}$ is proportional to $\mathbf{E} \cdot \mathbf{B}$, so, when we apply to the system parallel electric and magnetic fields, charge from one Weyl point starts to flow to the other Weyl point with opposite chirality, and it does not stop unless any other element puts the charges back to their original Weyl point, reaching a stationary situation [12].

As mentioned above, the calculation of (7.3) can be found in any standard textbook of quantum field theory. It is done in two ways: by directly computing the Feynman diagrams associated with the expectation value $\langle 0 | J_5^\mu J^\nu J^\rho | 0 \rangle$, or by analyzing the non-invariance of the fermionic measure in the partition function under the chiral transformation, method originally developed by Fujikawa [13, 14].

Here we will adopt a different approach, based on the chiral kinetic theory [15–17]. It is perhaps the most economic way to compute (7.3). It also provides a crystal clear reason of the *topological* origin of the chiral anomaly.

7.3 Chiral Kinetic Theory

Chiral kinetic theory is nothing but the standard semiclassical Boltzmann transport theory where the peculiarities of the Hamiltonian (7.1) are taken into account [15, 17]. To make this chapter as self-contained as possible, I will itemize this section in the following way. First, we will review some generalities of the Boltzmann transport equation, as it can be found everywhere. Second, I will comment on the semiclassical equations of motion for Weyl fermions. It will be clear here that, despite of being gapless, massless Weyl fermions are topologically non-trivial. This non-triviality will appear in front of our eyes. Then, with the Boltzmann equation in one hand, and the equations of motion in the other, we will compute the non-conservation of the chiral charge in presence of electromagnetic fields.

7.3.1 Boltzmann Equation

In classical statistical mechanics, the Boltzmann transport equation (aka kinetic equation) plays a central role. Imagine that we have a collection of several species of classical particles. We thus can define the configuration space made out of the real space and the space of momenta (or velocities, it is the same). Then, we can define some probability density, or distribution function f_s of finding a particle of type s at some infinitesimal region around the point $(\mathbf{x}, \mathbf{k})$. The time evolution of this probability density will depend on the external forces acting on the particles, given by the Newton law, $\dot{\mathbf{k}}_s = \mathbf{F}$, the diffusion processes, that is, changes in the velocities $\dot{\mathbf{x}}_s$ of the particles entering and getting out of this small region, and the possibility of collisions among all the types of particles or by external impurities within the region. All this allows us to write the Boltzmann kinetic equation [18]:

$$\frac{\partial f_s}{\partial t} + \dot{\mathbf{k}}_s \cdot \frac{\partial f_s}{\partial \mathbf{k}} + \dot{\mathbf{x}}_s \cdot \frac{\partial f_s}{\partial \mathbf{x}} = I_c[f_s, f_{s'}]. \tag{7.4}$$

The term describing how the particles collide among them or with impurities is written as $I_c[f_s, f_{s'}]$. Its precise form will depend a lot of the type of collisions. So, to solve (7.4), we need to know which types of collisions we have to consider, and we need to know the dynamics of our particles, that is, we need to find equations of motion for $\dot{\mathbf{k}}_s$ and $\dot{\mathbf{x}}_s$.

It is interesting to mention that, although being a description of kinetic processes in classical systems, the Boltzmann equation has been quite successfully used to describe transport properties in solids, where the electrons behave like a gas of quantum particles, and the natural description appears to be quantum mechanics. Quite generically, classical mechanics is recovered from quantum mechanics in the limit $\hbar \to 0$. In particular, when one considers transport phenomena, the natural way to compute them is by the Kubo formula, by computing averages of the density matrix $\rho_{ab} = |a\rangle \langle b|$. In the Kubo formula, in most of the cases, one can take the classical limit and compute the evolution of the diagonal elements of the density matrix ρ_{aa} (that will play the role of the distribution functions f_a) and progressively consider quantum corrections. However, we will not follow that route. We will follow a different route. We will assume that the classical Boltzmann description is valid, but we need to go from quantum states to some states that will describe the dynamics faithfully the dynamics of the quantum states but in the limit $\hbar \to 0$. In order to meet with the necessities of the kinetic equation, we need to find the way to simultaneously describe the position and the momentum (and their equations of motion) of the particles in the configuration space.

We will finish this section discussing the collision term $I_c[f_s, f_{s'}]$. We can argue the form of this term by considering that the probability to make a transition from some state $(\mathbf{k}, s)$ to a final state $(\mathbf{k}', s')$ has to do with the quantum-mechanical transition probability $W_{\mathbf{k}\mathbf{k}'}^{ss'}$ described in terms of the corresponding Bloch states, the probability of having the initial state $|\mathbf{k}, s\rangle$ occupied, that is, the distribution function $f_s(\mathbf{k})$, and the final state $|\mathbf{k}', s'\rangle$ to be empty, that is, $1 - f_{s'}(\mathbf{k}')$. Also, we have to consider the reversed process, $(\mathbf{k}', s') \to (\mathbf{k}, s)$, exchanging the role of $f_s(\mathbf{k})$ and $f_{s'}(\mathbf{k}')$. Since we have to consider all processes that are connected to $\mathbf{k}$, the collision term reads $((d\mathbf{k}) \equiv \frac{d^d \mathbf{k}}{(2\pi)^d})$:

$$I_c[f_s, f_{s'}] = \int (d\mathbf{k}) f_{s'}(\mathbf{k}')(1 - f_s(\mathbf{k})) W_{\mathbf{k}\mathbf{k}'}^{ss'} - f_s(\mathbf{k})(1 - f_{s'}(\mathbf{k}')) W_{\mathbf{k}'\mathbf{k}}^{ss'}. \tag{7.5}$$

The collision integral (7.5) makes the Boltzmann equation a nonlinear integro-differential equation, difficult to handle in general. We will see that, for impurity scattering, and under the principle of detailed balance, the collision integral $I_c[f_s, f_{s'}]$ can be simplified to make the Boltzmann equation a linear integro-differential equation, easier to handle.

Anticipating events, once we have solved the Boltzmann equation, we will need to compute expected values of transport quantities. For instance, we will make extensive use of the quasiparticle density ρ_s defined as

$$\rho_s = e \int (d\mathbf{k}) f_s, \tag{7.6}$$

or the particle current, $\mathbf{J}_s = e \int (d\mathbf{k}) f_s \dot{\mathbf{x}}_s$ ($\dot{\mathbf{x}}_s$ being the velocity), associated with the density ρ_s through the conservation law $\dot{\rho}_s = -\nabla \cdot \mathbf{J}_s$.

7.3.2 Semiclassical Equations of Motion

There are several ways to get a semiclassical description of the dynamics of electrons. Which one to use depends on particular preferences, but they all give the same information. One can get these equations of motion by performing the semiclassical approximation to the path integral [15], by computing the quasiclassical limit of the Bloch states within the WKB approximation [19], or using a time-dependent variational approach where the trial function is made of wavepacket states [20–22]. We will make a brief summary of the latter approach here.

In the standard variational approach, one chooses a trial wavefunction depending on a set of free parameters and computes the expected value of the Hamiltonian, trying to get the best upper bound to the ground state by finding the extremal values of these parameters. In the time-dependent variational approach, one chooses a wavefunction that depends on some set of parameters, but this time we make these parameters time-dependent. In this way, we compute the average not of the Hamiltonian but the Lagrangean.[1] We thus obtain some equations of motion for the parameters by looking for the extremal values of the action associated with the averaged Lagrangean.

---

[1] This is totally allowed since the Schrodinger equation can be obtained by a variational principle of a Lagrangean, as was shown by Dirac.

7 Anomalies and Kinetic Theory

Doing so, one does not directly obtain anything semiclassical. The semiclassical limit comes when one looks for an adequate trial wavefunction with parameters that have some sense in classical mechanics. Then, it seems natural the trial wavefunction to be a wavepacket made out of Bloch wavefunctions. Wavepackets are wavefunctions that have finite spread both in real and momentum spaces, in contrast to plane waves. They are useful because they naturally give smaller upper bounds for the Heisenberg uncertainly principle, $\Delta x \Delta k \geq \frac{\hbar}{2}$. The key point is to write the trial wavepacket having the appropriate phase (here we will largely follow the reasonings made in [18, 20]):

$$|\mathbf{x}_c, \mathbf{k}_c\rangle = \sum_{\mathbf{k}} w_{\mathbf{k}_c}(\mathbf{k}) e^{i\mathbf{k}\cdot\mathbf{x}_c - e\mathbf{A}(\mathbf{x}_c)\cdot\mathbf{x}} |u_{\mathbf{k}}\rangle, \quad (7.7)$$

where $|u_{\mathbf{k}}\rangle$ is the periodic part of the Bloch states for a given band, and the labels $(\mathbf{x}_c, \mathbf{k}_c)$ represent the position in real and momentum space around the wavepacket is constructed. We have also assumed that the electromagnetic vector field $\mathbf{A}(\mathbf{x}_c)$ is almost constant when compared with the spread of the wavepacket (i.e., we do not allow for inter-band transitions due to electromagnetism). Also, we need to assume that the amplitude $w_{\mathbf{k}_c}(\mathbf{k})$ is peaked around $\mathbf{k}_c$, *and* it is also centered around $\mathbf{r}_c$, so we need to assume that

$$w_{\mathbf{k}_c}(\mathbf{k}) = |w_{\mathbf{k}_c}(\mathbf{k})| e^{i(\mathbf{k}-\mathbf{k}_c)\cdot\mathscr{A}(\mathbf{k}_c)}, \quad (7.8)$$

where $\mathscr{A}(\mathbf{k}_c)$ turns out to be the Berry connection.[2]

Now, we can compute the averaged expression for the Lagrangean:

$$\mathscr{L} = \langle \mathbf{x}_c, \mathbf{k}_c | i\partial_t - H | \mathbf{x}_c, \mathbf{k}_c \rangle. \quad (7.9)$$

The effective Lagrangean thus reads

$$\mathscr{L} = -e\mathbf{A}(\mathbf{x}_c) \cdot \dot{\mathbf{x}}_c + \mathbf{k}_c \cdot \dot{\mathbf{x}}_c + \dot{\mathbf{k}}_c \cdot \mathscr{A}(\mathbf{k}_c) - \varepsilon_0(\mathbf{k}_c) + e\mathbf{m}_{\mathbf{k}_c} \cdot \mathbf{B}, \quad (7.10)$$

where we have assumed that the electromagnetic vector field $\mathbf{A}$ has a time-dependent portion giving an electric field $\mathbf{E}$ and another portion that gives a magnetic field $\mathbf{B}$. We have also used that $H|u_{\mathbf{k}}\rangle = \varepsilon_0^s(\mathbf{k})|u_{\mathbf{k}}\rangle$.

There is something worth to mention in the Lagrangean (7.10). Now, due to the presence of a non-vanishing Berry connection $\mathscr{A}$, the momentum $\mathbf{k}$ and $\mathbf{x}$ are not longer canonically conjugate variables. So there are different Lagrange equations for them:

$$\dot{\mathbf{k}}_c = e\mathbf{E} - e\dot{\mathbf{x}}_c \times \mathbf{B}, \quad (7.11)$$

$$\dot{\mathbf{x}}_c = \partial_{\mathbf{k}_c}\varepsilon(\mathbf{k}_c) - \dot{\mathbf{k}}_c \times \Omega_{\mathbf{k}_c}, \quad (7.12)$$

---

[2] The expression for $\mathscr{A}(\mathbf{k}_c)$ as $\mathscr{A}(\mathbf{k}_c) = i\langle u_{\mathbf{k}}|\nabla_{\mathbf{k}}|u_{\mathbf{k}}\rangle$, can be obtained by imposing that the wavepacket (7.7) is centered around $\mathbf{x}_c$: $\langle \mathbf{x}_c, \mathbf{k}_c | \mathbf{x} - \mathbf{x}_c | \mathbf{x}_c, \mathbf{k}_c \rangle = 0$.

where we have defined the Berry connection $\Omega_{\mathbf{k}_c} = \partial_{\mathbf{k}_c} \times \mathscr{A}(\mathbf{k}_c)$, and the dispersion relation $\varepsilon^s(\mathbf{k}_c) = \varepsilon_0^s(\mathbf{k}_c) - e\mathbf{m}_{\mathbf{k}_c}^s \cdot \mathbf{B}$ is modified due to the presence of the orbital magnetic moment $\mathbf{m}_{\mathbf{k}}^s$. These equations can be easily generalized for a number s of species of particles.

We can decouple (7.11) and (7.12) to get

$$D(\mathbf{k}_c)\dot{\mathbf{x}}_c = e\mathbf{E} + e\mathbf{v} \times \mathbf{B} + e^2(\mathbf{E} \cdot \mathbf{B})\Omega_{\mathbf{k}_c}, \qquad (7.13a)$$

$$D(\mathbf{k}_c)\dot{\mathbf{k}}_c = \mathbf{v} - e\Omega_{\mathbf{k}_c} \times \mathbf{E} + e(\mathbf{v} \cdot \Omega_{\mathbf{k}_c})\mathbf{B}. \qquad (7.13b)$$

The coefficient $D(\mathbf{k}_c) = 1 + e\mathbf{B} \cdot \Omega_{\mathbf{k}_c}$ is the modification of the phase space density of states. It appears because, under the effect of a magnetic field, and because $\mathbf{k}_c$ and $\mathbf{x}_c$ are no canonically conjugate variables, the Liouville theorem is not satisfied unless we redefine the change of volumes in phase space as $\Delta V \cdot D(\mathbf{k}_c)$ [23]. This is why in all what follows we will define all the averaged quantities as integrals in the momentum space weighted with $D(\mathbf{k}_c)$.

The reader might miss the presence of $\hbar$ in this semiclassical derivation. It is true, I have made $\hbar = 1$ all the time. The approximation to the classical world comes when we have computed the Lagrangean (7.10) with a wavepacket centered around $(\mathbf{k}_c, \mathbf{x}_c)$, and integrated over the *fast* variables $(\mathbf{k}, \mathbf{x})$, integrations that are in (7.7) and implicitly in (7.10), leaving us with a simultaneously well-defined pair $(\mathbf{k}_c, \mathbf{x}_c)$. From now on, we will omit the subscript c.

A final comment. The previous semiclassical treatment appears to be quite generic, and so far we have made no mention on anything specific on Weyl semimetals. It turns out that we have to be a little bit more careful when taking the semiclassical limit. The Berry connection is actually first order in $\hbar$, since is an information associated with the wavefunctions (inherently quantum-mechanical objects) even when we are working with definite trajectories in the configuration space. So we need to have a criterium to allow for objects like the Berry connection ($\mathscr{O}(\hbar)$) and neglect other effects, like quantum mechanical transitions to other bands, apart from the simultaneous definition of momentum and position. This extra criterion comes from scale considerations. In quantum electrodynamics, the inter-band transitions that induce the electron–hole pair production are blocked by the presence of a mass gap (the polarization tensor is analytic for frequencies satisfying $\omega^2 \leq 4m^2$). If on top of that, the Fermi level μ crosses the conduction bands, the range of frequencies that are blocked increases to 2μ, and the only dynamics comes from the processes at the Fermi surface, processes that are captured by the kinetic theory. This is why the kinetic theory can be understood as an effective field theory [24, 25]. In the particular case of Weyl fermions, we do not have a gap, so the only scale that blocks the inter-band transitions is the chemical potential, so it is all important to keep in mind that all that comes from the chiral kinetic theory will be strictly valid at finite chemical potential.

In any case, this is what one obtains when computes the Berry curvature for Weyl fermions:

7 Anomalies and Kinetic Theory

$$\Omega_{\mathbf{k}}^{s} = \frac{s}{2}\frac{\mathbf{k}}{k^3}, \quad (7.14)$$

where s is the chirality of each Weyl node. It looks like the field generated by a point particle in electrostatics. Using this analogy with electromagnetism, we can understand the topological protection in Weyl semimetals by considering that there is a Chern number associated with each Weyl node, which comes from applying the Gauss law to the Berry connection (7.14): The flux through any closed surface enclosing the monopole in (7.14) is the Chern number and it equal to the chirality of the monopole. There is another important quantity related to the wavefunctions, the orbital magnetic moment [26] $\mathbf{m}_{\mathbf{j}\mathbf{k}}^{s}$ [22]. This quantity enters in the problem by modifying the effective dispersion relation for Weyl fermions, and it is all important to correctly compute the total current in Weyl semimetals beyond the local limit, due to the appearance of magnetization currents [22, 27].

7.3.3 The Chiral Anomaly

A primer derivation. Now we are equipped with the Boltzmann equation, and we have seen that Weyl semimetals show their non-triviality through the presence of a Berry curvature $\Omega_{\mathbf{k}}^{s}$ and an orbital magnetic moment $\mathbf{m}_{\mathbf{k}}^{s}$ per Weyl point. As discussed above, we will deal only with the simple case of *elastic* scattering, and only in the case of elastic scattering satisfying the detailed balance condition:

$$W_{\mathbf{k}'\mathbf{k}} = W_{\mathbf{k}\mathbf{k}'}. \quad (7.15)$$

Under these conditions, the Boltzmann equation is written in terms of the linearized collision integral:

$$\dot{f}_s + \dot{\mathbf{k}}_s \cdot \partial_{\mathbf{k}} f_s + \dot{\mathbf{x}}_s \cdot \partial_{\mathbf{x}} f_s = \sum_{s'} \int (d\mathbf{k}') D_{s'}(\mathbf{k}') W_{\mathbf{k}\mathbf{k}'}^{ss'}(f_{s'} - f_s). \quad (7.16)$$

Let us go step by step and consider only intra-node scattering, that is, $W_{\mathbf{k}\mathbf{k}'}^{ss'} = W_{\mathbf{k}\mathbf{k}'}\delta_{ss'}$, where impurity scattering does not move electrons outside of any nodal point.

Considering homogeneous and *static* fields simplify matters a lot, because, since neither $\mathbf{E}$ nor $\mathbf{B}$ depend on $\mathbf{x}$, it is natural to assume that $\partial_{\mathbf{x}} f_s = 0$, and the Boltzmann equation reads

$$\dot{f}_s(\mathbf{k}) + \dot{\mathbf{k}}_s \cdot \partial_{\mathbf{k}} f_s(\mathbf{k}) = \int (d\mathbf{k}') D_s(\mathbf{k}') W_{\mathbf{k}\mathbf{k}'}(f_s(\mathbf{k}') - f_s(\mathbf{k})). \quad (7.17)$$

In the previous expression, we have made explicit the dependence of f_s (and D_s) with $\mathbf{k}$ to avoid any source of confusion.

We will use also time-independent electromagnetic fields, so the reader might ask why we keep the time dependence of f_s. The precise answer of this question will be given in the next section.

The rate of change with time of the charge density at each nodal point in presence of external electromagnetic fields is (making use of $\dot{\mathbf{B}} = 0$)

$$\dot{\rho}_s = e \int (d\mathbf{k}) D_s(\mathbf{k}) \dot{f}_s(\mathbf{k}). \tag{7.18}$$

Substituting $\dot{f}_s(\mathbf{k})$ by the rest of the expression (7.17) we simply obtain

$$\dot{\rho}_s = -e \int (d\mathbf{k}) D_s(\mathbf{k}) \dot{\mathbf{k}}_s \cdot \partial_\mathbf{k} f_s(\mathbf{k}), \tag{7.19}$$

The integral of the collision operator vanishes due to detailed balance condition (7.15). Now we can integrate by parts, throw away a surface term that vanishes if we consider an infinite momentum space, and get

$$\dot{\rho}_s = e \int (d\mathbf{k}) \partial_\mathbf{k} \cdot (D_s(\mathbf{k}) \dot{\mathbf{k}}_s) f_s(\mathbf{k}), \tag{7.20}$$

where we can use the EOM for $\dot{\mathbf{k}}_s$. The intermediate step is

$$\dot{\rho}_s = e \int (d\mathbf{k}) f_s(\mathbf{k}) \left(\mathbf{B} \cdot (\partial_\mathbf{k} \times \mathbf{v}_\mathbf{k}^s) + e^2 (\mathbf{E} \cdot \mathbf{B}) \partial_\mathbf{k} \cdot \Omega_\mathbf{k}^s \right). \tag{7.21}$$

The term $\partial_\mathbf{k} \times \mathbf{v}_\mathbf{k}^s$ is zero because $\mathbf{v}_\mathbf{k}^s$ is made out of derivatives of things: $\mathbf{v}_\mathbf{k}^s = \partial_\mathbf{k} \varepsilon(\mathbf{k}) - e \partial_\mathbf{k} (\mathbf{m}_\mathbf{k}^s \cdot \mathbf{B})$, so

$$\dot{\rho}_s = \frac{e^3}{8\pi^3} (\mathbf{E} \cdot \mathbf{B}) \int d^3\mathbf{k} f_s(\mathbf{k}) (\partial_\mathbf{k} \cdot \Omega_\mathbf{k}^s). \tag{7.22}$$

So far, we have been just (slightly) careful with the maths and gone step by step in deriving (7.22). Without entering in more details, we can see that there is something peculiar in (7.22). The rate of change of the electric charge at each nodal point is not certainly zero in presence of electric and magnetic fields. The intra-node scattering has nothing to do at this level, as expected, since we are not permitting the system to send particles to one nodal point to the other. And, more importantly, the coefficient is a momentum integral that crucially depends on the presence of a non-vanishing Berry curvature. Let us be more focused now and particularize this expression for Weyl semimetals.

We have seen before that the Berry curvature around each Weyl node takes the form of a *monopole* in momentum space.

$$\Omega_\mathbf{k}^s = \frac{s}{2} \frac{\mathbf{k}}{k^3}, \tag{7.23}$$

7 Anomalies and Kinetic Theory

being the *charge* of this monopole the chirality of the nodal point. This particular form of the monopole tells us that its divergence is just a Dirac delta, $\partial_\mathbf{k} \cdot \frac{\mathbf{k}}{k^3} = \delta^{(3)}(\mathbf{k})$, so

$$\dot{\rho}_s = \frac{e^3}{8\pi^3} (\mathbf{E} \cdot \mathbf{B}) \int d^3\mathbf{k} f_s(\mathbf{k}) \delta^{(3)}(\mathbf{k}) = s \frac{e^3}{4\pi^2} (\mathbf{E} \cdot \mathbf{B}) f_s(0). \quad (7.24)$$

We find a finite result, only written in terms of the (so far unknown) value of the non-equilibrium distribution function evaluated at the position of the Weyl node, and the chirality s of the node.

From (7.24), we can compute two things. The rate of change of the total electric charge $\rho = \rho_+ + \rho_-$, and the rate of change of the *difference*, which is what we call the chiral charge $\rho_5 = \rho_+ - \rho_-$:

$$\dot{\rho} = \dot{\rho}_+ + \dot{\rho}_- = \frac{e^3}{4\pi^2} (\mathbf{E} \cdot \mathbf{B}) (f_+(0) - f_-(0)) = 0, \quad (7.25a)$$

$$\dot{\rho}_5 = \dot{\rho}_+ - \dot{\rho}_- = \frac{e^3}{4\pi^2} (\mathbf{E} \cdot \mathbf{B}) (f_+(0) + f_-(0)). \quad (7.25b)$$

That $\dot{\rho} = 0$ is a necessity, that is, a non-vanishing rate $\dot{\rho}$ would mean that we do not have electromagnetic gauge invariance in our system (we do not want charges to pop up or disappear anywhere). This necessity tells us that $f_+(0) = f_-(0)$, without solving the Boltzmann equation. However, this necessity forces us to accept that the chiral density is *not conserved*:

$$\dot{\rho}_5 = \frac{e^3}{2\pi^2} (\mathbf{E} \cdot \mathbf{B}) f_+(0), \quad (7.26)$$

provided that $f_+(0)$ is not zero. Here it goes, the no conservation of the chiral charge [28].

This derivation is not definitively the end of the story. Some things that have been assumed are natural, like the vanishing of surface terms at the infinity when obtaining (7.20), and others that need more justification, like the non-vanishing value of $f_+(0)$ at the nodal point. Also, we have deliberately omitted the presence of inter-node scattering. This type of scattering will surely modify (7.26), since there is nothing wrong in sending particles from one nodal point to the other, and this process will also modify the chiral charge density. And lastly, we have to say something about the time dependence of f_s. This possible dependence might tell us that expression $\dot{\rho}_5$ has some time dependence through $f_+(0)$, but, according to other derivations of the anomaly, there is nothing time-dependent in the coefficient accompanying $\mathbf{E} \cdot \mathbf{B}$.

However, despite of all these shortcomings, this line of arguments is tremendously informative since it tells us that the anomaly is a consequence of the non-trivial topological structures associated with the Weyl Hamiltonian.

Detailed discussion. If we want to be more rigorous and discuss the details mentioned in the previous paragraphs, we need to treat the Boltzmann equation (7.16) more seriously. Again, we wont complicate too much the situation and keep the electromagnetic fields homogeneous and static, but i feel necessary to warn the reader that what follows is a lot of mathematical manipulations to transform the original Boltzmann equation, that is full of physics, into an integral equation that will enjoy mathematicians.[3]

First, let us discuss the scattering rate $W_{\mathbf{k}\mathbf{k}'}^{ss'}$ for the case of intra- and inter-node scattering. In what follows, we will assume that we have a doped system with two isotropic Weyl nodes at zero temperature. The Fermi level will cross the conduction band for both nodes.

The eigenstates and energies of the system are the solutions of the equation

$$H_s \psi_s = \varepsilon \psi_s, \tag{7.27}$$

with $\varepsilon = \pm v|\mathbf{k}| = \pm vk$. The conduction band corresponds to $\varepsilon = +vk$ for both chiralities. For $s = +$ we have

$$|+, +\rangle = \begin{pmatrix} \cos\frac{\theta}{2} \\ e^{i\phi} \sin\frac{\theta}{2} \end{pmatrix}, \tag{7.28}$$

and, for $s = -1$,

$$|+, -\rangle = \begin{pmatrix} \sin\frac{\theta}{2} \\ -e^{i\phi} \cos\frac{\theta}{2} \end{pmatrix}. \tag{7.29}$$

We have used spherical coordinates: $\mathbf{k} = k(\sin\theta \sin\phi, \sin\theta \cos\phi, \cos\theta)$. We can compute the scattering rate by using the Fermi golden rule:

$$W_{\mathbf{k}\mathbf{k}'}^{ss'} = \hat{W}_{\mathbf{k}\mathbf{k}'}^{ss'} \delta(\varepsilon_{\mathbf{k}} - \varepsilon_{\mathbf{k}'}) \equiv 2\pi n_r V_r^2 |\langle \mathbf{k}', s'| \sigma_r |\mathbf{k}, s\rangle|^2 \delta(\varepsilon_{\mathbf{k}} - \varepsilon_{\mathbf{k}'}), \tag{7.30}$$

where n is the impurity density and V is the parameter defining the type of scattering event. We will restrict ourselves to potential scattering, that is, $\sigma_r = \sigma_0$, the identity matrix. In this particular case, that is more than enough for our purposes, we will need the following squared matrix elements:

$$|\langle -', +|+, +\rangle|^2 = \frac{1}{2} \left(1 - \cos\theta \cos\theta' - \sin\theta \sin\theta' \cos(\phi - \phi')\right), \tag{7.31a}$$

$$|\langle +', +|+, +\rangle|^2 = \frac{1}{2} \left(1 + \cos\theta \cos\theta' + \sin\theta \sin\theta' \cos(\phi - \phi')\right). \tag{7.31b}$$

The previous expressions can be written in terms of the chiralities s and s' and the unit momentum vectors as

---

[3] All this section is inspired by [29].

7 Anomalies and Kinetic Theory

$$|\langle s', +| + s\rangle|^2 = \frac{1}{2}\left(1 + ss'\hat{\mathbf{k}} \cdot \hat{\mathbf{k}}'\right). \tag{7.32}$$

The Berry curvature takes the form of a monopole, where the monopole charge is the chirality:

$$\Omega_{\mathbf{k}}^s = s\frac{1}{2}\frac{\hat{\mathbf{k}}}{k^2}, \tag{7.33}$$

and the orbital magnetic moment reads

$$\mathbf{m}_{\mathbf{k}}^s = sv\frac{1}{2}\frac{\hat{\mathbf{k}}}{k}. \tag{7.34}$$

The orbital magnetic moment enters through a modification of the group velocity $\mathbf{v}_{\mathbf{k}}^s = \mathbf{v}_{\mathbf{k}}^0 - \partial_{\mathbf{k}}(\mathbf{m}_{\mathbf{k}}^s \cdot \mathbf{B})$. As usual, $\mathbf{v}_{\mathbf{k}}^0 = \frac{\partial \varepsilon_{\mathbf{k}}}{\partial \mathbf{k}} = v\hat{\mathbf{k}}$, that does not depend on the chirality. We leave the reader to verify all the previous relations.

Now, let us suppose that we do not consider that f_s does not depend on time:

$$\dot{\mathbf{k}}_s \cdot \partial_{\mathbf{k}} f_s = \sum_{s'} \int (d\mathbf{k}') D_{s'}(\mathbf{k}') W_{\mathbf{k}\mathbf{k}'}^{ss'}(f_{s'} - f_s). \tag{7.35}$$

Following the standard procedure, we will consider only small departures from equilibrium, so $f_s \simeq f_0 + f_{1s}$, where $f_0 = f_0(\varepsilon_{\mathbf{k}})$ is the equilibrium distribution function, and the function f_{1s} is assumed to be $\mathcal{O}(\mathbf{E})$. So, to fist order in the external electric field, we have, after multiplying by the density of states in the phase space $D_s(\mathbf{k}) = 1 + e\mathbf{B} \cdot \Omega_{\mathbf{k}}^s$:

$$D_s(\mathbf{k})\dot{\mathbf{k}}_s \cdot \mathbf{v}^0 \frac{\partial f_0}{\partial \varepsilon} = \sum_{s'} \int (d\mathbf{k}') D_s(\mathbf{k}) D_{s'}(\mathbf{k}') W_{\mathbf{k}\mathbf{k}'}^{ss'}(f_{1s'} - f_{1s}).$$

From now on, we will use the prime to refer both to the chirality s' and to the momentum components of this precise chirality:

$$D_s \dot{\mathbf{k}}_s \cdot \mathbf{v}^0 \frac{\partial f_0}{\partial \varepsilon} = \sum_{s'} \int (d\mathbf{k}') D_s D_{s'} W_{\mathbf{k}\mathbf{k}'}^{ss'}(f_{1s'} - f_{1s}). \tag{7.36}$$

The quasiparticle current *in the local limit* is

$$\mathbf{J}_s = e \int (d\mathbf{k}) D_s \dot{\mathbf{x}}_s f_s. \tag{7.37}$$

At zero temperature $T = 0$, the derivative of the Fermi distribution function is strongly peaked around the Fermi level μ:

$$\frac{\partial f_0}{\partial \varepsilon} \approx -\delta(\mu - \varepsilon_{\mathbf{k}}), \tag{7.38}$$

using this fact, we will solve (7.36) imposing the condition given by (7.38): $\delta(\varepsilon_{\mathbf{k}} - \varepsilon_{\mathbf{k}'}) = \delta(\mu - \varepsilon_{\mathbf{k}'})$, that implies $\varepsilon_{\mathbf{k}} = \mu$, and $\mathbf{k} = \mathbf{k}_F$, where $\mu = \varepsilon_{\mathbf{k}_F}$.

The result (7.26) tells us that it is worth if we assume $\mathbf{E}$ to be parallel to $\mathbf{B}$. We are dealing with an isotropic system, so, without loss of generality, we will assume $\mathbf{B} = B_3 \hat{\mathbf{z}}$. Let us simplify things more. It turns out to be more convenient to work, not with f_{1s}, but with a related function $g_s(\mathbf{k})$:

$$f_{1s} = eE_3 \frac{\partial f_0}{\partial \varepsilon} \frac{1}{D_s(\mathbf{k})} g_s(\mathbf{k}), \tag{7.39}$$

so the Boltzmann equation (7.36) now reads $((d\mathbf{k}') = \frac{1}{8\pi^3} d^3 k = \frac{1}{8\pi^3} dk k^2 d\theta \sin\theta d\phi)$

$$v_3^0 + eB_3(\Omega_r^s v_r^0) = \sum_{s'} 2\pi n_{s'} V_{s'}^2 \int (d\mathbf{k}') |\langle s'|s\rangle|^2 \delta(\mu - vk') \left(D_s(\mathbf{k}) g_{s'}(\mathbf{k}') - D_{s'}(\mathbf{k}') g_s(\mathbf{k})\right). \tag{7.40}$$

We have defined the scattering elements $V^{ss'} = V_0$ if $s = s'$ (intra-valley scattering), and $V^{ss'} = V_s$ if $s' \neq s$ (inter-valley scattering). Under these circumstances, no term in (7.40) depends on ϕ, so we can safely integrate this coordinate.

Substituting each quantity and remembering that we are computing everything at the Fermi level μ, (7.40) reads

$$vu + s\frac{v}{2k_F^2} eB_3 = \sum_{s'} \frac{n_{s'} V_{s'}^2}{2\pi} \frac{k_F^2}{v} \int_{-1}^{1} du' \frac{1}{2}(1 + ss'uu')(D_s(u)g_{s'}(u') - D_{s'}(u')g_s(u)), \tag{7.41}$$

after the change of variables $u = \cos\theta$. We can write the previous equation in terms of dimensionless quantities remembering that $eB_3 = 1/l_B^2$, and $k_F \equiv 1/l_F$. This gives the following dimensionless parameter $\alpha = l_F^2/2l_B^2$, and the characteristic mean free paths $1/l_{ss'} \equiv (n_{s'} V_{s'}^2/2\pi)(1/v^2 l_F^2)$. With these definitions, $D_s(u) = 1 + e\mathbf{B} \cdot \mathbf{\Omega}_{\mathbf{k}}^s = 1 + s\alpha u$, we can write the Boltzmann equation (7.41) in the following dimensionless fashion:

$$ul + s\alpha l = \sum_{s'} \frac{l}{l_{ss'}} \int_{-1}^{1} du' \frac{1}{2}(1 + ss'uu') \left((1 + s\alpha u)g_{s'}(u') - (1 + s'\alpha u')g_s(u)\right), \tag{7.42}$$

where $1/l = \sum_{s'} 1/l_{ss'}$ is the inverse of the total mean free path.

7 Anomalies and Kinetic Theory   191

Integrating over u' in the last term of the right-hand side of (7.42), we obtain

$$g_s(u) = -\frac{3(ul + s\alpha l)}{(3+su\alpha)} + \sum_{s'} \frac{l}{l_{ss'}} \int_{-1}^{1} du' \frac{3}{2}\frac{(1+ss'uu')(1+s\alpha u)}{(3+su\alpha)} g_{s'}(u'), \quad (7.43)$$

Mathematicians call this equation the inhomogeneous Fredholm equation of second kind.

Let us pause ourselves a little and take a breath. What we have done is to rearrange things to convert the Boltzmann equation, an integro-differential equation, into a *simpler* integral equation in terms of a dimensionless parameter α ($g_s(u)$ has units of length). As it stands, it is hopeless to try to solve the integral equation (7.43), so the reader could complain about all we have done. The first thing the reader should note is that the right-hand side of the integral equation (7.42) is not easy to cast into the form of a transport time: $-\frac{1}{\tau_0}(f_0 - f_s) - \frac{1}{\tau_{ss'}}(f_s - f_{s'})$. The reason for that is the explicit dependence of the scattering time with the matrix elements (7.31b). Normal systems usually show trivial state overlaps and zero Berry curvature, so in these cases it is a textbook matter to convert (7.42) into a simple algebraic equation. Also, there are systems like graphene, with zero Berry curvature, but a non-trivial state overlaps, that strongly influence the form of the Boltzmann equation, and the subsequent transport properties.

Despite of all the complexity of (7.43), the benefit of having a dimensionless parameter is that we can consider it small or large. The case of $\alpha \ll 1$ corresponds to magnetic lengths much larger than the Fermi wavelength (or very small magnetic fields), and yet, working within the limits imposed by the semiclassics. Also, we can track the information of the Berry curvature in the terms proportional to α.

Let us expand everything in powers of α:

$$g_s(u) = g_s^0(u) + g_s^1(u)\alpha + \cdots \quad (7.44a)$$

$$\mathscr{K}_{ss'}(u,u') \equiv \frac{3}{2}\frac{(1+ss'uu')(1+s\alpha u)}{(3+su\alpha)} = \mathscr{K}_{ss'}^0(u,u') + \mathscr{K}_{ss'}^1(u,u')\alpha + \cdots \quad (7.44b)$$

$$f(u) \equiv -\frac{3(ul+s\alpha l)}{(3+su\alpha)} = f_0(u) + f_1(u)\alpha + \cdots \quad (7.44c)$$

From (7.43), we obtain the following set of coupled equations:

$$\alpha^0 : g_s^0(u) = f_0(u) + \sum_{ss'} \frac{l}{l_{ss'}} \int du' \, \mathscr{K}_{ss'}^0(u,u') g_{s'}^0(u'), \quad (7.45)$$

$$\alpha^1 : g_s^1(u) = f_1(u) + \sum_{ss'} \frac{l}{l_{ss'}} \int du' \, \mathscr{K}_{ss'}^1(u,u') g_{s'}^0(u') + \sum_{ss'} \frac{l}{l_{ss'}} \int du' \, \mathscr{K}_{ss'}^0(u,u') g_{s'}^1(u'). \quad (7.46)$$

Once we have integral equations for each series component $g_s^n(u)$, we can apply the mathematical machinery presented in the appendix and transform these integral equations in simple algebraic equations.

We face two different situations. The first one is when only intra-node scattering is allowed, that is, we only allow for $s' = s$ scattering processes, so $1/l_{++} = l/l_{--} \equiv 1/l_0$ and $l_{+-} \equiv l_s \to \infty$, so $l \to l_0$. Intra-node scattering does not exchange particles between Weyl nodes so it is not difficult to understand that intra-node scattering will not enter into the expression of the chiral anomaly. However, there is something else that is unexpected from intuition. Let us see it.

We can write writing $g_s^0(u) = f_0(u) + \sum_n P_n^0(u) b_n^0$ and apply the methodology presented in the appendix to transform the integral equation (7.45) into a simple algebraic equation for the coefficients b_n^0:

$$\begin{bmatrix} a_{11}^0 & a_{12}^0 \\ a_{21}^0 & a_{22}^0 \end{bmatrix} \begin{bmatrix} b_1^0 \\ b_2^0 \end{bmatrix} \equiv \begin{bmatrix} 0 & 0 \\ 0 & \frac{2}{3} \end{bmatrix} \begin{bmatrix} b_1^0 \\ b_2^0 \end{bmatrix} = \begin{bmatrix} 0 \\ -\frac{2}{3}l \end{bmatrix} \equiv \begin{bmatrix} c_1^0 \\ c_2^0 \end{bmatrix}. \quad (7.47)$$

Quite simple-looking equation, but there is something that should call our attention. While we obtain $b_2^0 = -l$, the coefficient b_1^0 is undetermined (we also note that there is no dependence of the node label s so $g_+^0 = g_-^0$). The indeterminacy of b_1^0 is not an actual problem since it will not contribute to the current. Also, since we stablished that $b_{1,+}^0 = b_{1,-}^0$ charge conservation tells us that the total charge is the one we obtain in equilibrium, so any non-equilibrium contribution to the charge density would be zero, implying $b_{1,+}^0 = b_{1,-}^0 = 0$. But what if the coefficient c_1^0 were not zero? We would find a contradiction. Let us compute the expression for g_s^1.

Using the expression for g_s^0 in (7.46) and, again, writing $g_{1s}^1(u) = f_{1s}(u) + \sum_n P_n^1(u) b_{ns}^1$, we get the following algebraic equation after some little algebra:

$$\begin{bmatrix} 0 & 0 \\ 0 & \frac{2}{3} \end{bmatrix} \begin{bmatrix} b_{1s}^1 \\ b_{2s}^1 \end{bmatrix} = \begin{bmatrix} -2sl \\ 0 \end{bmatrix}. \quad (7.48)$$

Now, we obtain that $0 \cdot b_{1s}^1 = -2sl$ which is the contradiction we mentioned above. It says that we cannot find a solution to the Boltzmann equation to first order in α, or to first order in the magnetic field when intra-node scattering is considered alone. The procedure stops here since we need g_s^1 to compute g_s^2 to order α^2 and so on.

What we have actually found is that we cannot find an *static* solution of the Boltzmann equation if we only consider intra-node scattering. We could consider time-dependent electromagnetic fields $\mathbf{E} = \mathbf{E}e^{i\omega t}$, or $\mathbf{B} = \mathbf{B}e^{i\omega t}$. That introduces an extra term $i\frac{\omega l}{v} g_s(u)$ in the left-hand side of (7.42). The presence of the parameter $\delta = \frac{\omega l}{v}$ regularizes the expression for b_{1s}^1, so $b_{1s}^1 \sim i\frac{2sl}{\delta}$, and when the electromagnetic fields are time-dependent, we can safely obtain a finite solution of the Boltzmann equation, and eventually, $\dot{\rho}_s \sim E_3 B_3$.

However, transport experiments are not usually performed under time dependent electromagnetic fields, so we need an alternative to time dependent fields. This alternative is provided by the presence of inter-nodal scattering. To keep things simple, in what follows we will consider that there is no intra-node scattering, $l_0 \to \infty$ keeping l_s finite, so $l \to l_s$. Also, in the calculations we have to put $s' = -s$, so we will put $ss' = -1$ whenever necessary.

In this new scenario, the algebraic equation for g_s^0 is

$$\begin{bmatrix} 0 & 0 \\ 0 & \frac{4}{3} \end{bmatrix} \begin{bmatrix} b_{1s}^1 \\ b_{2s}^1 \end{bmatrix} = \begin{bmatrix} 0 \\ -\frac{2}{3}l_s \end{bmatrix}, \qquad (7.49)$$

so $g_s^0(u) = -\frac{3}{4}l_s u$. Doing the same for $g_s^1(u) = sg^1(u)$, the matrix equation now looks different than (7.48):

$$\begin{bmatrix} 2 & 0 \\ 0 & \frac{2}{3} \end{bmatrix} \begin{bmatrix} b_{1s}^1 \\ b_{2s}^1 \end{bmatrix} = -\frac{5}{3}l_s \begin{bmatrix} 1 \\ 0 \end{bmatrix}. \qquad (7.50)$$

The reason is that the kernel $\mathcal{K}_{+-}^0$ is different than $\mathcal{K}_{++}^0$, and at some intermediate point we have used that $ss' = -1$. Equation (7.50) now has a (unique) valid solution, so

$$g_s^1(u) = -sl_s\left(\frac{17}{12} - \frac{1}{2}u^2\right), \qquad (7.51)$$

implying that the presence of inter-node scattering now allows for a stationary solution of the Boltzmann equation.

Now that we have a solution of the Boltzmann equation to first order in the magnetic field, we can compute the non-equilibrium contribution to the electric density per node ρ_s:

$$\rho_s = \frac{e}{8\pi^3} \int d^3\mathbf{k} D_s(\mathbf{k}) f_{1s}. \qquad (7.52)$$

Using the expression (7.39) with $\frac{\partial f_0}{\partial \varepsilon} = -\delta(\mu - v|\mathbf{k}|)$ and integrating in k and ϕ we get

$$\rho_s = -\frac{e^2}{4\pi^2}\frac{\mu^2}{v^3}\int_{-1}^{1} du (g^0(u) + s\alpha g^1(u)). \qquad (7.53)$$

$g^0(u)$ is an odd function of u so it vanishes upon integration. Using the expression for $g^1(u)$ and remembering that the chiral density is $\rho_5 = \sum_s s\rho_s$ we finally get ($l_s = v\tau_s$)

$$\rho_5 = \frac{e^3}{2\pi^2}\frac{5\tau_s}{4}\mathbf{E} \cdot \mathbf{B}. \qquad (7.54)$$

This expression means that, under the effect of parallel electric and magnetic fields and due to inter-node scattering, a stationary non-equilibrium imbalance between the densities at different nodes appears in the system.

If we look at the field-theoretic expression (7.24), it is clear now how we have to modify it in order to accommodate the inter-node scattering:

$$\dot{\rho}_5 = \frac{e^3}{2\pi^2}\mathbf{E}\cdot\mathbf{B} - \frac{1}{\tau_5}\rho_5, \tag{7.55}$$

so the inter-node transport time τ_5 is just proportional to τ_s up to a numerical factor. Remember that τ_s is actually the inter-node *scattering* time, if we had solved the problem including both intra-node and inter-node scattering times, τ_5 would be a complicated expression of both times (the point is that we always need inter-node scattering, but once it is included, intra-node scattering enters as well).

We can also draw another important consequence of the existence of a stationary chiral imbalance ρ_5 (7.54). Another important consequence of the presence of the Berry curvature in the effective semiclassical dynamics in Weyl semimetals is the appearance of an electric current along the direction of the magnetic field $\mathbf{B}$, the so-called chiral magnetic effect [30]:

$$\mathbf{J}_{CME} = \frac{e^2}{2\pi^2}\mu_5\mathbf{B}. \tag{7.56}$$

This current appears due to a chiral imbalance, that is denoted by the chiral chemical potential $\mu_5 = \mu_+ - \mu_-$. While now there is consensus in that such chiral imbalance is not possible *in equilibrium*, we have shown that a chiral imbalance is totally allowed out of equilibrium due to the chiral anomaly (7.54), so, using standard knowledge of quantum statistical mechanics, we can relate μ_5 to ρ_5 to lowest order in the external electromagnetic fields, $\rho_5 \simeq \chi_5\mu_5$, where χ_5 is some sort of chiral susceptibility [7], so we can write (7.56) as

$$\mathbf{J} = \frac{e^4}{4\pi^4}\frac{\tau_5}{\chi_5}(\mathbf{E}\cdot\mathbf{B})\mathbf{B}. \tag{7.57}$$

We then conclude that when combining the chiral magnetic effect (7.56) with the chiral anomaly (7.54), we can define a longitudinal magnetoconductivity $\sigma(B)$ (notice that the current in (7.56) now is proportional to an electric field $\mathbf{E}$)

$$\sigma(B) = \frac{e^4}{4\pi^4}\frac{\tau_5}{\chi_5}B^2, \tag{7.58}$$

that is quadratic with the magnetic field and *positive* [31]. Well, we know that it is positive since it is easy to see that χ_5 is positive.

This a very remarkable result, and actually it is the result that has fueled all the experimental investigation in the last few years in the field of Weyl semimetals [9, 32]. It is remarkable for two reasons. First, because it is a longitudinal magnetoconductivity in an isotropic system. While it is possible to get a magnetoconductivity when the electric and magnetic fields are perpendicular in systems with isotropic dispersion relations, it is almost a theorem that we cannot get longitudinal magnetoconductivities for them. The result (7.58) is a beautiful counterexample of that.

Alternatively, we could had computed the expression of the distribution function at order α^2, g_s^2, using the same procedure as before. We leave it as an exercise to the reader.

7.4 Conclusions

In this chapter, we have discussed the chiral anomaly using the kinetic equation approach. This approach tells us that wisely including the Berry curvature into the equations of motion (i.e., quantum correction) we are able to obtain a version of this chiral anomaly. However, this is not the end of the story. It happens that there are more contributions to the chiral anomaly besides the product $\mathbf{E} \cdot \mathbf{B}$. There is a *gravitational* contribution to the chiral anomaly that leaves its fingerprint in transport measurements [33]. Also, the possibility of defining chiral gauge fields in condensed-matter-based Weyl semimetals [34, 35] allows for an interesting discussion of the covariant versus consistent forms of the chiral anomaly [36].

Moreover, there are more quantum anomalies present in Weyl semimetals leading to unconventional behavior of transport coefficients, like the conformal anomaly [37] which implies the non-conservation of the conformal invariance that classically appears in massless fields when one regularizes the interacting theory. This anomaly appears in the absence of any other formal scale in the problem, like mass or chemical potential, so it is hard to connect with the kinetic theory (i.e., an effective field theory at large chemical potentials or temperatures). There is still plenty of room at the bottom in the field of anomaly-related phenomena in Weyl semimetals.

Acknowledgements Almost all my knowledge of anomalies and transport in Weyl semimetals come from conversations with my colleagues and friends. Specially I would like to thank Maria A. H. Vozmediano, Karl Landstenier, Maxim Chernodoub, Yago Ferreiros, Fernando de Juan, and Adolfo G. Grushin. I also acknowledge financial support through the MINECO/AEI/FEDER, UE Grant No. FIS2015-73454-JIN, and the Comunidad de Madrid MAD2D-CM Program (S2013/MIT3007).

Appendix

In this appendix, I will explain some technicalities about the way to solve inhomogeneous Fredholm integral equations of second kind. General approaches, like the resolvent formalism, are based on iterative series, but when the integral kernel is separable (as the ones we find when we solve the Boltzmann equation in series of α), the problem can be reduced to an algebraic one.

Let us consider the following integral equation, where $g(x)$ is the function we want to find, and satisfies the following equation:

$$g(x) = f(x) + \int_a^b dx' K(x, x') g(x'). \tag{7.59}$$

The kernel $K(x, x')$ is separable when we can write it as a (finite) sum of products of functions of x and x':

$$K(x, x') = \sum_n P_n(x) Q_n(x'). \tag{7.60}$$

Putting this expression back into our equation, we get

$$g(x) = f(x) + \sum_n P_n(x) \int_a^b dx' Q_n(x') g(x'), \tag{7.61}$$

and write $b_n \equiv \int_a^b dx' Q_n(x') g(x')$, so

$$g(x) = f(x) + \sum_n P_n(x) b_n. \tag{7.62}$$

If we now multiply both sides by $Q_m(x)$ and integrate, we have

$$b_m = \int_a^b dx Q_m(x) f(x) + \sum_n \int_a^b dx Q_m(x) P_n(x) b_n. \tag{7.63}$$

If we define the parameters $c_m = \int_a^b dx Q_m(x) f(x)$ and $a_{mn} = \int_a^b dx Q_m(x) P_n(x)$, we get an algebraic set of equations

$$b_m = c_m + \sum_n a_{mn} b_n, \tag{7.64}$$

or

$$\sum_n (\delta_{mn} - a_{mn}) b_n = c_m. \tag{7.65}$$

Once we obtain the coefficients b_m, we can go back to (7.62) and plug them into the expression for $g(x)$. It is clear that if the matrix a_{mn} has at least one eigenvalue equal to one, the algebraic equation (and therefore the integral equation) has no solutions, provided that the corresponding element of the vector **c** is *nonzero*. This result, when properly stated, is known as the Fredholm alternative [38].

References

1. E. Noether, Invariant variation problems. Transp. Theor. Stat. Phys. **1**, 186–207 (1971)
2. S.L. Adler, Axial-vector vertex in spinor electrodynamics. Phys. Rev. **177**, 2426–2438 (1969)
3. J.S. Bell, R. Jackiw, A *pcac* puzzle: $\pi_0 \to \gamma\gamma$ in the σ-model. Il Nuovo Cimento A **1965–1970**(60), 47–61 (1969)
4. G.E. Volovik, *The Universe in a Helium Droplet* (Clarendon Press, Oxford, 2003)
5. N. Manton, The schwinger model and its axial anomaly. Ann. Phys. **159**, 220–251 (1985)
6. X.G. Wen, Chiral luttinger liquid and the edge excitations in the fractional quantum hall states. Phys. Rev. B **41**, 12838–12844 (1990)
7. K. Landsteiner, Notes on anomaly induced transport. Acta Phys. Pol., B **47**, 2617 (2016)
8. E.V. Gorbar, V.A. Miransky, I.A., Shovkovy, P.O. Sukhachov, Anomalous transport properties of Dirac and Weyl semimetals (2017), arXiv:1712.08947
9. N.P. Armitage, E.J. Mele, A. Vishwanath, Weyl and dirac semimetals in three-dimensional solids. Rev. Mod. Phys. **90**, 015001 (2018)
10. J.D. Jackson, *Classical Electrodynamics*, 3rd ed. (Wiley, 1998)
11. H. Nielsen, M. Ninomiya, A no-go theorem for regularizing chiral fermions. Phys. Lett. B **105**, 219–223 (1981)
12. H. Nielsen, M. Ninomiya, The adler-bell-jackiw anomaly and weyl fermions in a crystal. Phys. Lett. B **130**, 389–396 (1983)
13. K. Fujikawa, Path-integral measure for gauge-invariant fermion theories. Phys. Rev. Lett. **42**, 1195–1198 (1979)
14. K. Fujikawa, Path integral for gauge theories with fermions. Phys. Rev. D **21**, 2848–2858 (1980)
15. M.A. Stephanov, Y. Yin, Chiral kinetic theory. Phys. Rev. Lett. **109**, 162001 (2012)
16. D.T. Son, N. Yamamoto, Berry curvature, triangle anomalies, and the chiral magnetic effect in fermi liquids. Phys. Rev. Lett. **109**, 181602 (2012)
17. J.-W. Chen, J.-Y. Pang, S. Pu, Q. Wang, Kinetic equations for massive dirac fermions in electromagnetic field with non-abelian berry phase. Phys. Rev. D **89**, 094003 (2014)
18. M.P. Marder, *Condensed Matter Physics* (Wiley, 2015)
19. M. Stone, V. Dwivedi, T. Zhou, Berry phase, lorentz covariance, and anomalous velocity for dirac and weyl particles. Phys. Rev. D **91**, 025004 (2015)
20. G. Sundaram, Q. Niu, Wave-packet dynamics in slowly perturbed crystals: gradient corrections and berry-phase effects. Phys. Rev. B **59**, 14915–14925 (1999)
21. D. Culcer, Y. Yao, Q. Niu, Coherent wave-packet evolution in coupled bands. Phys. Rev. B **72**, 085110 (2005)
22. D. Xiao, M.-C. Chang, Q. Niu, Berry phase effects on electronic properties. Rev. Mod. Phys. **82**, 1959–2007 (2010)
23. D. Xiao, J. Shi, Q. Niu, Berry phase correction to electron density of states in solids. Phys. Rev. Lett. **95**, 137204 (2005)
24. C. Manuel, Hard dense loops in a cold non-abelian plasma. Phys. Rev. D **53**, 5866–5873 (1996)
25. D.K. Hong, Aspects of high density effective theory in qcd. Nucl. Phys. B **582**, 451–476 (2000)
26. D. Xiao, W. Yao, Q. Niu, Valley-contrasting physics in graphene: magnetic moment and topological transport. Phys. Rev. Lett. **99**, 236809 (2007)
27. J. Ma, D.A. Pesin, Chiral magnetic effect and natural optical activity in metals with or without weyl points. Phys. Rev. B **92**, 235205 (2015)
28. K.-S. Kim, H.-J. Kim, M. Sasaki, Boltzmann equation approach to anomalous transport in a weyl metal. Phys. Rev. B **89**, 195137 (2014)
29. G.M. Monteiro, A.G. Abanov, D.E. Kharzeev, Magnetotransport in dirac metals: chiral magnetic effect and quantum oscillations. Phys. Rev. B **92**, 165109 (2015)
30. K. Fukushima, D.E. Kharzeev, H.J. Warringa, Chiral magnetic effect. Phys. Rev. D **78**, 074033 (2008)
31. D.T. Son, B.Z. Spivak, Chiral anomaly and classical negative magnetoresistance of weyl metals. Phys. Rev. B **88**, 104412 (2013)

32. B. Yan, C. Felser, Topological materials: Weyl semimetals. Annu. Rev. Condens. Matter Phys. **8**, 337–354 (2017)
33. J. Gooth, et al., Experimental signatures of the mixed axial–gravitational anomaly in the weyl semimetal nbp. Nature **547**, 324 EP (2017)
34. C.-X. Liu, P. Ye, X.-L. Qi, Chiral gauge field and axial anomaly in a weyl semimetal. Phys. Rev. B **87**, 235306 (2013)
35. A. Cortijo, Y. Ferreiros, K. Landsteiner, M.A.H. Vozmediano, Elastic gauge fields in weyl semimetals. Phys. Rev. Lett. **115**, 177202 (2015)
36. K. Landsteiner, Anomalous transport of weyl fermions in weyl semimetals. Phys. Rev. B **89**, 075124 (2014)
37. M.N. Chernodub, Anomalous transport due to the conformal anomaly. Phys. Rev. Lett. **117**, 141601 (2016)
38. I. Fredholm, Sur une classe dequations fonctionnelles. Acta Math. **27**, 365–390 (1903)

Chapter 8
Topological Materials in Heusler Compounds

Yan Sun and Claudia Felser

Abstract As a class of tuneable materials, Heusler has grown into a family of more than 1000 compounds, synthesized from combinations of more than 40 elements. Recently, by incorporating heavy elements that can give rise to strong spin–orbit coupling (SOC), non-trivial topological phases of matter, such as topological insulators (TIs), have been discovered in Heusler materials. The interplay of symmetry, SOC and magnetic structure allows for the realization of a wide variety of topological phases through Berry curvature design. Weyl points and nodal lines can be manipulated by various external perturbations, which results in exotic properties such as the chiral anomaly, large anomalous, spin and topological Hall effects. The combination of a non-collinear magnetic structure and Berry curvature gives rise to a nonzero anomalous Hall effect, which was first observed in the antiferromagnets Mn_3Sn and Mn_3Ge. Besides this k-space Berry curvature, Heusler compounds with non-collinear magnetic structures also possess real-space topological states in the form of magnetic antiskyrmions, which have not yet been observed in other materials.

8.1 Topological Insulators in Heusler Compounds

Among the large variety of topological states found in the field of condensed-matter physics, the topological insulator (TI) is one of the most important classes. The first TI in HgTe/CdTe quantum wells were predicted by Bernevig et al. [1] in 2006 and experimentally verified by Koenig et al. [2] via the observation of a quantum spin Hall effect (SHE). In HgTe/CdTe quantum wells, the band inversion between the s-orbital-dominated Γ_6 state and the p-orbital-dominated Γ_8 state is the typical feature of the topological phase transition between normal and Z_2 TI. In 2010, similar electronic band structures were predicted in half-Heusler compounds by Chadov et al. [3], Lin [4] and Di Xiao et al. [5]. Similar to the binary zinc-blende

---

Y. Sun (✉) · C. Felser
Max Planck Institute for Chemical Physics of Solids, 01187 Dresden, Germany
e-mail: ysun@cpfs.mpg.de

C. Felser
e-mail: Claudia.Felser@cpfs.mpg.de

© Springer Nature Switzerland AG 2018
D. Bercioux et al. (eds.), *Topological Matter*, Springer Series in Solid-State Sciences 190, https://doi.org/10.1007/978-3-319-76388-0_8

semiconductors of HgTe and CdTe, the s-orbital-dominated Γ_6 state and p-orbital-dominated Γ_8 state also exist in a large number of half-Heusler compounds. Here the band gap and band order can be tuned by spin–orbit coupling (SOC), electronegativity difference of constituents and lattice constants.

Figure 8.2a shows the energy difference between Γ_6 and Γ_8 for all the relevant Heuslers (containing Sc, Y, La, Lu and Th) as a function of lattice constant. Each subgroup (e.g. Ln = Sc, Y, La, Lu) is marked by a certain colour. The compounds with E_{Γ_6}-E_{Γ_8} > 0 are trivial insulators, whereas those with E_{Γ_6}-E_{Γ_8} < 0 are the TI candidates. The latter group consists of zero-gap semiconductors with a doubly degenerate Γ_8 point at the Fermi energy. It follows that all existing Heuslers with zero bandgap at the Fermi energy under certain conditions will reveal the same type of band inversion as does HgTe. Indeed, the increase of the lattice constant reduces the hybridization and closes the nonzero bandgap. Combined with sufficiently strong SOC, it leads to a pronounced Γ_6-Γ_8 band inversion, which is the key to realize the TI state.

Figure 8.2b demonstrates the E_{Γ_6}-E_{Γ_8} difference as a function of the average SOC expressed by the average nuclear charge over the atoms in the unit cell by $\langle Z \rangle = (1/N) \sum_{i=1}^{N} Z(X_i)$, where N is 2 for binaries and 3 for ternaries. This seems to be a suitable order parameter, which sorts the materials almost along a straight line. The combinations of Pt with Bi in LnPtBi or Au with Pb in the LnAuPb series always lead to the inverted band structure. There is an additional advantage of Heusler materials: owing to the large number of compounds with different gap values, it is easy to construct a quantum well consisting of the trivial and topological parts with well-matching lattice constants, similar to the HgTe/CdTe quantum well. The appropriate pairs can be chosen from the candidates situated in the middle area of Fig. 8.2a along the same vertical line, because the transition from trivial to topological behaviour as a function of lattice constant seems to be fairly smooth on average. As a large family of tuneable materials, there are more than 50 Heusler compounds predicted to have on-trivial band order, and some of them have been experimentally verified via magnetotransport measurements [6] or angle-resolved photoemission spectroscopy (ARPES) [7–9].

8.2 Weyl Semimetal in Half-Heusler GdPtBi with External Field

The inverted band structure in Heusler compounds can be used to obtain a variety of other topological states; a typical example is the WSM. The half-Heusler compound GdPtBi (with N éel temperature TN = 9.2 K) has an electronic structure with inverted band order and a quadratic band touching at the Γ point. However, the f-electrons from the Gd ions provide the possibility of tuning the electronic structure via control of the spin orientation. With Zeeman splitting, the spin-up and spin-down states near the Fermi level shift oppositely in energy, and Weyl points are formed between the

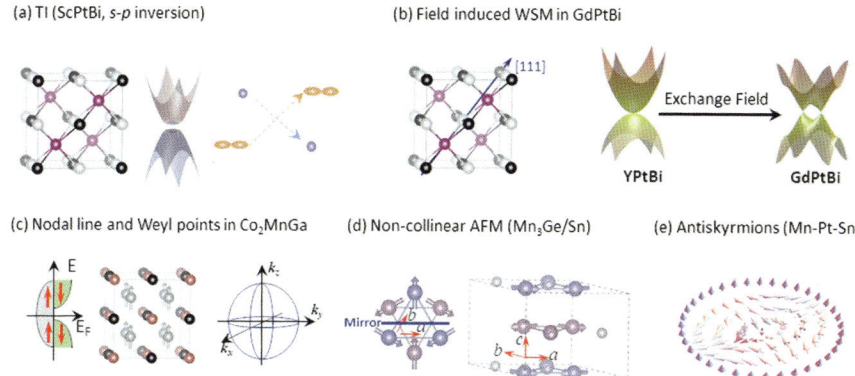

Fig. 8.1 Topological materials in Heusler compounds. **a** Topological insulators with s-p band inversion. **b** Crystal structure of half-Heusler GdPtBi, and band structure evolution from quadratic touching to the Weyl point with applied external magnetic field. **c** Schematic of density of states, and lattice structure for Half metal Co_2MnGa, with nodal line and Weyl points band structure. **d** Lattice and magnetic structure of non-collinear AFM Mn_3Ge/Sn. **e** Antiskyrmion structure in tetragonal Heusler alloy Mn-Pt-Sn

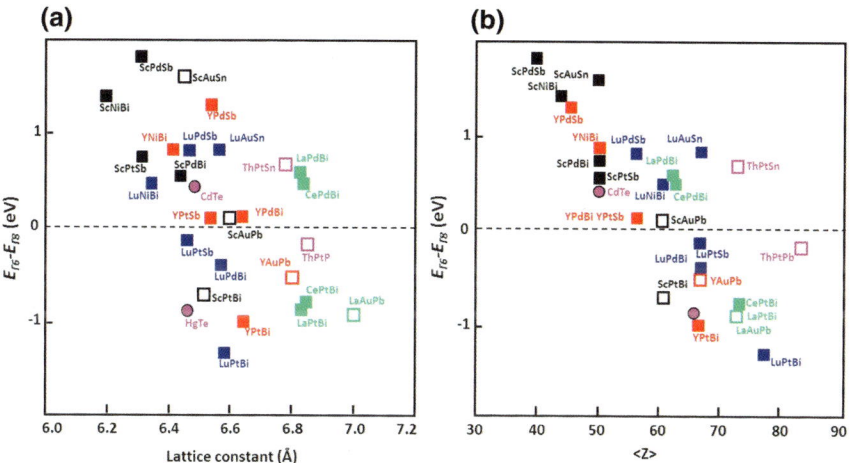

Fig. 8.2 $E_{\Gamma 6}$-$E_{\Gamma 8}$ difference calculated for various Heuslers at their experimental lattice constants. HgTe and CdTe binaries are shown for comparison. Open squares mark the systems not reported in the literate lattice constant. The borderline compounds (between trivial and topological) insulators (YPtSb, YPdBi, ScAuPb) are situated closer to the zero horizontal line. **b** $E_{\Gamma 06}$-$E_{\Gamma 08}$ difference as a function of the average SOC strength represented by the average nuclear charge. (From Nature Materials, 9, 541545 (2010), order Number: 4322551473137)

shifted spin-polarized bands, see Fig. 8.1b. The WSM state in GdPtBi was verified by the observation of different signatures of Weyl points, such as the chiral anomaly [10], unusual intrinsic AHE [11], non-trivial thermal effect [11], and strong planar Hall effect [12], as well as linear dependence of optical conductivity to temperature.

Owing to the defined chirality of each Weyl point, the charge carriers are pumped from one Weyl point to the other one with opposite chirality, when the magnetic field B is not perpendicular to electric field E. This breaks the conservation of Weyl fermions for a given chirality, which is the so-called chiral anomaly. The most important phenomenon induced by the chiral anomaly is the negative magnetoresistance (MR). As long as the magnetic field resides perpendicular to E, the negative MR disappears, implying that the contribution of negative MR originates only from the chiral anomaly of the Weyl points. Similar to electrical resistivity, Seebeck is thermal resistivity, where a thermal gradient is applied in place of the electrical gradient. In GdPtBi, variations similar to those of electrical resistivity are also observed in the thermal resistivity when a thermal gradient applied parallel to the applied magnetic field; this behaviour is also known as chiral anomaly because its origin is the same as that of Weyl [10]. Owing to the existence of Weyl points around the Fermi level, a sizable intrinsic AHC appears upon application of the magnetic field. Together with low charge carrier density and small longitudinal charge conductivity, the AHA can reach up to 10% in GdPtBi [11]. In addition to this chiral anomaly and AHC, GdPtBi exhibits an anomalously large value of $1.5\,m\Omega cm$ planar Hall resistivity at 2 K in a 9 T magnetic field, which is completely different from the Hall resistivity. Though the normal Hall signal is a function of the multiple of the sine and cosine of the applied field, a planar Hall signal is a function of cosine only [6]. This is another alternative way to detect the chiral anomaly in Weyl semimetals.

8.3 Tuneable Anomalous Hall Effect in Half-Metallic Topological Semimetal with Weyl Points and Nodal Lines

The concept of half-metallic ferromagnetism was first introduced by Groot et al. in 1983 [13]; in it, one spin channel is insulating or semiconducting and the other spin channel is metallic because of ferromagnetic decoupling [Fig. 8.1c]. Because of the tunability of SOC, this half-metallic behaviour plays an important role in the stability of topological semimetals in Heusler compounds, where the band crossings derive from bands with either the same or opposite spin polarization [14]. Recently, topological surface states were predicted by Wang et al. [15] and Chang et al. [16] in half-metallic Co_2-based full-Heusler alloys. The half-metallic electronic structure goes in hand with several useful properties: (i) the spin orientation can be easily altered by a small external magnetic field because most half-metallic ferromagnets are soft-magnets. (ii) The magnetic transition temperature is quite high, suitable for room temperature topo-spintronics applications. (iii) Heusler compounds offer tuneable band structures and symmetry elements by appropriate chemical substitution. Therefore, the Berry curvature distribution can be easily changed, and one can tune the anomalous Hall effect from zero to a very large value accordingly.

Considering Co_2MnGa as an example, in Fig. 8.1c, it was found mirror symmetry protected band crossings between the valence and conduction bands close to E_F. When the bands of opposite eigenvalues cross, a nodal line is formed. Three such nodal lines form around the point in the k_x, k_y and k_z planes and are protected by the mirror symmetries M_x, M_y, M_z of the $Fm\bar{3}m$ space group. Upon incorporating SOC, the electron spin is not a good quantum number any longer and the crystal symmetry changes depending on the direction of the magnetization. For example, if a sample is magnetized along the [001] direction, the M_x and M_y mirror symmetries are broken. Therefore, the nodal lines will open up unless there remain certain symmetries that protect the band crossings away from E_F. As a consequence, at least two Weyl points form along the k_z axis, leading to a finite AHC. For example, we calculate an intrinsic anomalous Hall conductivity of $\sim 1400\,\Omega^{-1}cm^{-1}$ for the fully stoichiometric Co_2MnGa compound. Depending on the details of the linear band crossings, such as the proximity of the nodal line to E_F and its dispersion, the AHC in topological Heuslers can range from $\sim 100\,\Omega^{-1}cm^{-1}$ in Co_2TiSn to $\sim 2000\,\Omega^{-1}cm^{-1}$ in Co_2MnAl [6, 17–19]. The high AHE in Co_2MnAl was already recognized in 2012 on the basis of Berry curvature calculation and agrees well with the experiment [20–22].

Maintaining the same number of valence electrons (NV) and reducing the crystal symmetry in such a way that inversion and mirror symmetries are broken. An easy example is Mn_2CoAl with an inverse-Heusler structure $F\bar{4}3m$ that shares the same NV = 26 of the full-Heusler compound Co_2TiSn. Interestingly, the compound belongs to a special class of materials, the spin-gapless semiconductor [23] . Here, the minority spin channel is insulating, similar to the half-metallic compounds, but the majority spin channel possesses a vanishingly small gap at Fermi level. Because of the non-centrosymmetric crystal structure, the mirror planes M_x, M_y and M_z of the full-Heusler no longer exist. Naturally, the nodal lines gap, and upon incorporating SOC, no Weyl points form. Hence, the band structure of Mn_2CoAl does not show any topologically protected crossings. For the spin-gapless compounds, the AHE shows an unusual behaviour. Though the materials can be highly magnetic (saturation magnetization $2\,\mu_B$/f.u. for Mn_2CoAl), the AHC nearly compensates around E_F, which is in contrast with the classical understanding that large magnetic moments always accompany a strong AHE. The predicted zero AHE was also found experimentally in Mn_2CoGa [24].

8.4 AHE in Non-collinear AFM with Weyl Points

For a long time, it was believed that an AHE cannot exist in AFM materials due to the zero net magnetic moment. However, it was recently revealed that the existence of the AHE relies only on the symmetry of the magnetic structure and corresponding Berry curvature distribution. Since the AHC can be understood as the integral of Berry curvature in k-space, and the Berry curvature is odd under time-reversal operation, the AHE can exist only in systems with broken time-reversal symmetry. In collinear AFMs, the combined symmetry of time reversal $\hat{T}$ and a space group operation $\hat{O}$ will change the sign of Berry curvature ($\Omega_i(\mathbf{k}) = -\Omega_i(\hat{T}\hat{O}\mathbf{k})$), leading to a vanishing AHC, despite the broken time-reversal symmetry due to the formation of local magnetic moments. However, in certain non-collinear AFMs, the symmetry to reverse the sign of the Berry curvature is absent, and a nonzero AHE can appear. A non-collinear AFM order was first demonstrated by Kren et al. in 1968 for the cubic compounds Mn_3Rh and Mn_3Pt [25]. A similar spin structure in the hexagonal series of Mn_3X (X = Ga, Ge, Sn) compounds was discovered by Kren et al., Nagamiya, Tomiyoshi et al., and Brown et al. [26–29]. These early investigations invoked the DM interaction to explain the observed triangular order by neutron diffraction experiments [30, 31]. In 1988, the first ab initio density functional calculations were reported by Kuebler et al. for the Cu_3Au structures of Mn_3Rh and Mn_3Pt, and they succeeded in explaining the observed non-collinear order [32]. By the same method, Sticht et al. dealt with the hexagonal Mn_3Sn, successfully obtaining and analysing the triangular magnetic ground-state structure [33]. Later, Sandratskii et al. showed that the DM interaction produces a weak ferromagnetism in Mn_3Sn [34]. The DM vector is oriented along the crystallographic c-axis and leads to a negative chirality of the spin structure.

The first nonzero AHE in non-clear AFMs was predicted in cubic Mn_3Ir by Chen et al. [35] However, its experimental realization has not yet been successful. Motivated by theoretical studies of the stability of cubic, tetragonal and hexagonal phases of Mn_3X (X = Ga, Sn, Ge) in connection with the Heusler family (see Fig. 8.1d), a new series of studies began, which led to the prediction of the AHE in hexagonal Mn_3Sn and Mn_3Ge by Kübler et al. [36, 37]. Soon after the predictions, the large AHE was experimentally verified in both Mn_3Sn and Mn_3Ge hexagonal antiferromagnets [38–40].

The AHC can be viewed as a vector in three dimensions, where the nonzero components are determined by symmetry. Both Mn_3Ge and Mn_3Sn exhibit a triangular antiferromagnetic structure with an ordering temperature above 365 K, and the magnetic structure is symmetric with respect to the glide mirror operation $[M_y|(0, 0, c/2)]$ [left panel in Fig. 8.1d]. Under this symmetry operation, the two components of Berry curvature Ω_x and Ω_z change sign, whereas Ω_y does not. As a consequence, σ_x and σ_z are forced to be zero, and only a nonzero σ_y survives. Therefore, a nonzero AHE can be obtained only when the magnetic field is applied perpendicular to c. The maximum AHC appears for the set-up with B⊥a (B//y). Since there is a weak net moment (∼0.01 μB/Mn) out of the a-b plane, a very small AHC was also detected in the situation with B//c, which is orders of magnitude smaller than that of the other configuration [41].

The strong anomalous Hall effect in Mn_3Sn and Mn_3Ge inspired the interest of the investigation for their band structure from topological point of view, and multiple Weyl points were observed [42]. Taking Mn_3Sn as the sample, there are six pairs of Weyl points in the first Brillouin zone, that can be classified into three groups according to their positions (noted as W1, W2 and W3), as indicated in Fig. 8.3. These Weyl points lie in the M_z plane (with W2 points being only slightly off this plane owing to the residual-moment-induced symmetry breaking) and slightly above the Fermi energy. Therefore, there are four copies for each of them according to the symmetry. A Weyl point (e.g. W1 in Fig. 8.3b, c) acts as a source or sink of the

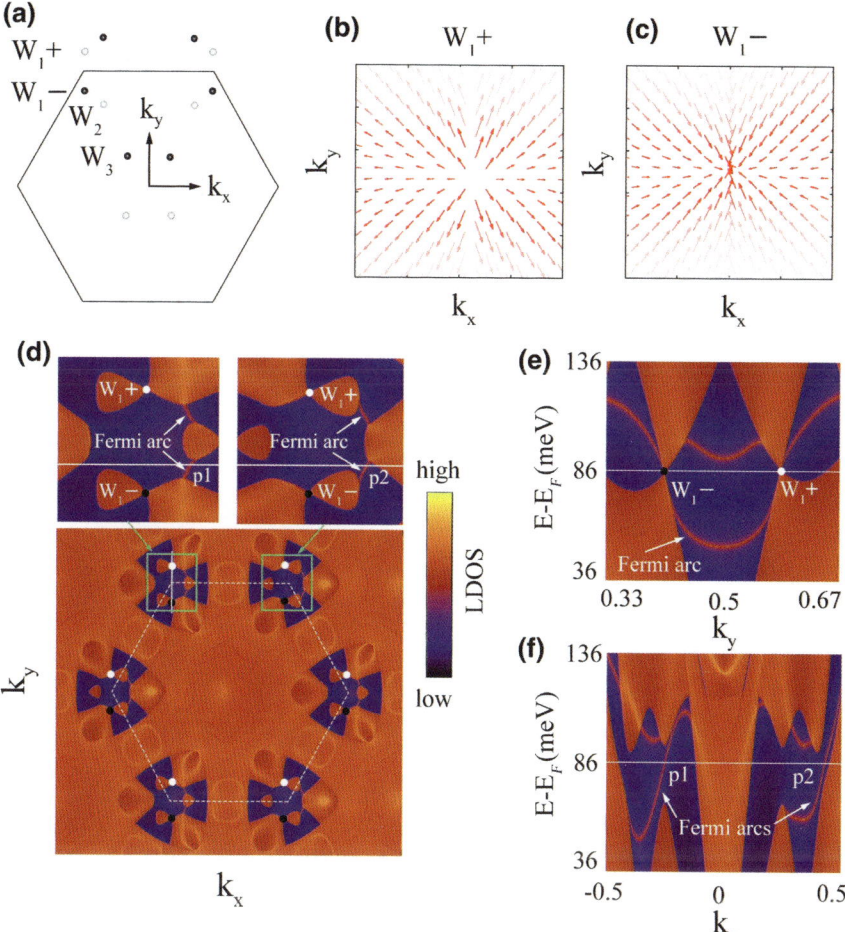

Fig. 8.3 Surface states of Mn_3Ge. **a** Distribution of Weyl points in momentum space. Black and white points represent Weyl points with − and + chirality, respectively. Larger points indicate two Weyl points ($\pm k_z$) projected into this plane. **b** and **c** monopole-like distribution of the Berry curvature near a W1 Weyl point. **d** Fermi surface at $E_F = 55$ meV crossing the W1 Weyl points. The colour represents the surface LDOS. Two pairs of W1 points are shown enlarged in the upper panels, where clear Fermi arcs exist. **e** Surface band structure along a line connecting a pair of W1 points with opposite chirality. **f** Surface band structure along the white horizontal line indicated in **b**. Here p1 and p2 are the chiral states corresponding to the Fermi arcs. (From New J. Phys. 19 015008 (2017))

Berry curvature, clearly showing the monopole feature with a definite chirality. The existence of Weyl points was experimentally verified by both surface ARPES and bulk magnetotransport measurements soon after the theoretical prediction [43].

The existence of Fermi arcs on the surface is one of the most significant consequences of Weyl points inside the three-dimensional bulk. We first investigate the surface states of Mn_3Sn that have a simple bulk band structure with fewer Weyl points. When projecting W2 and W3 Weyl points to the (001) surface, they overlap with other bulk bands that overwhelm the surface states. W1 Weyl points are visible on the Fermi surface. When the Fermi energy crosses them, W1 Weyl points appear as the touching points of neighbouring hole and electron pockets. Therefore, they are typical type-II Weyl points. Indeed, their energy dispersions demonstrate strongly tilted Weyl cones.

The Fermi surface of the surface band structure is shown in Fig. 8.3d for the Sn compound. In each corner of the surface Brillouin zone, a pair of W1 Weyl points exists with opposite chirality. Connecting such a pair of Weyl points, a long Fermi arc appears in both the Fermi surface (Fig. 8.3d) and the band structure (Fig. 8.3e). Although the projection of bulk bands exhibits pseudo-symmetry of a hexagonal lattice, the surface Fermi arcs do not. It is clear that the Fermi arcs originating from two neighbouring Weyl pairs, as shown in Fig. 8.3d do not exhibit M_x reflection, because the chirality of Weyl points apparently violates M_x symmetry. For a generic k_x-k_z plane between each pair of W1 Weyl points, the net Berry flux points in the $-k_y$ direction. As a consequence, the Fermi velocities of both Fermi arcs point in the $+k_x$ direction on the bottom surface (see Fig. 8.3f). These two right movers coincide with the nonzero net Berry flux, i.e. Chern number = 2.

8.5 Strong Anomalous Hall and Anomalous Nernst Effect in Compensated Ferrimagnets

Owing to the absence of a symmetry operation that inverses the sign of the Berry curvature in ferrimagnets, AHE and ANE are also allowed in compensated ferrimagnets with zero net magnetic moments. Because the charge carrier density is relatively small in most compensated ferrimagnets, the AHE is very weak and not easy to detect in transport measurements. However, if a compensated ferrimagnet possesses a special electronic band structure with a large Berry curvature, a strong AHE is expected. A typical example is the compensated ferrimagnetic Weyl semimetal. The integration of the Berry curvature around a Weyl point should provide a large Berry phase and therefore a strong AHE.

For convenience, we start to understand the AHE from a AFM model in the combination symmetry $\hat{T}\hat{O}$ of a glide operation to the centre of the unit cell and time reversal; see the left panel in Fig. 8.4a. A simple and effective way to remove this symmetry is by replacing the equivalent atoms lying on the other sublattice with a different element, see the right panel in Fig. 8.4a, which is just a compensated ferrimagnets (FiM), and a nonzero Berry phase from the whole BZ is allowed. Based on this guiding principle, both strong AHE and ANE were recently predicted in compensated ferrimagnetic Heusler WSM Ti$_2$MnX (X = Al, Ga and In) [44, 45].

These compounds have an inverse-Heusler lattice structure with space group $F\bar{4}3m$ (No. 216) (see Fig. 8.4b) [46]. Ti$_2$MnX (X = Al, Ga and In) have half-metallic ferrimagnetic structure, where magnetic moments are located at the

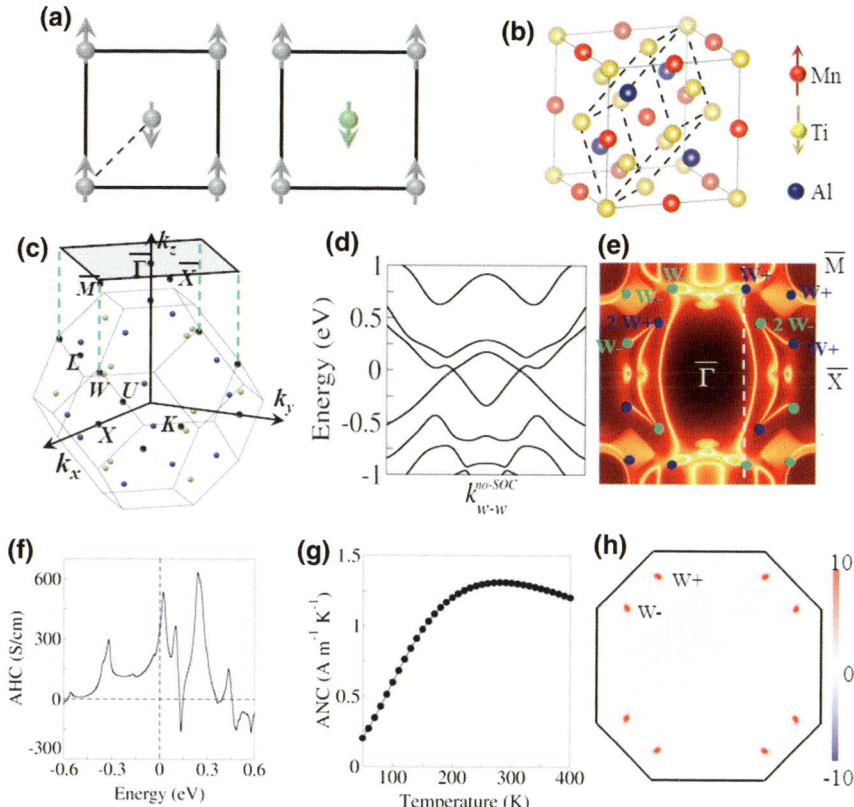

Fig. 8.4 Strong AHE and ANE in compensated ferrimagnets Ti_2MnAl. **a** Left: AFM structure with a time reversal + slide symmetry leading to a vanishing AHE. Right: Ferrimagnet with broken slide symmetry. **b** Inverted Heusler FCC crystal structure of Ti_2MnAl. The magnetic moments of Ti and Mn are all aligned along the (001) direction and compensate each other. **c** Brillouin zone of Ti_2MnAl with the location of the 12 pairs of Weyl points. **d** Energy dispersion along one pair of Weyl points. **e** Surface Fermi arc states with energy fixing at Weyl points. **f** AHC for Ti_2MnAl. A maximum linked to the Weyl points can be seen around 40 meV above the Fermi level. **g** The dependence of the ANC of the Ti_2MnAl at $E = E_F$ on temperature. **f** Berry curvature distribution in the $k_z = 0$ plane. The eight hot spots are just the positions of Weyl points. Colour bars are arbitrary units. (From Physical Review B, 97, 060406(R) (2018), Licence Number: RNP/18/APR/003014)

Ti ($\mu = 1.3(1.2)\mu B$) for first(second) atom) and Mn ($\mu = 2.5\mu B$) atoms. The net magnetic moment in Ti_2MnX vanishes because of the compensated magnetic sublattices formed by Ti and Mn. In total, there are 12 pairs of Weyl points. Their positions in the Brillouin zone are depicted in Fig. 8.4c, and they are located slightly above the Fermi level, as indicated in Fig. 8.4d.

The surface states in Ti_2MnX present very large Fermi arcs, as indicated in Fig. 8.4e for the example of Ti_2MnAl. By fixing the energy at the Weyl points, perfect Fermi arcs with tiny bulk states can clearly be seen in Fig. 8.4e. Dependent on the number of surface projected Weyl points, the number of Fermi arcs terminated at each Weyl points differs. Moreover, two long Fermi arcs extend around 75% of the reciprocal lattice vector. Therefore, the Weyl semimetal states in Ti_2MnX lead to the existence of isolated surface Fermi arcs, and the long Fermi arc around the charge neutral point should be easy to detect by surface techniques.

In bulk transport, Ti_2MnX also have large AHC and ANC. Taking Ti_2MnAl as an example, the intrinsic AHC is around 300 and 550 S/cm by fixing the energy at the charge neutrality point and Weyl points (see Fig. 8.4f), respectively. Replacing electric field by temperature-gradient, the ANC can reach up to 1.3 A $(m^{-1}K^{-1})$ at room temperature (see Fig. 8.4g), which is around five times large than that in non-collinear AFM Mn_3Sn. From the analysis of Berry curvature distribution in k−space, it is found that the large AHC and ANC are almost dominated by the Weyl points. Figure 8.4h shows the Berry curvature distribution in the $k_z = 0$ plane with four pairs of Weyl points very close to it. Except for the eight hot spots derived from the Weyl points, there are barely other contributions to the AHC. The other two high-symmetry planes $k_x = 0$ and $k_y = 0$ have almost the same Berry curvature distribution. Therefore, the AHE and ANE are topologically protected.

8.6 Antiskyrmions

Apart from a k-space topology, Heusler compounds with non-collinear spin structure also host real-space topological states such as magnetic skyrmions. Magnetic skyrmions are particle-like vortex spin textures surrounded by chiral boundaries that are separated from a region of reversed magnetization found in magnetic materials [47–50]. In this case, the topological skyrmion number is defined in real space. It measures the winding of the magnetization direction wrapped around the unit sphere and can take on integer values only. The mechanism of formation and stabilization of skyrmions can be understood as due to the competition of the ferromagnetic exchange and the relativistic DM interaction in non-centrosymmetric magnets. The typical size of a skyrmion can range from 1 to 100 nm, which enables the manipulation of many internal degrees of freedom. Owing to the magnetoelectric coupling, it is possible to control the skyrmions with an external electric field with low energy consumption.

Depending on the spin rotation, skyrmions can be classified into two fundamental types, Bloch skyrmions and N'eel skyrmions. Another type of skyrmion (antiskyrmion) was also proposed to exist, where the boundary domain walls alternate between the Bloch and Neel types as one traces around the boundary [51, 52]. The first two fundamental types of skyrmions were observed in B20 crystals and polar magnets with C_{nv} symmetry, respectively. However, despite a prediction of antiskyrmions in Co/Pt multilayers and B20 compounds, none have been experimentally verified. Very recently, by following the theoretical prediction and symmetry analysis, the first class of antiskyrmions has been observed in the inverse tetragonal acentric Mn-Pt-Sn Heusler compounds with D2d symmetry.

References

1. B.A. Bernevig, Quantum spin Hall effect and topological phase transition in HgTe quantum wells. Science **314**, 1757 (2006)
2. M. Knig, Quantum spin Hall insulator state in HgTe quantum wells. Science **318**(5851), 766–770 (2006)
3. S. Chadov, Tunable multifunctional topological insulators in ternary heusler compounds. Nat. Mater. **9**, 541 (2010)
4. H. Lin, Half-Heusler ternary compounds as new multifunctional experimental platforms for topological quantum phenomena. Nat. Mater. **9**, 546 (2010)
5. D. Xiao, Half-Heusler compounds as a new class of three-dimensional topological insulators. Phys. Rev. Lett. **105**, 096404 (2010)
6. C. Shekhar et al., Observation of chiral magneto-transport in RPtBi topological heusler compounds (2016), arXiv:1604.01641
7. Z.K. Liu, Observation of unusual topological surface states in half-Heusler compounds LnPtBi (Ln=Lu, Y). Nat. Commun. **7**, 12924 (2016)
8. J.A. Logan, Observation of a topologically non-trivial surface state in half-Heusler PtLuSb (001) thin films. Nat. Commun. **7**, 11993 (2016)
9. C. Liu, Metallic surface electronic state in half-Heusler compounds RPtBi (R = Lu, Dy, Gd). Phys. Rev. B **83**, 205133 (2011)
10. M. Hirschberger, The chiral anomaly and thermopower of Weyl fermions in the half-Heusler GdPtBi. Nat. Mater. **15**, 1161 (2016)
11. T. Suzuki, Large anomalous Hall effect in a half-Heusler antiferromagnet. Nat. Phys. **12**, 1119 (2016)
12. N. Kumar et al., Planar Hall effect in Weyl semimetal GdPtBi. **1711** (2017), arXiv:1711.04133
13. R.A. de Groot et al., Phys. Rev. Lett. **50**, 2024–2027 (1983)
14. M.I. Katsnelson, Half-metallic ferromagnets: from band structure to many-body effects. Rev. Mod. Phys. **80**, 315–378 (2008)
15. Z. Wang, Time-Reversal-Breaking Weyl fermions in magnetic heusler alloys. Phys. Rev. Lett. **117**, 236401 (2016)
16. G. Chang, Room-temperature magnetic topological Weyl fermion and nodal line semimetal states in half-metallic Heusler Co2TiX (X=Si, Ge, or Sn). Sci. Rep. **6**, 38839 (2016)
17. S. Wurmehl, Investigation of Co2FeSi: the heusler compound with highest curie temperature and magnetic moment. Appl. Phys. Lett. **88**, 032503 (2006)
18. H.C. Kandpal, Correlation in the transition-metal-based heusler compounds Co2MnSi and Co2FeSi. Phys. Rev. B **73**, 094422 (2006)
19. T. Jen-Chuan, High spin polarization of the anomalous Hall current in Co-based heusler compounds. New J. Phys. **15**, 033014 (2013)
20. E. Vilanova Vidal, Exploring Co2MnAl heusler compound for anomalous Hall effect sensors. Appl. Phys. Lett. **99**, 132509 (2011)
21. J. Kübler, Berry curvature and the anomalous Hall effect in heusler compounds. Phys. Rev. B **85**, 012405 (2012)
22. J. Kübler, C. Felser, Weyl points in the ferromagnetic Heusler compound Co2MnAl. EPL **114**, 47005 (2016)
23. S. Ouardi, Fecher et al., Realization of spin gapless semiconductors: the heusler compound Mn2CoAl. Phys. Rev. Lett. **110**, 100401 (2013)
24. K. Manna et al., From colossal to zero: controlling the anomalous Hall Effect in magnetic heusler compounds via berry curvature design (2017), arXiv:1712.10174
25. E. Krn, Magnetic structures and exchange interactions in the Mn-Pt system. Phys. Rev. **171**, 574–585 (1968)
26. E. Krn, Neutron diffraction study of Mn3Ga. Solid State Commun. **8**, 1653–1655 (1970)
27. T. Nagamiya, Triangular spin ordering in Mn3Sn and Mn3Ge. J. Phys. Soc. Jpn. **46**, 787–792 (1979)

28. S. Tomiyoshi, Magnetic structure and weak ferromagnetism of Mn3Sn studied by polarized neutron diffraction. J. Phys. Soc. Jpn. **51**, 2478–2486 (1982)
29. P.J. Brown, Determination of the magnetic structure of Mn3Sn using generalized neutron polarization analysis. J. Phys. Condens. Matter **2**, 9409 (1990)
30. I. Dzyaloshinsky, A thermodynamic theory of weak ferromagnetism of antiferromagnetics. J. Phys. Chem. Solids **4**, 241–255 (1958)
31. T. Moriya, Anisotropic superexchange interaction and weak ferromagnetism. Phys. Rev. **120**, 91–98 (1960)
32. J. Kübler, Local spin-density functional theory of noncollinear magnetism (invited). J. Appl. Phys. **63**, 3482–3486 (1988)
33. J. Sticht, Non-collinear itinerant magnetism: the case of Mn3Sn. J. Phys. Condens. Matter **1**, 8155 (1989)
34. L.M. Sandratskii, Role of orbital polarization in weak ferromagnetism. Phys. Rev. Lett. **76**, 4963–4966 (1996)
35. H. Chen, Anomalous Hall effect arising from noncollinear antiferromagnetism. Phys. Rev. Lett. **112**, 017205 (2014)
36. D. Zhang, First-principles study of the structural stability of cubic, tetragonal and hexagonal phases in Mn3Z (Z = Ga, Sn and Ge) heusler compounds. J. Phys. Condens. Matter **25**, 206006 (2013)
37. J. Kübler, Non-collinear antiferromagnets and the anomalous Hall effect. EPL **108**, 67001 (2014)
38. S. Nakatsuji, Large anomalous Hall effect in a non-collinear antiferromagnet at room temperature. Nature **527**, 212–215 (2015)
39. A.K. Nayak et al., Large anomalous Hall effect driven by a nonvanishing Berry curvature in the noncolinear antiferromagnet Mn3Ge. Sci. Adv. **2** (2016)
40. N. Kiyohara, Giant anomalous Hall effect in the chiral antiferromagnet Mn3Ge. Phys. Rev. Appl. **5**, 064009 (2016)
41. J. Zelezn, Spin-Polarized current in noncollinear antiferromagnets. Phys. Rev. Lett. **119**, 187204 (2017)
42. H. Yang et al., Topological Weyl semimetals in the chiral antiferromagnetic materials Mn3Ge and Mn3Sn. New J. Phys. **19**, 015008 (2017)
43. K. Kuroda, Evidence for magnetic Weyl fermions in a correlated metal. Nat. Mater. **16**, 1090 (2017)
44. W. Shi et al. (2017), arXiv:1801.03273
45. J. Noky et al., Strong anomalous nernst effect in collinear magnetic Weyl semimetals without net magnetic moments (2018), arXiv:1803.03439
46. W. Feng et al., Phys. Status Solidi RRL **11**, 641 (2015)
47. A.K. Nayak, Magnetic antiskyrmions above room temperature in tetragonal heusler materials. Nature **548**, 561 (2017)
48. A. Fert, Magnetic skyrmions: advances in physics and potential applications. Nat. Rev. Mater. **2**, 17031 (2017)
49. N. Nagaosa, Topological properties and dynamics of magnetic skyrmions. Nat. Nanotechnol. **8**, 899 (2013)
50. C. Felser, Skyrmions. Angew. Chem. Int. Ed. **52**, 1631–1634 (2013)
51. W. Koshibae, Theory of antiskyrmions in magnets. Nat. Commun. **7**, 10542 (2016)
52. M. Hoffmann, Antiskyrmions stabilized at interfaces by anisotropic dzyaloshinskii-moriya interactions. Nat. Commun. **8**, 308 (2017)

Chapter 9
Topological Materials and Solid-State Chemistry—Finding and Characterizing New Topological Materials

L. M. Schoop and A. Topp

Abstract In this chapter, we will start by introducing some basic concepts of solid-state chemistry and how they can help us identify new topological materials. We give a short overview of common crystal growth methods and the most significant characterization techniques available to identify topological properties. Finally, we summarize this knowledge in a step-by-step procedure that will guide us from the idea to a real compound. The aim of this chapter is to give physics students a guide for implementing simple chemical principles in their search for new topological materials, as well as giving a basic introduction to the steps necessary to experimentally verify the electronic structure of a material.

9.1 The Role of Solid-State Chemistry in the Search for Topological Materials

The field of topology is currently developing at an extremely fast rate. Exciting predictions of new physical phenomena that can arise in topological matter appear frequently [1–6], and once a material that shows the desired electronic structure is discovered, experimental evidence of the proposed features usually appears rapidly [7–11]. Finding a material candidate that fulfills the requirements for a new prediction is often the bottleneck in this process. Predicting and developing new topological materials is an interdisciplinary endeavor between physics and chemistry. On a fundamental level, a material's crystal structure and the types of bonds within the structure

---

L. M. Schoop (✉)
Department of Chemistry, Princeton University, Princeton, NJ 08544, USA
e-mail: lschoop@princeton.edu

A. Topp
Max-Planck-Institut für Festkörperforschung, Heisenbergstraße 1,
70569 Stuttgart, Germany
e-mail: a.topp@fkf.mpg.de

are connected to its electronic structure and physical properties. A challenge for the advancement of the field of topology is not only to predict desired physical properties but also to relate these properties to structural motifs that allow researchers to link materials to their properties.

9.2 Simple Rules from Solid-State Chemistry

Historically, the field of solid-state chemistry focused primarily on the synthesis of materials and their structural and chemical properties. Often times, not much attention was directed to their physical properties. Still, the field of solid-state chemistry developed a deep understanding of the structure of matter and created rules for material's stabilities in different structure types. These rules depend on certain characteristics of the elements used, such as their size or electronegativity. The rules help to understand materials and their crystal structure and grant some intuition for predicting possible unknown ones.

9.2.1 Counting Electrons in Solids

One of the most basic chemical principles states that most molecules follow the 8 (or 18, if d electrons are included)-electron rule [12]. This number arises from the desire of each element to have a filled electron shell (with 8 or 18 valances electrons, all s, d, and p orbitals or bands are filled). If a shell is only partially filled, bonds are formed to gain a stable, closed shell state. This concept explains the most basic ideas of chemistry, for example, why oxygen forms a diatomic species and argon does not. The same concept can be (less rigorously) expanded to solid materials [13]. Closed shell systems, which are often found in crystalline solids, are either insulators, semiconductors, or semimetals, while open shell systems tend to be metals. Just by counting valence electrons, we can quickly make predictions about general aspects of the electronic structure. Note that one might be tempted to think that any material with $2n$ electrons could be insulating, since in this case there could technically be only filled bands. However, since individual bands commonly overlap due to their bandwidth, open shell systems often show a metallic state.

For counting the electrons of an extended crystalline solid, one can count the number of electrons per formula unit. The number of valence electrons an element has is determined by its group in the periodic table. Therefore, one can just add up the number of the valence electrons from each element. It needs to be kept in mind that each element ideally prefers a filled shell. In order to achieve a closed shell state for all elements composing the material, electrons are transferred between different elements. Depending on its electronegativity, an element can be an acceptor or donor. Electron acceptors require a few electrons to fill their shell, while electron donors need to give up a few to empty theirs. The examples below should clarify the electron

9 Topological Materials and Solid-State Chemistry ...

transfer in ionic compounds.

Example 1 (rock salt NaCl):

$$1 \text{ (Na)} + 7 \text{ (Cl)} = 8 \text{ (filled } s \text{ and } p \text{ shell)}$$

Na gives its extra electron to Cl to fill the shell, forming NaCl, an insulating ionic solid.

Example 2 (Heusler compound ScPtSb):

$$3 \text{ (Sc)} + 10 \text{ (Pt)} + 5 \text{ (Sb)} = 18 \text{ (filled } s, \ p \text{ and } d \text{ shell)}$$

ScPtSb is an intermetallic compound where some bonds cannot be considered fully ionic. As a result, the electron transfer is less obvious and oxidation states have to be taken into account. Still, counting the valence electrons indicates a filled valence shell and ScPtSb is indeed a semiconductor [14].

Example 3 (CaF_2):

$$2 \text{ (Ca)} + 2 \cdot 7 \text{ (F)} = 16 \rightarrow 8 \text{ per F}$$

If the composition is not just simply 1:1, the total electron count must be divided by the number of acceptors. In the above example, Ca gives up two electrons, one for every F; thus all three atoms will have filled shells.

Sometimes, simple electron counting seems not to work. Some semiconductors have a formal electron count that differs from 8 or 18. An example is given below.

Example 4 (LaAuSb):

$$3 \text{ (La)} + 11 \text{ (Au)} + 5 \text{ (Sb)} = 19 \rightarrow \text{ metallic?}$$

Phases that have a 1:1:1 composition and contain a rather electropositive element such as La are usually always 18 electron phases and semiconducting or semimetallic. So why does the 19 electron compound LaAuSb exist? Unlike suggested by the electron count, LaAuSb is a semimetal [15]. A closer look at its crystal structure reveals that two Au atoms are in close proximity. Thus, one electron per formula unit is located between the Au atoms and an Au–Au bond is formed. Therefore, one electron per formula unit has to be subtracted from the count, resulting in a stable, semimetallic phase with 18 electrons per formula sum. A further example of a material where simple electron counting would wrongly suggest metallic conductivity is

$CaSi_2$, which is known to be a semiconductor [16].

Example 5 ($CaSi_2$):

$$2 \, (Ca) + 2 \cdot 4 \, (Si) = 10 \rightarrow 5 \text{ per Si}$$

The electron count is not capturing that every Si atom is bonded to three other Si atoms in the $CaSi_2$ crystal structure. The bonded Si atoms share one electron each, resulting in a count of 8 electrons per Si. Phases such as $CaSi_2$ are often referred to as Zintl phases. The Zintl concept is a description for closed shell (semiconducting) phases that contain covalent bonds in the crystal structure [17]. It distinguishes between polyanions and polycations; the Au–Au bond in LaAuSb is an example of the former and the Si_2^{2-} network in $CaSi_2$ is an example of the latter. Another example for bonds affecting the electron count is elemental silicon. It has four electrons and adopts a crystal structure where each Si atom shares four bonds with further Si atoms. Even though we only count 4 electrons per formula unit, silicon reaches a closed shell state by adding covalent bonds.

For Zintl phases, the electron count can be used to find the expected number of bonds via the following equation:

$$b(X - X) = 8 \text{ (or 18)} - VEC(X), \tag{9.1}$$

where b is the number of covalent bonds between identical atoms X and VEC is the valence electron count for that atom X. For example, this formula can explain why silicon and diamond have a band gap.

Example 6 (bond order of Diamond):

$$b(C - C) = 8 - 4 \, (C) = 4$$

In the diamond structure, each carbon is bonded to 4 other carbon atoms. Each C formally has a filled valence shell, making diamond an insulator.

Similarly, (9.1) can be used to predict the number of bonds in $CaSi_2$.

Example 7 (bond order of $CaSi_2$):

$$b(Si - Si) = 8 - \frac{2 \, (Ca) + 2 \cdot 4 \, (Si)}{2} = 3$$

As described above, the equation reveals that each Si must be bonded to three further Si to maintain a charge-balanced electron count.

We can conclude that electron counting allows to distinguish a metal from an insulator/semiconductor/semimetal, merely based on the formula sum. The examples above highlight that it is important to understand the crystal structure of a material to count its electrons correctly. Still, electron counting is only one part in predicting new materials. In addition to its electronic stability, one has to consider the sizes of the different atoms composing the structure, to see if the proposed crystal structure can actually stable.

9.2.2 Size of the Elements

The crystal structure adopted by a material largely depends on the size of the elements it is composed of. The ratio of the cation and anion radii defines the possible coordinations of a cation by anions. Based on simple geometric arguments, only certain coordination geometries are possible, which is summarized in the Pauling rules [19]. For that matter, we assume that ions are hard spheres that touch each other. The resulting coordination geometries for different ionic radii ratios are summarized in (9.2).

$$r_c/r_a > 0.732 \rightarrow \text{cubic coordination}$$
$$0.424 < r_c/r_a < 0.732 \rightarrow \text{octahedral coordination}$$
$$r_c/r_a < 0.424 \rightarrow \text{tetrahedral coordination} \quad (9.2)$$

Here r_c is the cationic radius and r_a is the anionic radius. The Pauling rules can help to test if a made-up structure can exist. These rules are widely followed by known oxide, fluoride, and other strongly ionic compounds. Systems that violate them tend to be unstable, to the point that they cannot exist.

Until here, we assumed bonds to be either purely ionic or covalent. In reality, the notion of ionic and covalent bonds describes two ends of a spectrum. Note that metallic bonds are a special case within this spectrum.

9.2.3 Bonding Type

As mentioned earlier, the type of bonds between atoms ranges over a wide spectrum. From purely ionic, where electrons are transferred from one atom to another, to purely covalent, where electrons are shared between atoms. Metallic bonds have a special status. Here, the electrons are not constrained to the location of one bond but are delocalized over the entirety of the metallic system. A common way to picture this is describing atoms as "positive ions in a sea of free electrons." This delocalization of electrons is responsible for the high thermal and electrical conductivity of metals. We now see that the type of a bond determines not only the crystal structure but also the physical properties of a material.

To find out what kinds of bonds are present in a material, one has to take a look at the electronegativity difference between the elements. If the difference is large, the bonds will be strongly ionic, resulting in salt-like compounds that always have to be charge-balanced. Such compounds are closed shell systems, which are usually strongly insulating. If the electronegativity difference is small, covalent or metallic bonds form a closed or open shell system. Closed shell systems formed by covalent or metallic bonds are semiconductors or semimetals. Open shell systems are metals with often high electric and thermal conductivity and metallic shine.

Bonding types can have a significant influence on the crystal structure. For example, while binary ionic compounds with a 1:1 composition usually adopt the cubic NaCl structure, binary compounds with a small electronegativity difference prefer the hexagonal NiAs structure [12].

9.2.4 A Database for Inorganic Crystalline Compounds

A great resource of experimentally determined crystal structures is the inorganic crystal structure database (ICSD). It contains information on the crystal structures of all known inorganic compounds, including pure elements and alloys [20]. It allows for a quick identification of the crystal structures of a known material with a given element composition. Nowadays, it is commonly used for data mining approaches to identify new topological materials. Because ICSD gathered information of published structures since 1913 and the field of crystal structure characterization has advanced drastically in the last 100 years, older entries can be outdated. Improved synthesis and measurement methods might reveal that the real crystal structure is slightly or even dramatically different from the reported one. Consulting the ICSD is only a starting point and does not replace a thorough characterization of your sample.

9.2.5 Linking Structures to Properties

Solid-state chemists have discovered and characterized a plethora of different compounds, but for a large fraction of these, the electronic properties have not yet been characterized. Sometimes, a structural motif can be linked to a physical property. A famous example of a repeating structural motif in different compounds that share the same physical property is the Fe-based superconductors. These compounds all contain edge-sharing Fe–As tetrahedra in their crystal structure [21, 22]. It stands to reason that similar structural motifs can be linked to topological properties of materials, which we will try to show in this chapter.

9.3 Topological Materials

Among the first, and most famous realizations of topological materials were HgTe quantum wells [23, 24] and the 3D topological insulators based on $Bi_{1-x}Sb_x$ and Bi_2Se_3 [25–27]. Only after the discovery of materials that experimentally showed

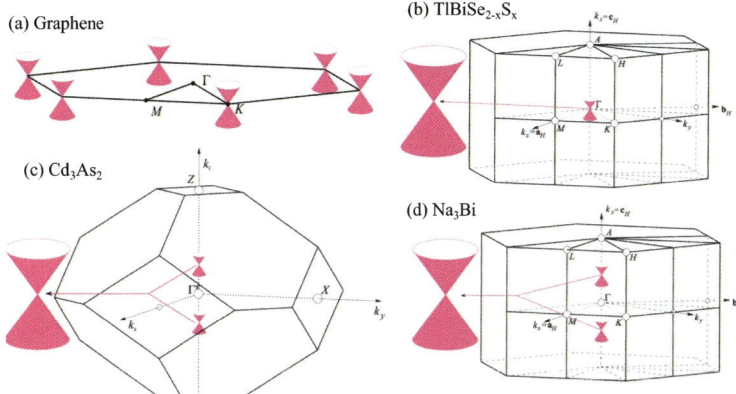

Fig. 9.1 Position of Dirac cones in the BZ in the Dirac semimetals **a** Graphene, **b** TlBiSe$_{2-x}$S$_x$, **c** Cd$_3$As$_2$ and **d** Na$_3$Bi. Reprinted figure with permission from [32]. Copyright 2018 by the American Physical Society

topological surface/edge states, the field grew to what it is today. Since then, the area of interest has expanded to 3D Dirac, Weyl, and Nodal line semimetals, which have been experimentally confirmed. For the advancement of the field, the discovery of materials that exhibit the desired electronic structure is essential. In the following section, we will show how we can use the chemical concepts introduced in Sect. 9.2 to identify new topological semimetals.

9.3.1 3D Analogs of Graphene—3D Dirac Semimetals

Graphene was the first Dirac semimetal that could be experimentally realized, and as a consequence, a new research field emerged [28]. Long before graphene was synthesized, it was theoretically predicted to be a zero-gap semiconductor [29]. Graphene's electronic structure consists of a conduction and valence band that cross at the K and K' points in the Brillouin zone (BZ), resulting in fourfold degenerate points at the Fermi level. These points are called Dirac points, since the low-energy excitation quasi-particles in such a linearly dispersing band structure behave like massless fermions, following the Dirac equation in high-energy physics [30]. The slope of the bands determines the Fermi velocity of the quasi-particles. The Fermi velocity relates to the speed of light for particles in high-energy physics. Thus, the linear band dispersion results in exotic properties such as a very high carrier mobility and extreme magnetoresistance [10, 31].

Graphene, as a 2D material, proved difficult to be implemented in practical applications, such as ultrahigh-frequency transistors. This led to a search for 3D analogs of graphene [33], resulting in the discovery of 3D Dirac semimetals in 2014 [7]. Figure 9.1 shows the BZ of graphene and some 3D Dirac semimetals.

All materials contain fourfold degenerate Dirac points in their electronic structure. Figure 9.2 shows the k dispersion of the electronic structures for a few examples of 3D Dirac semimetals; the Dirac cone is highlighted with circles.

Comparing the known 3D Dirac semimetals, one notices that they all have a few features in common:

(a) they are mostly charge-balanced compounds,
(b) they all crystallize in highly symmetric space groups,
(c) they all have a low DOS at the Fermi level.

These similarities can give us some guidelines on where to search for more 3D Dirac semimetals. In order to fulfill (a), we can count electrons as explained in Sect. 9.2.1. (b) results from group theory and the fact that crossings are only allowed in double groups that feature at least two irreducible representations. In the presence of spin–orbit coupling (SOC), our search is thus limited to cubic, hexagonal, or tetragonal compounds. These symmetry constraints have been discussed several times before [32, 34]. In the following, we want to focus on (c) and how we can influence the DOS at the Fermi level.

A "better" 3D Dirac semimetal has an electronic structure with a "clean" Dirac cone at the Fermi level, meaning the cone is not convoluted with further states. Of all the examples of 3D Dirac semimetals in Fig. 9.2 (taken from [32]), only BaAgBi (d) shows a Dirac crossing without additional bands in its vicinity. SrAgBi (e) shows an additional band crossing in the vicinity of the Dirac cone, while YbAuSb (f) has, in addition to other bands interfering at the Fermi level, its Dirac cone located below the Fermi level.

The "perfect" 3D Dirac semimetal has a zero density of states at the Fermi level. In general, the DOS at E_F allows for assessing a material's merit as a potential 3D Dirac semimetal. Gibson et al. tried to quantify this property by investigating the family of hexagonal ZrBeSi-type compounds [32]. To understand what affects the amount of states crossing the Fermi level, the authors plotted the DOS at E_F versus the total atomic number Z_{tot} divided by the electronegativity difference ΔE_N, where Z_{tot} is a measure for the SOC strength (Fig. 9.3a). Only above a certain threshold of $Z_{tot}/\Delta E_N$, a Dirac cone appears in the band structure. This indicates that the ratio of SOC to electronegativity difference can be a measure for the appearances of band inversions.

If the DOS is plotted against Z_{tot} (shown in Fig. 9.3b), one can see that a high Z_{tot} (and thus high SOC) results in a lower DOS at E_F and thus cleaner/more isolated Dirac cones. Therefore, there is a link between SOC strength divided by electronegativity difference and a material's potential for being a Dirac semimetal.

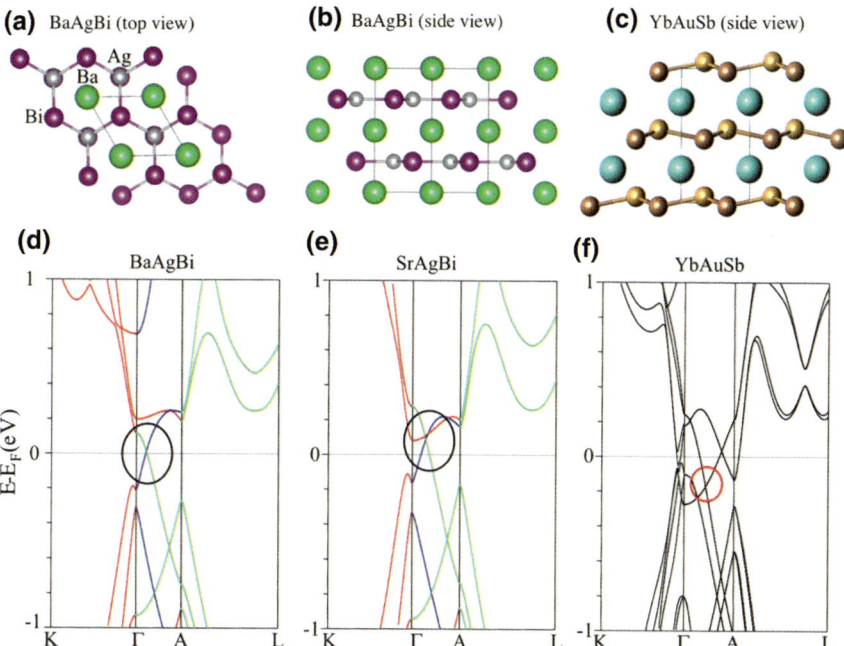

Fig. 9.2 Crystal and band structure of the Dirac materials BaAgBi, SrAgBi, and YbAuSb. The Dirac cones are highlighted with circles. Reprinted figure with permission from [32]. Copyright 2018 by the American Physical Society

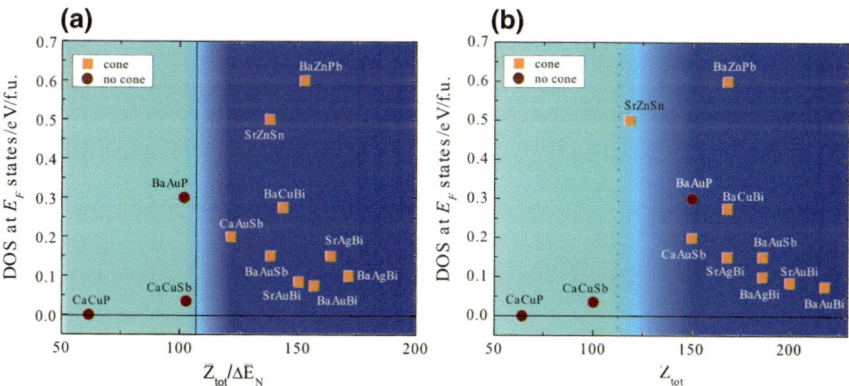

Fig. 9.3 Comparison of compounds with 18 electrons crystallizing in the ZrBeSi-type structure. Above a threshold of $Z_{tot}/\Delta E_N \approx 110$, Dirac cones appear in the band structure. The crossing becomes cleaner with increasing Z_{tot}. Reprinted figure with permission from [32]. Copyright 2018 by the American Physical Society

9.3.2 Weyl Semimetals

Weyl crossings can appear in materials that lack either inversion or time-reversal symmetry [35]. Unlike in inversion and time-reversal symmetric materials, where all bands always have to be doubly degenerate, this is not required if either of the symmetries is broken. This allows for twofold degenerate crossing points, i.e., Weyl points. In contrast to compounds that lack time-reversal symmetry, Weyl crossings of inversion asymmetric compounds are often found away from high-symmetry lines, which makes them hard to find. In Sect. 9.5.4.2, we will show an example of how magnetic order and the resulting absence of time-reversal symmetry can be used to obtain Weyl crossings.

9.3.3 Nodal Line Semimetals

Nodal line semimetals are materials, whose Fermi surface consists of a line or loop (rather than a point) of fourfold degenerate crossing points [36–38]. One can imagine a nodal line, if one pictures two parabolic bands that overlap, and the crossing is not gapped in any k space direction. Thus nodal lines are very rare, since SOC usually causes the band crossing to gap along some k vectors. Line node materials are thus usually found in materials composed of light elements that crystallize in highly symmetric space groups.

9.4 Nonsymmorphic Symmetries

3D Dirac semimetals are limited to highly symmetric space groups since SOC gaps the band crossings in the absence of a C_3, C_4 or C_6 rotation axis [32]. A way to circumvent this limitation can be found in crystals that contain nonsymmorphic symmetry elements [4, 39, 40]. Nonsymmorphic symmetry elements are mirror planes or rotation axes that are combined with a translational symmetry element. This combination yields glide planes and screw axes, which are a common appearance in space groups (157 of the 230 space groups are nonsymmorphic). An example for nonsymmorphic symmetry elements is shown in Fig. 9.4a. Due to the translational part of the symmetry element, the unit cell is multiplied, which directly results in a back-folding of the BZ in reciprocal space (see Fig. 9.4b, c). This band folding results in a forced band degeneracy at high-symmetry points. The degeneracy is enforced by group theory and SOC has no effect on it. The idea that nonsymmorphic symmetry enforces degeneracies has been used for one of the earliest predictions of 3D Dirac semimetals [39].

Besides fourfold Dirac-type crossings, nonsymmorphic symmetry can also cause higher-fold degeneracies. These have been suggested to result in so-called new fermions that have no counterpart in high-energy physics but can exist in condensed matter [4].

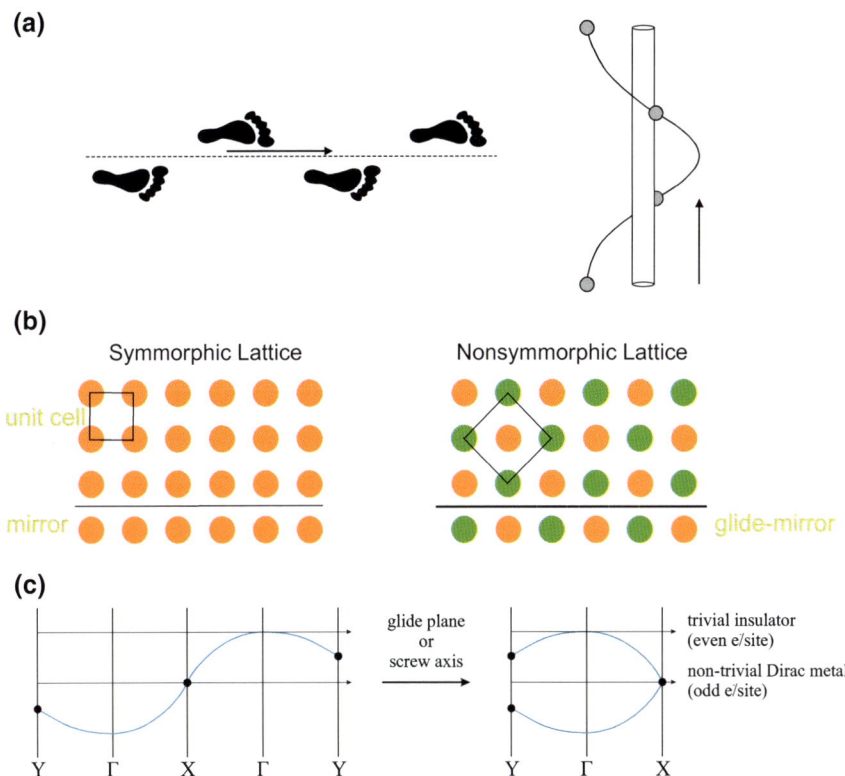

Fig. 9.4 a Examples of nonsymmorphic symmetry elements: The footprints resemble a glide plane; a screw axis is sketched beside. The arrow indicates the translational part of the symmetry element, which must be a fraction of a unit cell. **b** The translational part of the nonsymmorphic symmetry doubles the unit cell (now containing two atoms instead of only one in the symmorphic case). **c** Starting from the symmorphic case, the nonsymmorphic symmetry element back-folds the bands at a high-symmetry point (here X). Panel **c** adapted and reprinted with permission from [32]. Copyright 2018 by the American Physical Society

9.4.1 The Problem with the Half-Filled Band

A problem for realizing nonsymmorphic materials is that the Fermi level is usually not located at the degenerate crossing points (see, e.g., [4]). If a compound is charge-balanced, the folded bands are completely filled, thus yielding a trivial, insulating compound [32] (see Fig. 9.4c). In order for the Fermi level to be located at the degenerate point, the band would have to be half filled. However, such an odd electron count per formula unit is usually chemically unstable. Materials that formally have half-filled bands usually either undergo Peierl's distortions (form charge density waves) or become Mott insulators [41–43].

It is therefore challenging to predict materials that should have nonsymmorphically enforced degenerate points at the Fermi level, and if we are not careful, it is easy to predict a new material that is chemically unstable. One example is the compound "BiO_2" [39]. This material was predicted based on cristobalite-type SiO_2. In the cristobalite structure, each Si is coordinated tetrahedrally by four oxygens. The resulting SiO_4 tetrahedra are connected via their corners, forming a three-dimensional network. SiO_2 is charge-balanced; each O receives two electrons from the Si atoms, resulting in closed shell O^{2-} and Si^{4+} ions. SiO_2 is thus an insulator with a fairly large band gap. Since it crystallizes in a nonsymmorphic space group, it exhibits enforced band degeneracies well below and above the Fermi level. The original proposal to exchange Si with Bi lies near, since Bi has one electron more than Si resulting in a half-filled Bi 6s band, pinning the Fermi level to the degenerate point. In addition, the increased SOC in Bi would cause the bands to split along the high-symmetry lines, thus forming a symmetric 3D Dirac cone. Sadly, the compound "BiO_2" is not chemically stable. The first issue lies in the ratio of the ionic radii. As mentioned above (see (9.2)), the ratio of atomic radii can be used to predict the coordination geometry of the cations in the crystal structure. The ionic radii of all elements were tabulated and can be looked up online[1] [44]. In hypothetical BiO_2, Bi would be coordinated by four oxygen atoms in a tetrahedral fashion. In SiO_2, this coordination is favored since the Si^{4+} cation is very small. As can be seen in (9.2), a tetrahedral coordination is only found for a very small ratio of the cationic to the anionic radius. Since Bi is located much lower in the periodic table, Bi^{4+} is much larger than Si^{4+}. If we would look up the radius of a Bi^{4+} ion with a coordination sphere of four, we would encounter a problem—what we are looking for is not listed in the database. This arises due to two different problems. For once, Bi does not exist in the oxidation state +4, a problem we will come back to later. Second, Bi is too large, no matter how positively charged it is, to fit in a tetrahedral environment. Even if we consider the smallest tabulated radius for Bi, the ionic radius ratio we obtain is $r_{Bi}/r_O = 0.53$, which is too large to allow for a tetrahedral coordination. Thus, we can conclude that the cristobalite-type structure will not be adopted by the elements Bi and O.

We already briefly mentioned earlier that one of the problems we face in hypothetical BiO_2 is that the 4+ oxidation state does not exist in Bi. This results from the inert pair effect, which is a relativistic effect and a result of the lanthanide contraction and causes the separation of 6s and 6p electrons in energy. Thus Bi favors the 3+ oxidation state. A hypothetical Bi^{4+} would separate in Bi^{3+} and Bi^{5+} since having a single electron in the 6s shell is unfavorable.

An example for this kind of charge redistribution is found in $BaBiO_3$, a compound that became famous because it becomes superconducting at fairly high temperatures, if doped away from its insulating state [45–47]. If we count electrons for $BaBiO_3$, we might falsely conclude that Bi would be in the oxidation state 4+. In this case, $BaBiO_3$ would be metallic. Early density functional theory (DFT) calculations [48], based on an incorrect structural model, predicted the compound to be metallic, which was in contrast to the measured insulating properties [49]. A more careful structural analysis

---

[1] http://abulafia.mt.ic.ac.uk/shannon/ptable.php.

revealed that in the actual crystal structure, the unit cell is doubled to accompany alternating Bi^{3+} and Bi^{5+} ions [47]. If this structure is plugged into a DFT code, the insulating properties are predicted correctly. Thus we can conclude that even if BiO_2 would exist, it would likely also double its unit cell to accompany two different sized Bi cations (3+ and 5+). This way, Bi would avoid its half-filled band and the Fermi level would not be at a degenerate crossing point. Pure electron counting can thus be misleading and it is important to take bonding, ionic sizes, and common oxidation states into account, when predicting materials.

Does this mean that there is no hope for finding nonsymmorphic materials that have clean bulk band degeneracies at the Fermi level? In order to think about how we could stabilize a half-filled band, it might be worth looking at how organic chemists stabilize radicals, which are unpaired electrons in a molecule. Molecules that contain half-filled orbitals face the same kind of stability issues. A radical usually rapidly rips off atoms from other surrounding molecules, initiating a chain reaction, which is often used for polymerizing materials and making plastic. Nevertheless, stable radicals (stable at room temperature at least) can exist. Radicals can be stabilized if the electron can be delocalized over many atoms, for example in conjugated carbon systems. The solid-state equivalent, of such a conjugated carbon network, is cluster compounds. A cluster is composed of several metal atoms that are covalently bonded to each other. Thus they can have molecular orbitals that have the potential to delocalize an electron over the cluster. Nonsymmorphic cluster compounds can, therefore, potentially contain half-filled bands that are stabilized by the cluster, in their electronic structure [32]. Although there are many cluster compounds known [50], only a few are found in nonsymmorphic space groups. An example is $Tl_2Mo_6Se_6$, which was shown to host nonsymmorphically protected Dirac crossings very close to the Fermi level [32].

9.5 The Cycle of Material Development

Predicting a new material on the basis of physically motivated parameters is only one step in a cycle necessary to develop new materials. It usually starts with an idea, what kind of material or property is wished for. The next step would be to design a suitable material based on the chemical principles that have been discussed so far. Next, the electronic structure needs to be calculated to see whether the desired electronic structure is exhibited by the hypothetical material. The following step, the synthesis of the desired material, is the most time-consuming step and will be discussed in detail below. It is very important to subsequently characterize the grown material to evaluate the growing process and make sure that the material really crystallizes in the structure that was assumed for the electronic structure calculation. Finally, the physical properties can be measured, and if the desired properties are observed, the cycle is complete. For the last step, a number of spectroscopic methods are available. Here we will focus on angle-resolved photoemission spectroscopy (ARPES), since it directly maps the band structure and allows for the comparison with the calculations.

We describe this process as a cycle, since the realization of one compound usually allows for extending the idea toward a whole family of isostructural and isoelectronic materials.

9.5.1 Synthesis Methods

To grow a desired material in single crystalline form can be challenging at times. In order to probe the band structure with ARPES, a single crystal of several mm size is usually needed, without any grain boundaries.

There are a variety of different crystal growth methods available that can be combined and fine-tuned, to obtain crystals of very specific compositions and high purity. In this section, we will discuss the four most common crystal growth methods, while providing material examples that can be grown with each method. Furthermore, we will discuss scenarios where certain methods produce crystals of worse quality than others. The aim of this section is to provide the reader with a general feeling of which method should be considered first, when trying to grow a new material in single crystalline form.

9.5.1.1 Vapor Transport

The vapor transport method is the easiest way to obtain a sizable single crystal [51, 52]. In order to perform a vapor transport reaction, the solid reactants are sealed with a gaseous transport agent into a reaction vessel, which is typically a quartz tube. When heated to a certain temperature, the transport reagent will react with the solid starting materials to form a volatile intermediate product. By applying a temperature gradient, crystals can form at the colder side of the tube where the intermediate product decomposes. The temperature gradient and transport agent have to be chosen accordingly to the desired product. The most common transport agents are elemental halides since many transition metal halides (that would be formed in situ) are labile enough to decompose at the cold end. Therefore, they can be transported to the cold end of the tube and then decomposed to form the final product. Out of the elemental halides, I_2, which is solid at room temperature and atmospheric pressure, but vaporizes at only slightly elevated temperatures, is the easiest to handle. I_2 transports transition metals such as Zr, Hf, Nb and Ta well, but sometimes, Br_2 can achieve better results. Br_2 is a liquid at ambient conditions and very corrosive and volatile, which makes it more challenging to handle. When both, I_2 and Br_2, fail as a transport agent, Cl_2 might be the way to go. Cl_2 is a hazardous gas though, and its use should be avoided if possible. It is possible to substitute Cl_2 with solid compounds that decompose to release Cl_2 gas at elevated temperatures, such as $SeCl_4$ or $TeCl_4$, to overcome this obstacle. Further, NH_4Cl (where HCl is the active transport agent) can be used for growing SnS_2 or oxysulfides, for example. The vapor transport technique can be improved by adding a small seed crystal on the cold side of the tube.

That way, extremely pure and large crystals with defined orientations can be grown, since the reactants directly crystallize on the seed crystal [53].

If there are competing phases that can appear instead of the desired product, the purity of the crystal can be compromised. Since vapor transport requires a temperature gradient, the temperature window in the reaction is often too wide to just favor one phase. An example is Cd_3As_2, where four different phases with this composition exist and only the high-temperature phase is a 3D Dirac semimetal. This is the reason why only flux-grown crystals, where the temperature can be controlled more accurately, show ultrahigh mobility [10, 54].

9.5.1.2 Flux Growth

The flux growth method might be the preferred method for Cd_3As_2, but it comes with its own challenges and disadvantages. For this method a molten metal or salt (the flux) is used as a solvent to grow crystals from [55]. The flux can be one of the (low-melting) elements of the target compound (self-flux) or it can be a foreign element. For example, Cd_3As_2 can be grown from a Cd or a Bi flux [10, 56]. In a flux growth synthesis, the reactants are placed in a sealed reaction vessel that contains a filter and is placed upright in the furnace. The vessel is heated to a temperature where the flux is molten and all reactants are dissolved in the flux. The mixture is then cooled very slowly so that the desired phase can crystallize from the flux. Finally, at a temperature where the flux is still liquid (which is usually several hundred degrees C), the tube will be removed from the furnace and placed upside down in a centrifuge to separate the flux from the crystals [57, 58]. The crystals will be in the filter. While this procedure ensures a single-phase product, the centrifugation at high temperature can introduce defects itself. Nevertheless, high-quality crystals of Cd_3As_2, WTe_2, $ZrTe_5$, Na_3Bi, and Fe-based superconductors are grown by this method. If a salt flux is used, the centrifugation step might not be necessary if the product is stable in water, which often dissolves the flux. Salt fluxes are more commonly used to grow oxide materials, however.

9.5.1.3 Bridgman–Stockbarger Method

The Bridgman method can be used to grow extremely large, high-quality single crystals [59], but it only works for phases that melt congruently; i.e., the phase shows a straight line down to low temperatures at the phase diagram. In order to grow crystals via the Bridgman method, stoichiometric amounts of the elements have to be filled in evacuated reaction vessels and then are heated until melting. The sample is subsequently cooled very slowly while the hot temperature zone is slowly moved along the reaction vessel. Most famously, Bi_2Se_3 and other tetradymites are grown this way [60, 61].

9.5.1.4 Floating Zone Method

Lastly, it is possible to grow crystals with the floating zone method [62], which requires expensive equipment which might not be available for everyone. For this method, polycrystalline feed rods are melted locally, often with the use of expensive mirror furnaces that focus light on a single spot. The hot zone is then moved slowly, which can yield extremely large and pure single crystals. Since the ratio of impurities in the liquid phase is higher compared to the solid phase, defects will diffuse into the liquid phase at the melting boundary, leaving behind a purified crystal. Laser floating zone furnaces, where a laser is used as a heat source, can have a very small melting zone. This allows for the growth of single crystals of very high melting materials in a controlled way. Most commonly, oxides and single crystals relevant for optical applications are grown that way, but this method can also give access to high-quality single crystals of many other materials [63, 64].

9.5.2 Measuring the Electronic Structure of Materials—ARPES

After we successfully grew the desired material in single crystalline form, we need to verify the predicted electronic structure experimentally. Angle-resolved photoemission spectroscopy (ARPES) offers an easy way to directly measure the band structure of a crystalline solid. However, a clean and flat surface that has a size in the order of 100 μm is required to get a sufficiently strong signal, and thus fairly large single crystals are needed for this experiment. Newly developed methods such as nano-ARPES can measure smaller crystals, but there are currently only a few beam lines available for this method. In order to study a clean surface, crystals are usually either cleaved in ultrahigh vacuum or heated in situ to remove adsorbents.

For a more thorough insight into the principles of ARPES, we would refer the interested reader to Stefan Hüfner's book "Photoelectron Spectroscopy" [65], since here we will just give a very basic introduction. ARPES is based on the photoelectric effect, stating that photons can release electrons from a material, if their photon energy $\hbar\omega$ is higher than the binding energy E_B of the electrons plus the work function Φ (the energy needed to remove the electrons from the solid and eject them into the vacuum) [66, 67]. In order to fulfill conservation of energy, the excess energy must be transformed into kinetic energy E_{kin} of the released electrons:

$$\begin{aligned} E_{kin} &= \hbar\omega - \Phi - E_B \quad , \text{or} \\ E_i &= E_{kin} - \hbar\omega + \Phi \\ &= E_{kin} - E_F, \end{aligned} \quad (9.3)$$

where E_i is the initial state energy (which is the negative binding energy) and E_F is the Fermi energy.

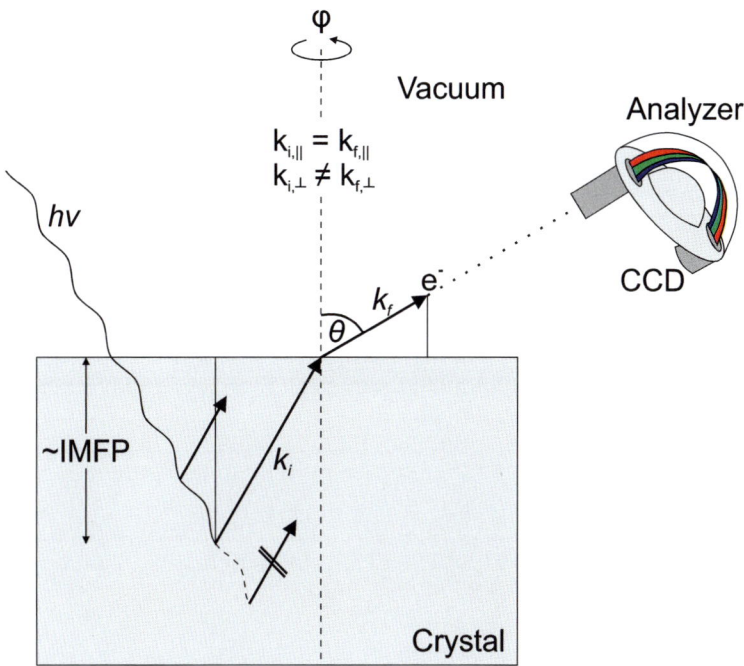

Fig. 9.5 Schematic illustration of the photoemission process and the ARPES setup. k and real space are shown in the same picture

Figure 9.5 shows the setup of a typical ARPES experiment. Note that k and real space are shown in the same picture here. The monochromatic photons can penetrate deeply into the bulk, but electrons can only be released from the first few monolayers of a crystal, since the electrons are inelastically scattered on their way to the surface. The inelastic mean free path (IMFP) gives a good estimate for the distance an electron can travel without being scattered inelastically in the crystal. After reaching the surfaces, electrons will be ejected to the vacuum and be available for band imaging. Electrons that are scattered inelastically below the surface, but still reach it with sufficient energy to overcome the work function, are responsible for the Shirley background discussed later. When the electrons reach the surface of the crystal they are refracted, and only the parallel part of the wave vector is preserved. Since their momentum needs to be conserved, the following equation holds (the moment of the photon can be ignored since it is tiny compared to one of the electrons):

$$k_{i,\|} = k_{f,\|} = \frac{\sqrt{2m}}{\hbar} \sqrt{E_{\text{kin}}} \sin \theta. \tag{9.4}$$

Here, θ is the angle between normal emission ($k_\| = 0$) and the analyzer. To scan the full angle range of θ, the sample is usually rotated, but there are also setups that rotate the whole analyzer. Rotating the analyzer has the advantage that the beam spot

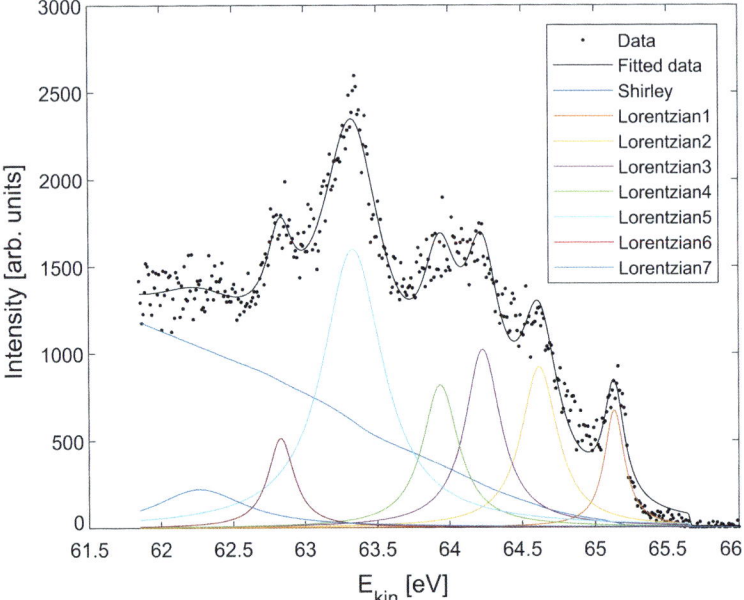

Fig. 9.6 Individual ARPES EDC for $\hbar\omega = 70\,\text{eV}$. The black fit curve consists of seven Lorentzian curves, multiplied by the Fermi–Dirac function, and the Shirley background

is kept constant. Rotation of the sample on the contrary is much easier to realize, but requires adjustment of the x and y positions, if the sample is not perfectly centered, since in this case, the beam spot will move depending on θ.

After being ejected from the crystal, the electrons travel through the vacuum to the analyzer where their kinetic energy is measured. From (9.3) to (9.4), it is apparent that the dispersion $E_i(k_\parallel)$ can be obtained by measuring E_{kin} while varying θ. For a constant θ angle, the analyzer provides an energy distribution curve (EDC) for a limited energy region. If a wider energy region is required, a number of these "fixed mode" spectra are accumulated, while the energy window is slowly changed ("swept mode").

Such an EDC is shown in Fig. 9.6 for a photon energy of 70 eV. Above the Fermi level, at 65.6 eV, the intensity reduces to almost zero, since there the bands are empty and no electrons can be emitted. If intensity above E_F is visible, it can be caused by higher-order resonances from the monochromator, which release electrons from photons with a multitude of the photon energy. In general, an EDC is fitted by a Fermi–Dirac distribution at the Fermi level, which is located at $\hbar\omega - \Phi$ ($\Phi \approx 4.5\,\text{eV}$, depending on the analyzer). The measured peaks have a Lorentzian shape and are superimposed by the Shirley background. In first approximation, this background can be corrected with:

$$I_{\text{Shirley}}(E) = c \int_E^\infty dE'\, I(E'), \tag{9.5}$$

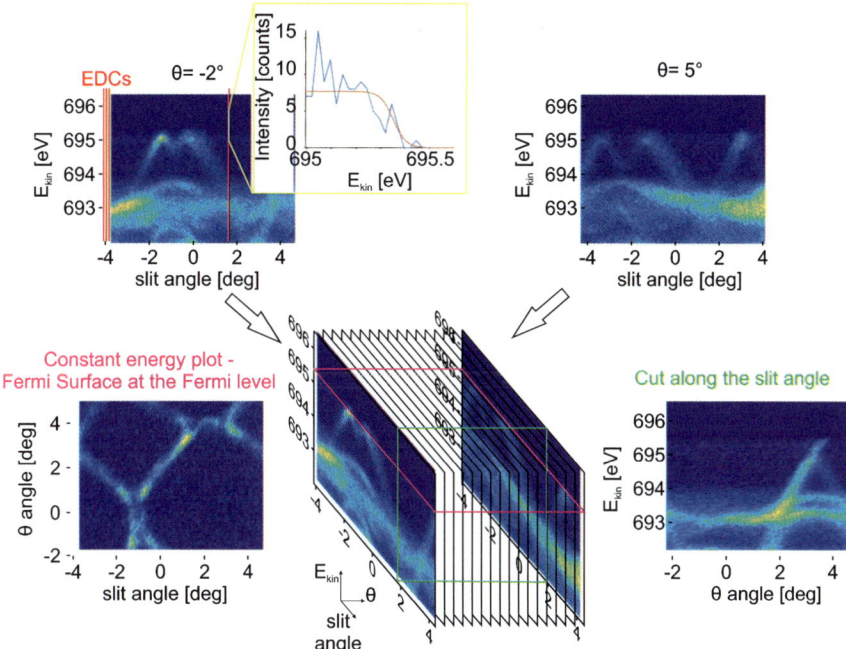

Fig. 9.7 Process of acquiring APRES data on the example of ZrSiS at $\hbar\omega = 700$ eV. The constant θ slices are stacked up to form a three-dimensional data cube that contains all information about the surface BZ

which models the contribution of inelastically scattered electrons by using a specific form of the inelastic energy-loss cross section [68].

Modern analyzers contain a CCD chip, so that the full slit angle α can be measured simultaneously. Therefore, for each θ, a two-dimensional picture of the band structure, consisting of individual EDCs, can be measured directly (EDC curves are indicated as red lines in Fig. 9.7). Changing θ and adding up the two-dimensional spectra result in a data cube as shown in Fig. 9.7. By cutting this cube along a constant α angle or constant energy, different directions of the band structure can be plotted. For example, cutting the cube along the energy direction at E_F results in the Fermi surface. Note that a constant angle does not necessarily correspond to a constant $k_{\parallel}$, if measured away from that high-symmetry line containing Γ. It is therefore important to convert the data cube into k space to show the actual dispersion $E(k)$.

Two of the conversion equations were already derived earlier ((9.3) and (9.4)), and the last one follows from simple geometric considerations:

$$E_i = E_{kin} - E_F$$
$$k_x = \underbrace{\frac{\sqrt{2m}}{\hbar}}_{\approx 0.512} \sqrt{E_{kin}} \sin\alpha \cos\theta$$
$$k_y = \frac{\sqrt{2m}}{\hbar} \sqrt{E_{kin}} \sin\theta \quad (9.6)$$

It is important to consider that crystals can usually never be glued perfectly flat to the stage, which causes an offset in α and θ that needs to be corrected before (9.6) can be applied. The Γ point of the first BZ should be located at $(\alpha, \theta) = (0, 0)$ and can thus be easily identified, since it is the only high-symmetry point that remains at a constant (α, θ) position in all constant energy slices.

Furthermore, since sine and cosine operations are part of the k conversion in (9.6), it is usually not possible to directly obtain $k_x = \text{const}$ or $k_x = \text{const}$ lines from the data, but rather an interpolation between the measured points is necessary.

Lastly, the φ angle in Fig. 9.5 can be used to align the α and θ angles along high-symmetry lines before a data cube is recorded. Since this alignment is rarely perfect, a rotation of the converted data cube can align the k_x and k_y-direction along high-symmetry lines:

$$\begin{pmatrix} k_x \\ k_y \end{pmatrix} = \begin{pmatrix} \cos\varphi & -\sin\varphi \\ \sin\varphi & \cos\varphi \end{pmatrix} \begin{pmatrix} k_x \\ k_y \end{pmatrix} \quad (9.7)$$

After discussing the parallel components of the k vector, let us now consider the perpendicular direction $k_\perp = k_z$, which is not preserved at the surface. k_z is usually obtained by tuning the photon energy, while remaining at a constant (k_x, k_y) position. The Γ point is predestined for such a measurement, since it also remains constant in angle space when changing the photon energy. For a free-electron-like final state approximation, $k_{i,\perp} = k_z$ can be obtained by

$$k_z = \frac{\sqrt{2m}}{\hbar} \sqrt{E_{kin} \cos^2\theta + V_0}, \quad (9.8)$$

where V_0 is determined by fitting to a periodic behavior of the initial state energy of the bands $E_i(\hbar\omega)$.

9.5.3 Example—The Nonsymmorphic Square-Net Compound ZrSiS

After having explained how material candidates can be identified and experimentally verified, we want to proceed with giving some examples of how the cycle (Sect. 9.5) works in real life. The Dirac line node material ZrSiS [18] will be used as an example here. Since ZrSiS crystallizes in a nonsymmorphic space group ($P4/nmm$, no. 129),

9 Topological Materials and Solid-State Chemistry …

we will additionally discuss how the theoretical idea of band degeneracies enforced by nonsymmorphic symmetry (see Sect. 9.4) can be observed in experiment.

The Idea—Where to Look?

When trying to identify a new topological material, the big question is where to start. If we don't want to screen the complete ICSD, we need to have some starting point, some idea of what kind of crystal structure or space group might host the material we desire. Many materials that have been identified to be of interest for their topologically nontrivial behavior, have been known to exist for a long time. The compound ZrSiS was, for example, presented in a paper entitled "Square Nets of Main Group Elements in Solid-State Materials" from 1986 by Tremel and Hoffmann [69] (although it was known to exist even before [70]). Its crystal structure contains a square-net arrangement of Si atoms. The paper by Tremel and Hoffmann focuses on the commonly observed Peierl's distortions in square-net compounds. The instability of the square-net was justified by a nonzero DOS at E_F that is found in the electronic structure, if the square-net is intact, while a distortion causes a band gap to open. The nonzero DOS is caused by linearly dispersing bands that cross at E_F. While these crossings were not connected to Dirac physics in the original paper, they are visible in the calculated electronic structure plots shown in the manuscript. We can infer that a square-net arrangement of atoms might frequently result in an electronic structure featuring Dirac crossings. We thus identified a structural motif that can result in Dirac crossings and, as it turns out, has a high potential even outside of the ZrSiS-type family of compounds [71].

Design a Suitable Material

As mentioned above, the crystal structure of ZrSiS shows exactly this type of structural motif, the square-net arrangement of atoms. According to the published crystal

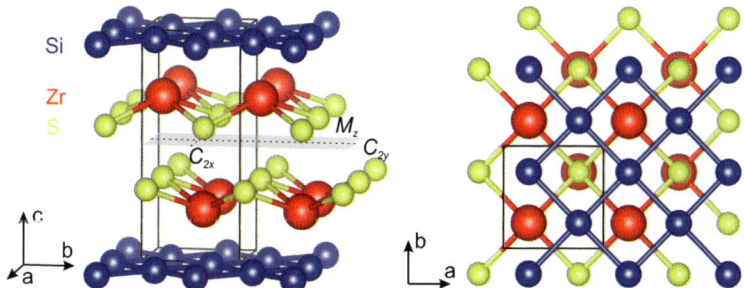

Fig. 9.8 Crystal structure of ZrSiS. The structure consists of layers of square-nets. In the Si square-net, the atoms are located close enough to be bonded to each other. The material cleaves between the two S layers along the z-direction, which is why the material cleaves between these layers. Since the structure is nonsymmorphic, the glide plane $\{M_z|\frac{1}{2}\frac{1}{2}0\}$ as well as the screw axes $\{C_{2x}|\frac{1}{2}00\}$ and $\{C_{2y}|0\frac{1}{2}0\}$ double the amount of atoms in the unit cell along x and y. The glide plane and the screw axes are indicated in the left panel. The right panel shows a top view of the structure where the glide plane can be seen easily

structure the square net is not distroted in ZrSiS. There are actually quite a few materials known, where the square-net does not distort, and thus, there might be many more candidates of interest. For simplicity, we will focus on ZrSiS for now. The crystal structure of ZrSiS is presented in Fig. 9.8. Each of the three elements is arranged in a square-net fashion and in the case of the Si square-net; the distance between the atoms is short enough to form chemical bonds. The short bond distance is important for the electronic structure, since it will result in highly disperse bands, which increase the likelihood of a band inversion.

Another convenient feature of ZrSiS is its nontoxic nature as well as the low cost of the elements it is composed of. It is very stable in air and water and is, therefore, the perfect candidate for an application-focused research. Furthermore, its natural cleavage plane between the S layers provides easy access to electronic structure investigations with ARPES and other surface-sensitive methods.

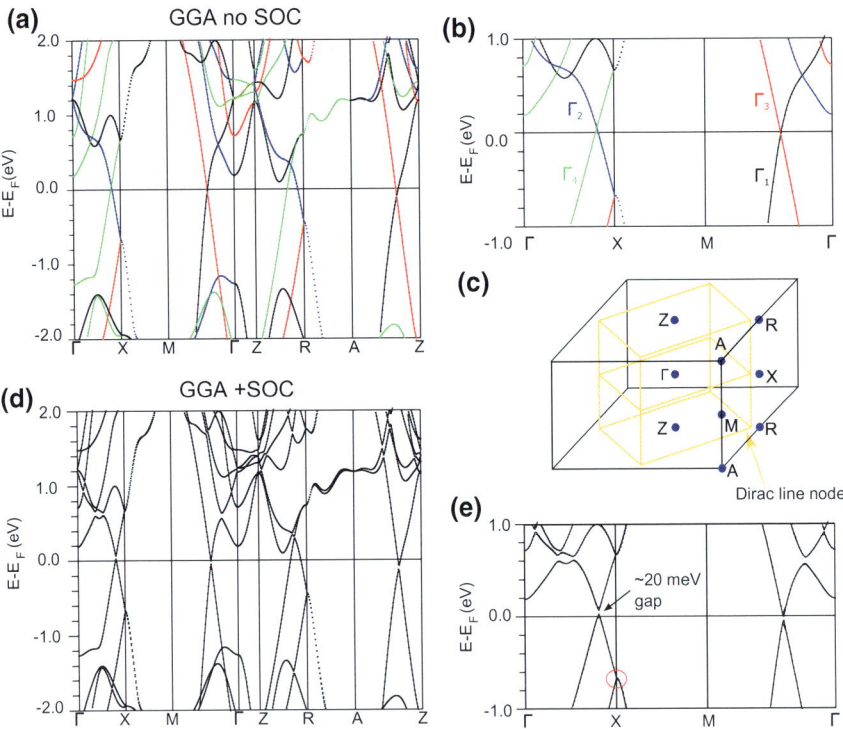

Fig. 9.9 Calculated band structure of ZrSiS. **a** and **b** present the electronic structure without SOC. The irreducible representations of the bands are shown in different colors. The path of the resulting Dirac line node is shown in the 3D BZ in (**c**). **d** and **e** show the band structure along the same high-symmetry lines, calculated with SOC. The Dirac line node is slightly gapped, but the degeneracies at the high-symmetry points X and M (and R and A, respectively) remain protected by the nonsymmorphic symmetry. Figure reprinted from [18]

Calculate the Electronic Structure

But before growing the material, we should confirm the initial idea that the square-net motif in ZrSiS results in an electronic structure with Dirac crossings. Figure 9.9 shows the calculated electronic structure of ZrSiS with and without SOC interaction included. We can see several Dirac crossings, and the bands are linearly dispersed over a wide energy range. Since there are multiple Dirac crossings visible in panel (a), a more thorough analysis shows that the crossings extend along a line in the BZ, shown in yellow in panel (c). The line node adopts a diamond-shaped Fermi surface in the $k_z = 0$ and $k_z = \pi$ planes. The diamonds are connected along k_z to form a cage-like structure of a line node. Therefore, ZrSiS is a line node material and not a 3D Dirac semimetal such as Cd_3As_2. If SOC is considered in the calculations, a small gap is induced as shown in panel (e). In ZrSiS, all elements are relatively light, which causes the gap to be small (recently, the gap was measured optically to be smaller than 30 meV [72]). Besides the nodal line, the electronic structure of ZrSiS also shows band degeneracies enforced by nonsymmorphic symmetry; they can be found at the high-symmetry points X, M, R, and A. These points are located below and above the Fermi level, and they will not directly contribute to transport measurements. In the calculated band structure plots, it can be seen that SOC does not affect these nonsymmorphic crossings, as expected. Without (or with very weak SOC), the nonsymmorphic degeneracies are extended along the complete XM (RA, respectively) high-symmetry line. Thus, the nonsymmorphic crossings are not shaped like a cone, but very anisotropically.

Synthesizing the Desired Material

After confirming with DFT that a square-net structure can result in a Dirac cone, we now have to grow a single crystal of the material to confirm the prediction experimentally. ZrSiS is a chalcogenide compound, and chalcogenides can often be grown with vapor transport, which is why this method should be tried first. Additionally, Zr is known to react to volatile ZrI_4 in the presence of I_2 [73], which is a further hint for a successful vapor transport reaction. Indeed, ZrSiS can be grown by the vapor transport method using I_2 as a transport agent. A temperature gradient of 200 K (from 1100 °C to 900 °C) yields crystals of several mm size (both shown in Fig. 9.10a).

Structural Characterization

It is of crucial importance to confirm the published crystal structure before proceeding with an experiment. The crystal structure of ZrSiS was solved in the 1960s [70], where diffraction data and fitting software was of much lower quality. If only a slight distortion of the square-net was missed, our predicted electronic structure will be incorrect. In addition, we should confirm that the crystals are of high quality, so that we can expect them to show the typical behavior of topological semimetals in transport experiments. The crystal structure can be confirmed with diffraction experiments. Here, we show precession electron diffraction (PED) patterns (taken from [18]) for different lattice planes (Fig. 9.10b). Simulations, based on the previously published ZrSiS crystal structure of Fig. 9.8, are shown in white next to the measurements—a very good agreement between measured and simulated data can be

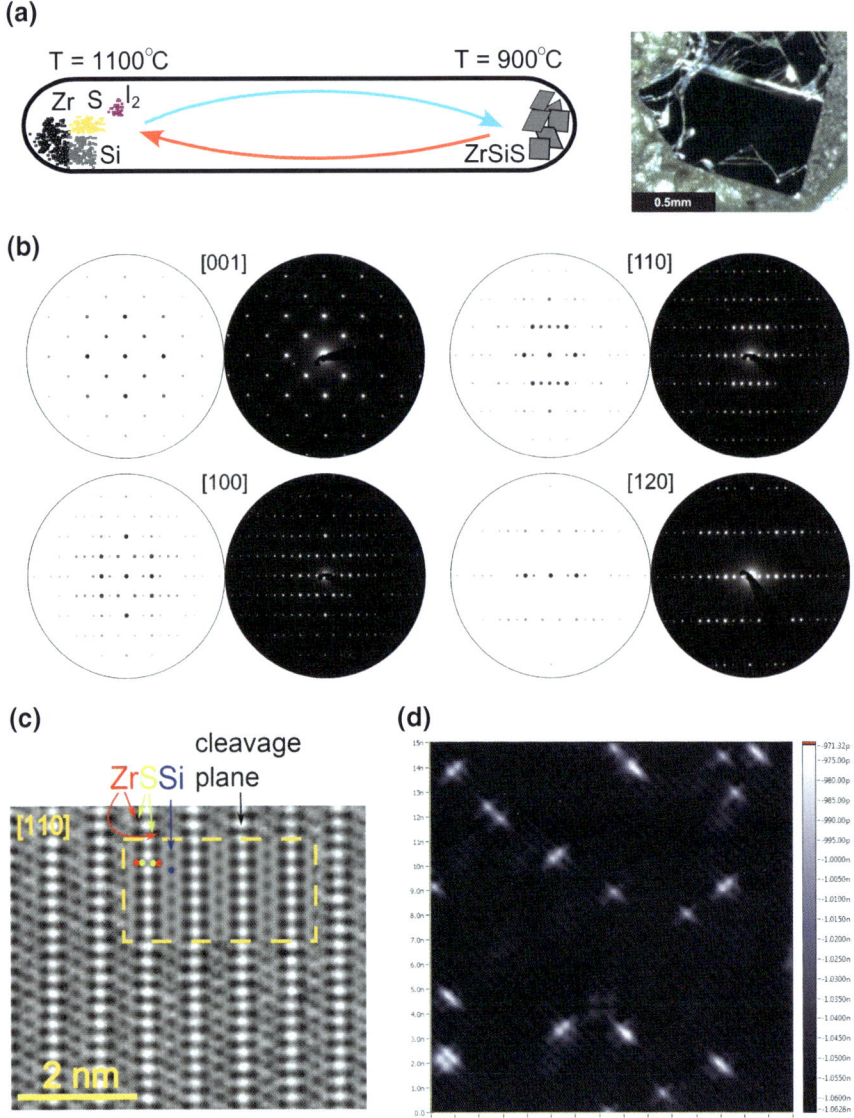

Fig. 9.10 Crystal growth and characterization of ZrSiS single crystals. **a** ZrSiS is an ideal example for a vapor transport crystal growth. The crystal dimensions reach centimeters in width and length, while the thickness is rarely more than one or two millimeters. **b** PED patterns along different zone axes show the very good agreement between simulated (white) and experimental data (black). **c** HRTEM of the (110) plane measured on a powdered sample. The simulation as well as atomic positions are superimposed. **d** STM picture of the ZrSiS surface. Defects appear in white. Panel **b** and **c** are reprinted from [18]

observed. Further diffraction experiments (single crystal and high-resolution powder diffraction) also confirmed the published structure [18], and we can thus be confident that the square-net structure is correct.

There are further methods that can be used to examine the crystal quality of the samples. For example, high-resolution transmission electron microscopy (HRTEM) allows to resolve individual atoms in a ZrSiS crystal. Images of ZrSiS, taken with a HRTEM, are shown in Fig. 9.10c [18]. A simulated image is superimposed in the dashed area, and it is visible that both the individual atoms and the cleavage plane can be clearly resolved. Another method that can be used is scanning tunneling microscopy (STM), which allows to estimate the amount of surface defects in a crystal. Such an image is shown in Fig. 9.10d; the defects appear much brighter than the atoms. From this image, one can estimate that there is about one defect for every 50 unit cells, which is a relatively low concentration. Note that this image was taken on a crystal with a relatively low quality and that the synthesis conditions were improved thereafter.

Electronic Structure Verification via ARPES

ZrSiS is perfectly suited for cleaving due to its layered structure is perfectly suited for cleaving. Therefore, ARPES will be suitable for the experimental determination of the electronic structure.

Figure 9.11a shows the measured band dispersion along $\overline{\Gamma}\,\overline{X}$, using a photon energy of 21.2 eV (He I). The data shown here is taken from [18, 74]. Several bands cross the Fermi level and meet slightly above E_F. The dashed lines in panel (a) serve as a guide to the eye. Comparing these bands to Fig. 9.9b, one can conclude that they correspond to the bands forming the Dirac line node. However, there are further bands appearing in the experiment that were not predicted by DFT. In Fig. 9.11a, one such a band crosses the bulk bands just short of the $\overline{X}$ point. Since ARPES is a surface-sensitive technique, this band is most likely surface-derived, which is why it won't appear in the calculated bulk band structure. For now, it is sufficient to label the additional states as surface states. We will discuss their origin in more detail below. Figure 9.11b shows a cut parallel to $\overline{\Gamma}\,\overline{X}$, where the surface bands follow the dispersion of the bulk bands, while they appear to cross them along the $\overline{\Gamma}\,\overline{X}$ line. Panel (c) shows a cut perpendicular to $\overline{\Gamma}\,\overline{X}$ along $\overline{M}\,\overline{X}\,\overline{M}$. The upper part of the band structure, starting at $E_i = -0.4\,\mathrm{eV}$, is again, not reproduced in the calculations of Fig. 9.9b, indicating that these are surface-derived states. Panel (d) shows a parallel cut of $\overline{M}\,\overline{X}\,\overline{M}$ where the bulk and surface states touch. These panels were gathered along the original measurement direction, which explains their higher resolution compared to the panels (a) and (b).

A constant energy plot at E_F (Fermi surface) is shown in panel (e) for $\hbar\omega = 21.2\,\mathrm{eV}$ and in panel (f) for $\hbar\omega = 700\,\mathrm{eV}$. Low photon energies (panel (e)) allow for a much higher resolution, but limit the excerpt of the BZ, usually to the surrounding of a single high-symmetry point. Higher photon energies (panel (f)) map a much bigger part of the BZ, but limit the resolution. The Dirac line node forms a diamond-shaped Fermi surface around the $\overline{\Gamma}$ point and does not touch any high-symmetry points. Since the Dirac line node is slightly above the Fermi level in the crystal measured

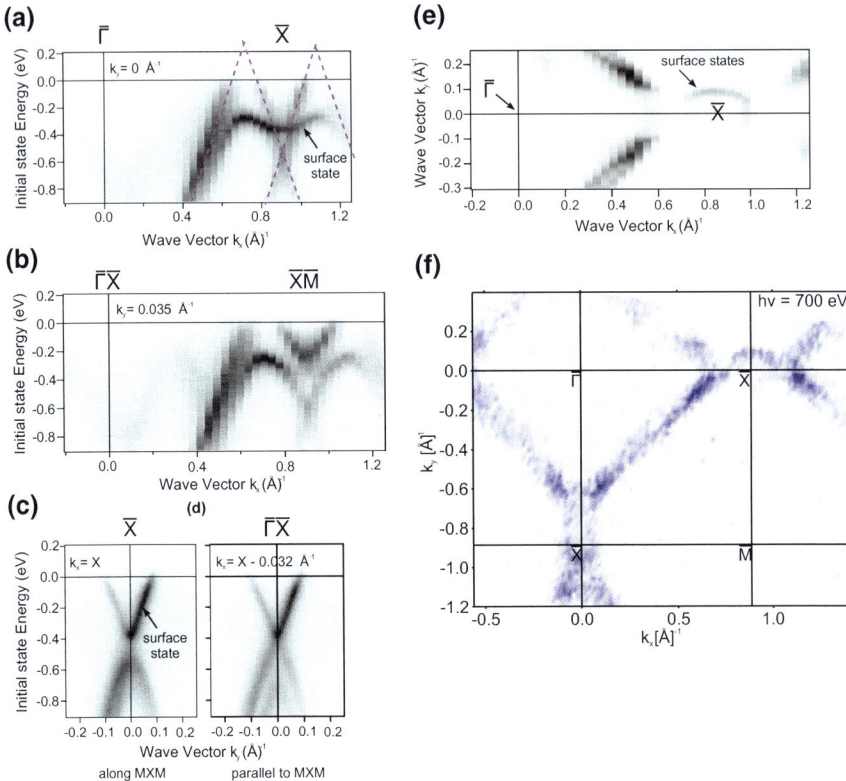

Fig. 9.11 Measured ARPES dispersion of ZrSiS for $\hbar\omega = 21.2$ eV along **a** $\overline{\Gamma}\overline{X}$, **b** parallel to $\overline{\Gamma}\overline{X}$, **c** along $\overline{M}\overline{X}\overline{M}$, **d** parallel to $\overline{M}\overline{X}\overline{M}$. The expected nodal line features predicted by bulk DFT calculations are marked by the dashed lines. Additionally, a surface state appears in the band structure. **e** Fermi surface measured for a photon energy of $\hbar\omega = 21.2$ eV. The low photon energy leads to a high-resolution around the X point. The surface states are represented by the ring-like structure around $\overline{X}$. **f** Fermi surface for $\hbar\omega = 700$ eV. The higher photon energy allows for the measurement of a larger part of the BZ. The diamond-like structure around $\overline{\Gamma}$ shows the nodal line. Panels **a**–**e** are reprinted from [18]; panel **f** is reprinted from [74]

here, the diamond consists of two branches. The ring-like structure around the $\overline{X}$ points is not expected to appear according to the bulk band structure calculations and can be attributed to the surface states.

Besides the Dirac line node, the band degeneracy enforced by nonsymmorphic symmetry is also visible in the data shown in panel (a). The data clearly resolve two bands crossing in an X-shape at the $\overline{X}$ point.

In order to verify that the measured electronic structure is in agreement with the calculated one, it is necessary to model the surface in the calculation. This is possible by modeling a slab containing a limited number of unit cells in the *c*-direction, while retaining the continuous crystal in the *ab*-plane (Fig. 9.12a). If the

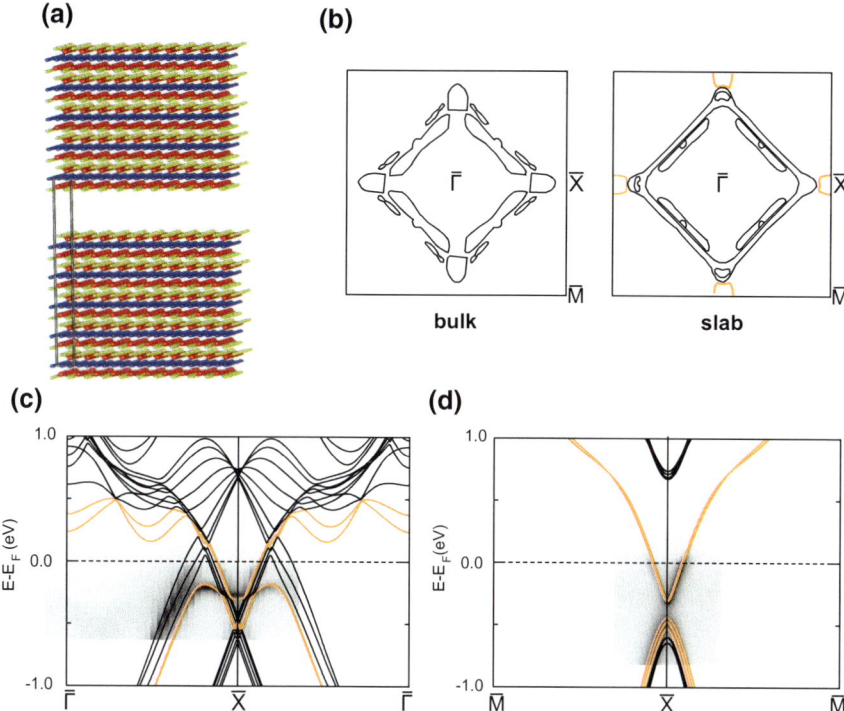

Fig. 9.12 Surface state calculations of ZrSiS. **a** Periodic crystal structure used to simulate the semi-infinite ZrSiS surface. While the *a*- and *b*-direction are continuous, the *c*-direction is interrupted by a vacuum layer, thus simulating a five-unit-cell-thick crystal with vacuum on both sides. **b** Bulk and slab calculation of the Fermi surface. The ring-like structure around the X points (shown in orange) is surface-derived. **c–d** ZrSiS slab calculations superimposed on the measured ARPES data (**c**) along $\overline{\Gamma}\,\overline{X}\,\overline{\Gamma}$, **d** along $\overline{M}\,\overline{X}\,\overline{M}$. Surface-derived bands are shown in orange and were identified by comparing them with the bulk band calculations from Fig. 9.9. Figures **b–d** reprinted from [18]

slab is thick enough, it simulates the surface–vacuum interface, next to the bulk. Such slab calculations (with a slab thickness of five unit cells) are shown in Fig. 9.12. In general, the agreement between the predicted (surface) electronic structure and the experimentally measured spectra is very high [18].

The remaining question is why such prominent surface states appear in ZrSiS and what their origin is. One might be tempted to conclude that these states are of topological origin. However, ZrSiS has a $\mathbb{Z}_2$ invariant of 0(001) (analogous to ZrSiO [75]), which means it is a weak topological insulator with the (001) surface being a dark surface. The surface states can therefore not be classified as topological surface states. A surface in general reduces the symmetry of any bulk system. For example, in the presence of large SOC, Rashba-type surface states appear on the surface of inversion symmetric crystals, since inversion symmetry is not preserved at the surface. A similar behavior is observable at the surface of ZrSiS, but here it is the nonsymmorphic symmetry that is not conserved. In an infinite crystal, all crossings at X are forced to be fourfold degenerate within the space group P4/nmm. At the (001) surface, all involved nonsymmorphic symmetries are broken and the surface wallpaper group is reduced to *P4mm* (no. 99). The bands at the surface are therefore not enforced to degenerate at X, which causes them to "float" freely in the surface layer [74].

Since it was the task of the ARPES experiment to confirm the calculated electronic structure, an important lesson can be learned from this paragraph. If only some of the experimental bands are in accordance with the DFT bulk calculations, but these fit fairly well, it is possible that surface-derived bands could interfere with pure bulk bands. This is of course only important for surface-sensitive measurements such as ARPES. In such a case, it is advisable to compare the data to a slab calculation and consider all the known derivations of surface states as a potential origin.

9.5.4 Beyond ZrSiS

As mentioned previously, the identification of a new topological material often opens the door to study a whole family of new compounds that all have a common structural motif, here the square-net. Thus, we can now explore more compounds containing square-nets and try to use chemical concepts to tweak the band structure in the desired way. For example, the nonsymmorphically protected, SOC-resistant, fourfold degeneracies at X and M (R and A, respectively) would be very interesting, if they weren't so anisotropic and located closer to the Fermi level. We can thus ask the question whether we can tune the electronic structure in such a way that we move the nonsymmorphic degeneracy close to the Fermi level and/or lift the degeneracy along the XM line so that the fourfold crossing only persists at high-symmetry points forming a 2D Dirac cone.

9.5.4.1 Influencing the Electronic Structure with Chemical Strain

As explained earlier, it is difficult to pin the energy of the nonsymmorphic crossings to the Fermi level, because this would require a half-filled band. It is, however, possible to tune the position of the Fermi level by strain. "Chemical" strain can be easily applied by replacing atoms in the crystal structure with atoms of a different size (while staying within the limits of the Pauling rules). In the ZrSiS-type family of compounds, chemical strain can be used to increase the *c*-axis almost independently of the *ab*-plane [78]. This is related to the square-net structure. As long as Si is forming the square-net, the Si-Si bond distance determines the size of the square-net and thus the length of the *a*- and *b*-axis. Replacing, for example, S with a larger element such as Te, causes only the *c*-axis to increase in length. This results in uniaxial strain since the relation between the *a*- and *c*-direction is varied. In order to understand if such an uniaxial strain affects the electronic structure of the compounds in question,

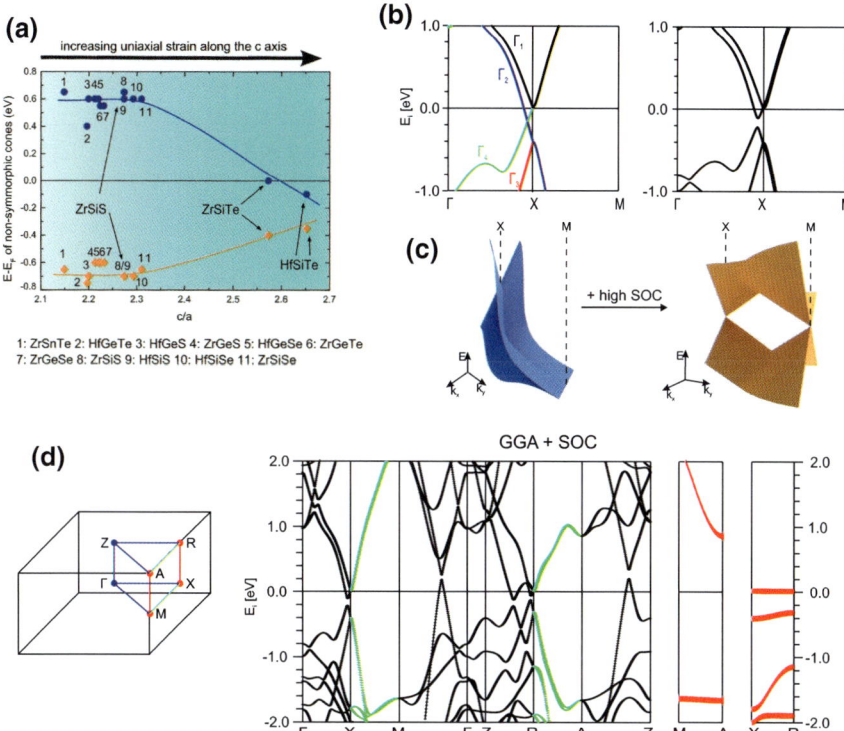

Fig. 9.13 a Energy positions of the nonsymmorphic points at X depending on the lattice parameter ratio c/a of different ZrSiS-like compounds. For ZrSiTe, the upper crossing lies at the Fermi level. **b** Magnification of the band structure of ZrSiTe around the Fermi level at the X point. The bands along ΓX have the same irreducible representation with SOC and gap, while the X point stays fourfold degenerate. **c** Schematic picture of the band dispersion along XM. Only with sufficiently high SOC, a real 2D Dirac cone can be achieved. **d** BZ of ZrSiTe next to the bulk band calculations along all high-symmetry lines including SOC. The high-symmetry lines highlighted in red are enforced to be fourfold degenerate by nonsymmorphic symmetry, while the bands along the green high-symmetry line can lift their fourfold degeneracy depending on the strength of SOC. Panels **a**, **b** and **d** are reprinted from [76]; panel **c** is reprinted from [77], Copyright 2018, with permission from Elsevier

Topp et al. plotted the position of the nonsymmorphically protected crossings at the X point against the c/a ratio [76]. The results are shown in Fig. 9.13a. As discussed in the last section, ZrSiS (number 8 in the figure) shows the nonsymmorphically protected degeneracies below and above the Fermi level. In the case of ZrSiTe, the upper degeneracy is located very close to the Fermi level. It is important to note here that the Fermi level is not pinned to the nonsymmorphic point, but is located there coincidentally due to the c/a ratio. Figure 9.13b shows the calculated band structure of ZrSiTe along Γ X M with and without SOC. The nonsymmorphic crossings are resilient against a gapping induced by SOC. Since the semimetallic nature of ZrSiTe is not filling enforced, other bands have to cross the Fermi level,

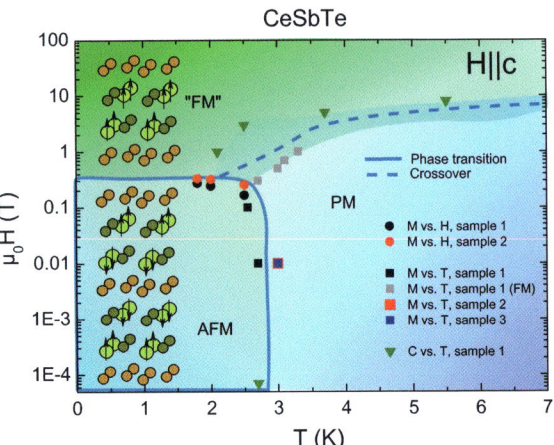

Fig. 9.14 Magnetic phase diagram of CeSbTe. Insets show the magnetic structures. Each magnetic phase changes the symmetry conditions, influencing the band structure. Figure taken from [80]

and the nonsymmorphic degeneracy at the Fermi level is stabilized by electron and hole pockets (visible in Fig. 9.13d). ARPES measurements have verified the expected band structure and revealed additional surface states, similarly as occurred in ZrSiS [76]. Despite the fact that the additional states at the Fermi level will contribute to transport experiments, ZrSiTe is nevertheless so far the only experimentally verified example of a nonsymmorphic material with the protected degeneracy at the Fermi level.

Another noteworthy observation concerning the band structure of ZrSiTe is the effect of the, in comparison to ZrSiS, increased SOC. Since Te ($Z = 52$) is much heavier than S ($Z = 16$), the extended degeneracy along XM is slightly lifted. This behavior is visible in panel (b). A schematic view of the influence of SOC can be seen in Fig. 9.13c. In ZrSiTe, the effect of SOC is still too small to significantly reduce the anisotropy of the "cone" at the X point, but this suggests that increasing the atomic mass of the involved elements pushes the electronic structure in the desired direction. In order to have an isotropic 2D Dirac cone at X and M, all of the involved elements need to be replaced by heavier counterparts, while the material should remain isostructural and isoelectronic to ZrSiS. An example for such an extremely heavy analog is CeSbTe. It could be theoretically and experimentally shown that the nonsymmorphic cone is much more isotropic in this compound [77]. The same argument holds for two-dimensional monolayers of the ZrSiS family; for example, monolayers HfGeTe were predicted to feature isotropic 2D Dirac cones [79].

9.5.4.2 Introducing Magnetism to Square-Net Materials

The material CeSbTe has a much more exciting property than enhanced SOC; the presence of the magnetic lanthanide Ce suggests the possibility to implement magnetism into the electronic structure of ZrSiS-type materials. Magnetic order can break time-reversal symmetry (if ferromagnetic) or influence the symmetry in other ways (if antiferromagnetic). This can change the electronic structure since it affects symmetry-protected crossings. Lanthanides that contain a partially filled f shell often cause cooperative magnetism in compounds. In CeSbTe, the unpaired $4f$ electron of the Ce^{3+} ions causes the material to be ordered antiferromagnetically below $T_N = 2.7$ K. If a small field of about 0.25 T is applied, the compound undergoes a metamagnetic transition into a fully polarized state (Fig. 9.14). The rich magnetic phase diagram allows to access several different magnetic phases in the same compound resulting in several topological states including Weyl crossings and new fermions [80]. Thus, magnetism can be a tool to switch between different topological properties.

CeSbTe is a good example for the importance of verifying the crystal structure of candidate materials. While this material was originally reported to crystallize in an orthorhombic structure [81] and thus disporting the square-net, it was recently been shown to be isostructural to ZrSiS [80].

9.6 Conclusion

In this chapter, we introduced chemical concepts and how they can help in finding new topological materials. We discussed common synthesis methods for topological materials and explained the basic concepts of ARPES, which is the most common method to experimentally verify topological materials. We then explained the full process of identifying, synthesizing, and verifying a new material on the example of ZrSiS and gave an outlook on the interesting physics arising in this family of compounds.

Acknowledgements The authors thank Andreas W. Rost for providing STM data on ZrSiS and Christian R. Ast for the helpful discussions on ARPES-related matters.

References

1. Y. Ando, J. Phys. Soc. Jpn. **82**(10), 102001 (2013)
2. Y. Sun, Y. Zhang, C. Felser, B. Yan, Phys. Rev. Lett. **117**(14), 146403 (2016)
3. C. Beenakker, L. Kouwenhoven, Nat. Phys. **12**(7), 618 (2016)
4. B. Bradlyn, J. Cano, Z. Wang, M. Vergniory, C. Felser, R. Cava, B. Bernevig, Science **353**(6299) (2016)
5. M. Zubkov, Mod. Phys. Lett. A **33**(07/n08), 1850047 (2018). https://www.worldscientific.com/doi/abs/10.1142/S0217732318500475
6. S.A. Yang, *Spin*, vol. 6 (World Scientific, 2016), p. 1640003
7. S. Borisenko, Q. Gibson, D. Evtushinsky, V. Zabolotnyy, B. Büchner, R.J. Cava, Phys. Rev. Lett. **113**(2), 027603 (2014)
8. Z. Liu, B. Zhou, Y. Zhang, Z. Wang, H. Weng, D. Prabhakaran, S. Mo, Z. Shen, Z. Fang, X. Dai, Z. Hussain, Y. Chen, Science **343**(6173), 864 (2014)

9. P.J. Moll, N.L. Nair, T. Helm, A.C. Potter, I. Kimchi, A. Vishwanath, J.G. Analytis, Nature **535**(7611), 266 (2016)
10. T. Liang, Q. Gibson, M. Ali, M. Liu, R. Cava, N. Ong, Nat. Mater. **14**(3), 280 (2015)
11. J. Xiong, S. Kushwaha, T. Liang, J. Krizan, M. Hirschberger, W. Wang, R. Cava, N. Ong, Science **350**, 413 (2015)
12. U. Müller, *Anorganische Strukturchemie* (Springer, Berlin, 2006)
13. T. Graf, C. Felser, S.S. Parkin, Progr. Solid State Chem. **39**(1), 1 (2011)
14. S. Chadov, X. Qi, J. Kübler, G.H. Fecher, C. Felser, S.C. Zhang, Nat. Mater. **9**(7), 541 (2010)
15. E.M. Seibel, L.M. Schoop, W. Xie, Q.D. Gibson, J.B. Webb, M.K. Fuccillo, J.W. Krizan, R.J. Cava, J. Am. Chem. Soc. **137**(3), 1282 (2015)
16. W. Pearson, Acta Crystallogr. **17**(1), 1 (1964)
17. S.M. Kauzlarich, *Chemistry, Structure, and Bonding of Zintl Phases and Ions*, vol. 6 (VCH New York, 1996)
18. L. Schoop, M. Ali, C. Straßer, A. Topp, A. Varykhalov, D. Marchenko, V. Duppel, S. Parkin, B. Lotsch, C. Ast, Nat. Commun. **7**, 11696 (2016)
19. L. Pauling, J. Am. Chem. Soc. **51**(4), 1010 (1929)
20. A. Belsky, M. Hellenbrandt, V.L. Karen, P. Luksch, Acta Crystallogr. Sect. B Struct. Sci. **58**(3), 364 (2002)
21. J. Paglione, R.L. Greene, Nat. Phys. **6**(9), 645 (2010)
22. I.I. Mazin, Nature **464**(7286), 183 (2010)
23. B.A. Bernevig, T.L. Hughes, S.C. Zhang, Science **314**(5806), 1757 (2006)
24. M. König, S. Wiedmann, C. Brüne, A. Roth, H. Buhmann, L.W. Molenkamp, X.L. Qi, S.C. Zhang, Science **318**(5851), 766 (2007)
25. L. Fu, C.L. Kane, Phys. Rev. B **76**(4), 045302 (2007)
26. D. Hsieh, D. Qian, L. Wray, Y. Xia, Y.S. Hor, R.J. Cava, M.Z. Hasan, Nature **452**(7190), 970 (2008)
27. Y. Xia, D. Qian, D. Hsieh, L. Wray, A. Pal, H. Lin, A. Bansil, D. Grauer, Y. Hor, R. Cava, M. Hasan, Nat. Phys. **5**(6), 398 (2009)
28. K.S. Novoselov, A.K. Geim, S.V. Morozov, D. Jiang, Y. Zhang, S.V. Dubonos, I.V. Grigorieva, A.A. Firsov, Science **306**(5696), 666 (2004)
29. P.R. Wallace, Phys. Rev. **71**(9), 622 (1947)
30. K.S. Novoselov, A.K. Geim, S. Morozov, D. Jiang, M. Katsnelson, I. Grigorieva, S. Dubonos, A. Firsov, Nature **438**(7065), 197 (2005)
31. L. Banszerus, M. Schmitz, S. Engels, J. Dauber, M. Oellers, F. Haupt, K. Watanabe, T. Taniguchi, B. Beschoten, C. Stampfer, Sci. Adv. **1**(6), e1500222 (2015)
32. Q. Gibson, L. Schoop, L. Muechler, L. Xie, M. Hirschberger, N. Ong, R. Car, R. Cava, Phys. Rev. B **91**(20), 205128 (2015)
33. J.L. Manes, Phys. Rev. B **85**(15), 155118 (2012)
34. L.M. Schoop, F. Pielnhofer, B.V. Lotsch, Chem. Mater. **30**(10), 3155–3176 (2018) https://pubs.acs.org/doi/abs/10.1021/acs.chemmater.7b05133
35. O. Vafek, A. Vishwanath, Annu. Rev. Condens. Matter Phys. **5**(1), 83 (2014)
36. C. Fang, Y. Chen, H.Y. Kee, L. Fu, Phys. Rev. B **92**(8), 081201 (2015)
37. A. Burkov, M. Hook, L. Balents, Phys. Rev. B **84**(23), 235126 (2011)
38. Y. Kim, B.J. Wieder, C.L. Kane, A.M. Rappe, Phys. Rev. Lett. **115**, 036806 (2015)
39. S. Young, S. Zaheer, J. Teo, C. Kane, E. Mele, A. Rappe, Phys. Rev. Lett. **108**(14), 140405 (2012)
40. S. Young, C. Kane, Phys. Rev. Lett. **115**(12), 126803 (2015)
41. R. Hoffmann, Angew. Chem. Int. Ed. Engl. **26**(9), 846 (1987)
42. S. Dixit, S. Mazumdar, Phys. Rev. B **29**(4), 1824 (1984)
43. N. Mott, J. Solid State Chem. **88**(1), 5 (1990)
44. R.D. Shannon, Acta Crystallogr. Sect. A Cryst. Phys. Diffr. Theor. Gen. Crystallogr. **32**(5), 751 (1976)
45. A.W. Sleight, J. Gillson, P. Bierstedt, Solid State Commun. **17**(1), 27 (1975)

46. R. Cava, B. Batlogg, J. Krajewski, R. Farrow, L. Rupp, A. White, K. Short, W. Peck, T. Kometani, Nature **332**(6167), 814 (1988)
47. B. Baumert, J. Supercond. **8**(1), 175 (1995)
48. D. Korotin, V. Kukolev, A. Kozhevnikov, D. Novoselov, V. Anisimov, J. Phys. Condens. Matter **24**(41), 415603 (2012)
49. G. Ruani, A. Pal, C. Taliani, R. Zamboni, X. Wei, L. Chen, Z.V. Vardeny, Synth. Met. **43**(3), 3977 (1991)
50. R. Chevrel, P. Gougeon, M. Potel, M. Sergent, J. Solid State Chem. **57**(1), 25 (1985)
51. H. Schäfer, *Chemical Transport Reactions* (Elsevier, Amsterdam, 2016)
52. C. Goodman, *Crystal Growth* (Springer, Berlin, 1974)
53. J.R. Panella, B.A. Trump, G.G. Marcus, T.M. McQueen, Cryst. Growth Des. **17**(9), 4944 (2017). https://doi.org/10.1021/acs.cgd.7b00879
54. R. Sankar, M. Neupane, S.Y. Xu, C. Butler, I. Zeljkovic, I.P. Muthuselvam, F.T. Huang, S.T. Guo, S.K. Karna, M.W. Chu et al., Sci. Rep. **5** (2015)
55. I. Fisher, M. Shapiro, J. Analytis, Philos. Mag. **92**(19–21), 2401 (2012)
56. M.N. Ali, Q. Gibson, S. Jeon, B.B. Zhou, A. Yazdani, R. Cava, Inorg. Chem. **53**(8), 4062 (2014)
57. P.C. Canfield, I.R. Fisher, J. Cryst. Growth **225**(2), 155 (2001)
58. Z. Fisk, J. Remeika, *Handbook on the Physics and Chemistry of Rare Earths* (1989), p. 53
59. P.W. Bridgman. Crystals and their manufacture. US Patent 1,793,672, 1931
60. D. Hsieh, Y. Xia, D. Qian, L. Wray, F. Meier, J. Dil, J. Osterwalder, L. Patthey, A. Fedorov, H. Lin, Phys. Rev. Lett. **103**(14), 146401 (2009)
61. S. Kushwaha, Q. Gibson, J. Xiong, I. Pletikosic, A. Weber, A. Fedorov, N. Ong, T. Valla, R. Cava, J. Appl. Phys. **115**(14), 143708 (2014)
62. H. Riemann, A. Luedge, Crystal Growth of Si for Solar Cells, pp. 41–53 (2009)
63. A. Yakub, Float zone silicon solar panels - 60% less expensive and 25% more efficient. https://www.altenergymag.com/article/2015/08/float-zone-silicon-solar-panels--60-less-expensive-and-25-more-efficient/21139. Accessed: 10 Jan 2018
64. P. Nabokin, D. Souptel, A. Balbashov, J. Cryst. Growth **250**(3), 397 (2003)
65. S. Hüfner, *Photoelectron Spectroscopy*, 2nd edn. (Springer, Berlin, 1996)
66. H. Hertz, Ann. der Phys. **267**(8), 983 (1887)
67. A. Einstein, Ann. der Phys. **322**(6), 132 (1905)
68. J. Végh, J. Electron Spectrosc. Relat. Phenom. **151**(3), 159 (2006)
69. W. Tremel, R. Hoffmann, J. Am. Chem. Soc. **109**(1), 124 (1987)
70. F. Jellinek, H. Hahn, Naturwissenschaften **49**(5), 103 (1962)
71. K.A. Benavides, I.W. Oswald, J.Y. Chan, Acc. Chem. Res. (2017)
72. M. Schilling, L. Schoop, B. Lotsch, M. Dressel, A. Pronin, Phys. Rev. Lett. **119**(18), 187401 (2017)
73. M. Balooch, D. Olander, J. Electrochem. Soc. **130**(1), 151 (1983)
74. A. Topp, R. Queiroz, A. Grüneis, L. Müchler, A.W. Rost, A. Varykhalov, D. Marchenko, M. Krivenkov, F. Rodolakis, J.L. McChesney, Phys. Rev. X **7**(4), 041073 (2017)
75. Q. Xu, Z. Song, S. Nie, H. Weng, Z. Fang, X. Dai, Phys. Rev. B **92**(20), 205310 (2015)
76. A. Topp, J.M. Lippmann, A. Varykhalov, V. Duppel, B.V. Lotsch, C.R. Ast, L.M. Schoop, New J. Phys. **18**(12), 125014 (2016)
77. A. Topp, M.G. Vergniory, M. Krivenkov, A. Varykhalov, F. Rodolakis, J.L. McChesney, B.V. Lotsch, C.R. Ast, L.M. Schoop, J. Phys. Chem. Solids (2017)
78. A. Klein Haneveld, F. Jellinek, Rec. Trav. Chim. Pays-Bas **83**(8), 776 (1964)
79. S. Guan, Y. Liu, Z.M. Yu, S.S. Wang, Y. Yao, S.A. Yang, Phys. Rev. Mater. **1**(5), 054003 (2017)
80. L.M. Schoop, A. Topp, J. Lippmann, F. Orlandi, L. Müchler, M.G. Vergniory, Y. Sun, A.W. Rost, V. Duppel, M. Krivenkov et al., Sci. Adv. **4**(2), eaar2317 (2018)
81. Y.C. Wang, K.M. Poduska, R. Hoffmann, F.J. DiSalvo, J. Alloy. Compd. **314**(1), 132 (2001)

Chapter 10
Momentum and Real-Space Study of Topological Semimetals and Topological Defects

Haim Beidenkopf

Abstract We draw a phenomenological analogy between the topological defect of a screw dislocation and the electronic Weyl semimetal topology class including their bulk and surface manifestations. In the bulk, both can be assigned a chirality which can be calculated from the crystallographic curvature of the screw dislocation or the Berry curvature of the Weyl bands. On the surface, the chiral screw dislocations give rise to open-contour surface modes in the form of a crystallographic step edge uniquely emanating from the screw termination. The bulk Weyl nodes induce surface Fermi-arc states that uniquely terminate at the surface projection of the bulk Weyl node. We use scanning tunneling microscopy to visualize the surface manifestation of both topological structures. The surface topology of the screw dislocation is visualized in the surface topography. The surface momentum-space topology of the Weyl semimetal is visualized and characterized spectroscopically using quasi-particle interference.

10.1 Introduction

Symmetry and topology are the fundamental building blocks in the crystallographic description of materials and defects in them [1, 2]. The crystallographic symmetries determine to a large extent also the electronic band structure of the materials, which in turn sets many of their physical properties [3]. Recently, it was realized that electronic band structures can also be classified by their topology, resulting in novel exotic electronic properties [4–11]. Bulk-boundary correspondence assures the formation of surface states that cannot be realized but as the surface termination of a topologically classified bulk. These include surface Dirac-like dispersions, Majorana

---

H. Beidenkopf (✉)
Condensed Matter Physics Department, Weizmann Institute of Science,
7610001 Rehovot, Israel
e-mail: haim.beidenkopf@weizmann.ac.il

© Springer Nature Switzerland AG 2018
D. Bercioux et al. (eds.), *Topological Matter*, Springer Series in Solid-State Sciences 190, https://doi.org/10.1007/978-3-319-76388-0_10

modes, and Fermi-arc states. Topological defects also exhibit similar bulk-boundary correspondence which results in surface crystallographic structures that can be found only at surface terminations of such bulk defects. Here, we investigate the properties of topological screw dislocations and the topological Weyl semimetal TaAs and draw an intriguing analogy between their topological characters.

10.2 Topological Screw Dislocations

A screw dislocation, sketched in Fig. 10.1a, is a topological crystallographic defect [1, 2]. It can be obtained by cutting the crystal through a half-infinite plane up to a certain axis, sliding it by an integer number, n, of primitive vectors, $\mathbf{u}$, along that axis and stitching it back together. The line defect obtained is characterized by its locus within the bulk crystal and by its topological charge quantified by the Burger's vector. The length of the Burger's vector is set by the integer number, n, of lattice constants along which the crystal is translated, and its direction by the direction of that translation, $\mathbf{u}$. Accordingly, the Burger's vector $\mathbf{b}_n = n\mathbf{u}$ defines a certain handedness or chirality about the screw axis. In an infinite crystal without boundaries, pairs of screw dislocation lines with opposite chirality can be generated in the bulk as shown in Fig. 10.1b. On the same footing, pairs of screw dislocations with opposite chirality can be annihilated. Accordingly, the bulk of a crystal can be characterised topologically by the integer number of pairs of screw dislocation lines excited in it.

A screw dislocation deforms the crystal about its axis and correspondingly induces a local strain field. The displacement of the atoms about the screw line is purely along the direction of the Burger's vector. The atoms are not displaced within the plane perpendicular to the screw axis. The amount of displacement is given by displacement as a function of the angle φ is given by the displacement field $\mathbf{u}(\varphi) = b\varphi/2\pi$ (see Fig. 10.1a). Its spatial gradients give the strain tensor. In the case of a screw dislocation, a pure shear strain, $\mathbf{\Omega} = \nabla \times \mathbf{u}(\mathbf{r})$, is induced. It decays radially away from the screw axis as the strain distributes over increasingly larger circumferences as $\Omega = b/2\pi r$. We thus find that the topological charge carried by the screw dislocation can be obtained through $b_n = \oint d\mathbf{l} \cdot \mathbf{\Omega}_n$. Physically, this means that by measuring the local curvature of a layer along a closed path one can deduce whether a screw dislocation threads this contour or not.

We now introduce a surface termination to the crystal that is normal to the screw axis. A topographic image of the (111) surface of a single crystal of Cu measured with a scanning tunneling microscope (STM) is shown in Fig.10.1c. The topography is nothing but the displacement field $\mathbf{u}(r)$ exposed on the crystal surface. A termination of a screw dislocation at this surface is identified with an arrow. Indeed, by integrating the local curvature of the topography along a closed path the magnitude of the Burger's vector b associated with the screw dislocation located somewhere within such path is recovered. As contour C1 in Fig. 10.1c denotes, creating a step edge by removing a partial monolayer off the surface will necessarily add an even number

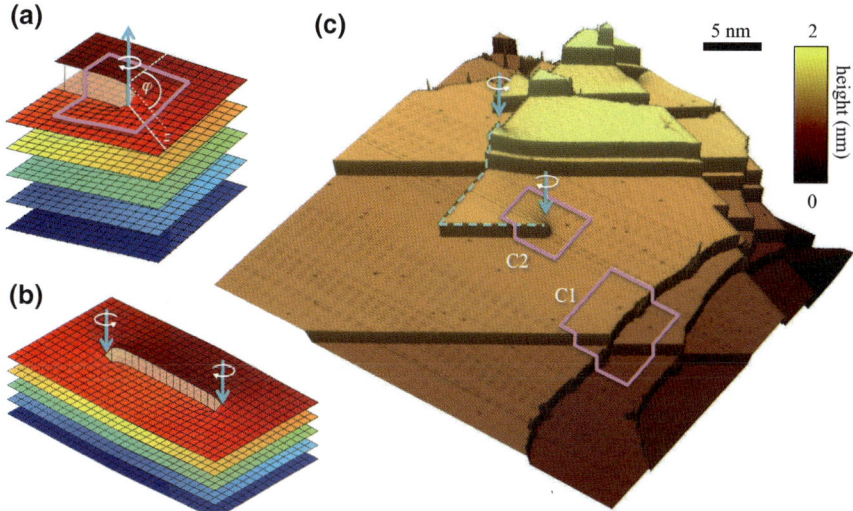

Fig. 10.1 Screw dislocations in Cu(111). **a** A sketch of a screw dislocation in a cubic crystal. A step edge ends at the surface termination of the screw dislocation line. **b** An 'open contour' crystallographic step edge between surface terminations of two screw dislocation with opposite chiralities. **c** A topographic image of Cu(111) surface shows several crystallographic step edges and two screw dislocations of opposite chirality indicated by the surface curvature around them (marked in arrows) that terminate within the imaged field of view. Crossing an odd number of step edges along a closed contour on the surface necessarily indicates the termination of a screw dislocation within that contour

of points where the closed contour crosses those step edges. The only way to create an odd number of step edges crossings along a closed contour is by termination of a screw dislocation within that contour, demonstrated by the contour C2 in Fig. 10.1c. A screw dislocation must terminate at a step edge that emanates from it. In that sense, the open-contour step edge can be regarded as the surface mode associated with the surface termination of the bulk screw dislocation line. Accordingly, the surface of the crystal exhibits an even–odd effect which allows to determine the existence of a screw dislocation line within a closed contour through the parity of the number of step edges crossed.

In the semi-infinite crystal, if one follows the open-contour step edge away from the screw dislocation it emanated from, a termination at a second dislocation line of opposite chirality must appear (in a finite crystal, it can terminate at the side boundary of the sample). Indeed, by following the step edge that emanates from a screw dislocation in Fig. 10.1c, one arrives at a second dislocation of opposite chirality (both are marked with an arrow). The connectivity of the open-contour step edges among the pairs of screw dislocations is not a protected property nor is their specific contour. A closed-contour step edge (edges of a monolayer thick island) can always be added and cross the open-contour ones thus obscure the connectivity beyond recognition. Still, the local parity of step-edge crossings along arbitrary closed contours is con-

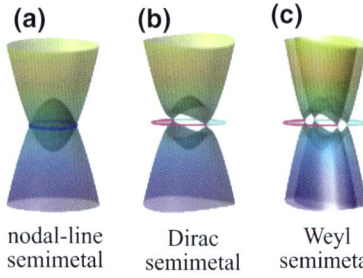

Fig. 10.2 Topological semimetals. a Nodal line semimetal **b** Dirac semimetal **c** Weyl semimetal

served. The continuation of the open-contour step edges is not completely removed, but rather split by the bulk dislocation line to the opposite surface of the crystal. Correspondence between the bulk dislocation line and the surface step-edge modes embodies also the topological protection the screw dislocation exhibits since and open-contour step edge cannot be smoothly eliminated from the surface by local addition or subtraction of atomic monolayers.

10.3 Topological Weyl Semimetals and Their Analogy to Screw Dislocations

We now demonstrate the close analogy between the bulk and surface phenomena induced by a bulk screw dislocation and those that occur in the bulk and on the surface of a Weyl semimetal. A fundamental difference is that for screw dislocation lines the topological structure is found in real space, while in Weyl semimetals and topological electronic phases in general, the topological winding occurs in the bulk momentum space and its surface projection.

We start with the bulk band structure of a Weyl semimetal [12–17]. This intriguing topological phase is formed when two bands with distinct-conserved quantum numbers interpenetrate, thus forming a semimetallic nodal ring of degenerate states with Dirac dispersion (Fig. 10.2a). This nodal line can be gapped out by breaking of either time reversal symmetry or inversion symmetry [13]. Once any of these symmetries has been broken, it may happen so that the nodal line will gap out but at a discrete set of points along it (Fig. 10.2c). In such a case, the electrons will exhibit linear dispersion along all three dimensions with a well-defined chirality, hence Weyl-like dispersion. Such a gapping mechanism assures pairwise formation of Weyl cones of opposite chirality. In the special case that the nodal line is shrunk down to a single touching point or that symmetry allows two Weyl nodes with opposite chirality to overlap a Dirac semimetal forms (Fig. 10.2b). Its doubly degenerate three-dimensional Dirac band comprises of the Weyl band pair of opposite chirality. As in the nodal line case, the Weyl bands can be split off by breaking inversion or time reversal symmetry. In a more abstract way, Weyl cones of opposite chirality can be pairwise excited (or be pairwise annihilated) within a gapped bulk spectrum similar to the pairwise creation (or annihilation) of screw dislocation lines.

The low-energy physics of a Weyl semimetal is generally described by the Weyl Hamiltonian: $H = \pm v\mathbf{k} \cdot \sigma \pm \mathbf{b}/2$, where $\mathbf{k}$ is the crystal momentum about the Weyl nodes located $\mathbf{b}$ apart in the Brillouin zone, σ is a vector of Pauli matrices acting in spin space, and v is the Fermi velocity taken here to be isotropic in all directions for simplicity. It corresponds to electrons with linear dispersion $E = v|\mathbf{k}|$ with a chiral spin texture captured by their spinor wave function: $u(\mathbf{k}) = u(\cos\frac{\theta_k}{2}, -\sin\frac{\theta_k}{2}e^{i\varphi_k})$ where θ_k and φ_k are polar angles of $\mathbf{k}$. Based on that one can compute the Berry connection, defined as $\mathbf{A}_n(k) = -i\langle u_n(k)|\nabla_k|u_n(k)\rangle$, which acts like a vector potential in momentum space. For instance, it adds up to the dynamical phase an electron accumulates along a closed trajectory in momentum space by contributing a Berry phase $\gamma_n = \oint d\mathbf{l}_k \cdot \mathbf{A}_n(k) = \int d\mathbf{s}_k \cdot \mathbf{\Omega}_n(k)$. In the second equation, we have used Stoke's theorem and defined the Berry curvature (also referred to as the Berry flux) as $\mathbf{\Omega}_n(k) = \nabla \times \mathbf{A}_n(k)$. The Berry phase, γ_n, the electron gains as it traverses a closed contour in momentum space is accordingly set by the integrated Berry flux $\mathbf{\Omega}_n$ threading that contour.

By substituting the low-energy Weyl electron's wave function, $u_n(\mathbf{k})$, described above, one finds a monopole-like behavior in momentum space $\mathbf{\Omega}_n(\mathbf{k}) = \pm \mathbf{k}/2k^3$ about the Weyl nodes at momenta $\pm \mathbf{b}/2$. This shows that the Weyl nodes are sources and drains of Berry curvature in analogy to the screw dislocations being sources and drains of the real-space curvature that induces a radially decaying shear strain. Integrating the Berry curvature over a closed sphere in momentum space containing a single Weyl node will then yield the quantized chirality charge of the Weyl node, $C_n = \gamma_n/2\pi$, which is the topological index akin to the Burger's vector.

So far, we have discussed the properties of bulk Weyl nodes and their analogy to crystallographic topological defects. We conclude the analogy with comparing the surface manifestations of the two. The surface manifestation of a Weyl semimetal is the formation of Fermi-arc states. These are open-contour surface modes that terminate at the surface projection of the Weyl nodes. Such a dispersion cannot occur but as the surface termination of a topological Weyl bulk. Fermi arcs can be thought of as the collection of the chiral edge modes that form at the edges of consecutive two-dimensional quantum Hall slices pierced by the quantized Berry flux. Such phenomenology is directly analogue to the open step edge that is inevitably found at the surface ends of the bulk screw dislocations. Indeed, the only way to form a step edge on the surface that terminates at a point is by terminating a screw dislocation line at the surface. We thus find that bulk screw dislocations and bulk Weyl nodes exhibit a similar bulk-boundary correspondence in which unique surface modes are induced that exhibit open contours either in real-space surface or in the surface momentum space, respectively. In both cases, the exact contour of the surface mode is not unique as trivial closed-contour modes can be added and modify it, but its unique terminations are topologically assured. The bulk and surface analogy of screw dislocations and Weyl semimetals is summarized in Table 10.1.

Table 10.1 Phenomenology of Topological defects versus Topological semimetals

	Property	Screw dislocation	Weyl semimetal
Bulk	Space	Real	Momentum
	Topological index	Burger's vector: $\mathbf{b}_n = n\mathbf{u}$	Chern number: $C_n = n$
	Curvature	Elastic: $\Omega = b/2\pi r$	Berry: $\Omega_n(k) = \pm 1/2k^2$
	Momopole	Shear strain	Berry flux
	Curvature-index	$b_n = \oint d\mathbf{l} \cdot \boldsymbol{\Omega}_n$	$C_n = \frac{1}{2\pi} \int d\mathbf{s} \cdot \boldsymbol{\Omega}_n(k)$
Surface	Topological surface mode	Open-contour step edge	Fermi-arc - Open-contour surface band
	Non locality (Topological protection)	Open-contour step edges on opposite sample surfaces connected through bulk screw dislocations	Open-contour Fermi arcs on opposite sample surfaces connected through bulk Weyl bands

10.4 The Topological Weyl Semimetal TaAs

After demonstrating the real-space surface phenomenology of screw dislocations in STM, we move on to study the surface manifestation of Bulk Weyl semimetals as it appears in momentum space. To investigate the energy-momentum structure of the electronic wave functions, we measure the quasi-particle interference (QPI) patterns electrons embed in the local density of states as they scatter off impurities and crystallographic step edges [18–21]. The modulated local density of states entails the electronic wavelengths involved in the scattering processes that can be identified by Fourier analysis. The QPI method has been used to study the existence of Fermi-arc states and their unique connectivity to the bulk Weyl cones in several material systems. These include TaAs [22, 23] that we discuss here, and NbP [24]. The type-two Weyl semimetal MoTe$_2$, in which the Weyl cone dispersion is tilted such that the electron and hole cones overlap in energy, was also probed [25], and the related Dirac semimetal Cd$_3$As$_2$ was studied using Landau level spectroscopy [26].

We have used the method of QPI to uniquely characterize the structure of the wave function of the topological Fermi-arc surface states in the Weyl semimetal TaAs [23]. We further used it to distinguish the Fermi arcs from non-topological surface states that coexist on the sample surface. Unlike the simplistic case described above, TaAs is a Weyl semimetal with a total of 24 bulk Weyl nodes which on the (001) surface project to 16 surface Weyl nodes with 8 Fermi arcs connecting them shown in DFT calculation in Fig. 10.3a [27–31]. In addition, there are trivial surface states induced by the dangling bonds on the exposed (001) surface [24]. This richness poses both a challenge of distinguishing the topological bands from the trivial ones but also an opportunity to compare the properties of the two kinds [22]. The topography of the cleaved surface is shown in Fig. 10.3b. We find a perfectly ordered square lattice (zoomed in image shown in the inset) with a low concentration of atom vacancies. By comparing the measured dI/dV spectrum over that surface, shown by solid line in Fig. 10.3c, to the calculated one for Ta versus As (red versus

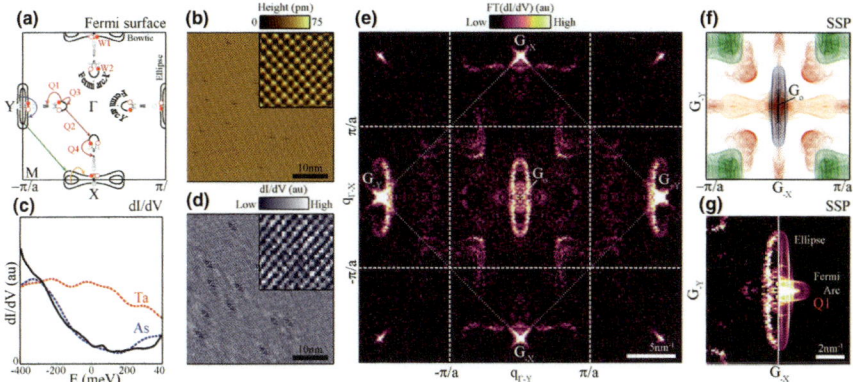

Fig. 10.3 QPI in TaAs. a The calculated Fermi surface of TaAs (110) surface comprising of both trivial bands and topological Fermi-arc states. The arrows mark possible scattering processes with the surface band structure. **b** Topographic image of the surface. **c** Comparison of the measured dI/dV spectrum (solid line) with calculated spectrum of As- versus Ta-terminated surfaces (blue versus red lines, respectively) suggests the cleave exposes the As layer. **d** dI/dV map at the Fermi energy finds complex QPI patterns around each As vacancy imaged in **b**. **e** Two-dimensional Fourier transformation of **d**. **f** calculated JDOS based on **a**. **g** The leaf-like QPI pattern that peaks beyond the central ellipse corresponds to scattering between the Fermi arc along Γ-Y and an adjacent trivial band

blue dashed lines, respectively) termination, we identify the cleaved surface as As terminated. Accordingly, the deficiencies are As vacancies. These scatter the surface electrons and give rise to the standing wave pattern captured by the dI/dV map in Fig. 10.3d.

Fourier analysis, presented in Fig. 10.3e, resolves the elaborate structure of scattering wave vectors that participate in the formation of that standing wave pattern. The brightest spots, marked with Γ_x and Γ_y, are the atomic Bragg peaks. Based on their location, we can divide the QPI pattern to scattering wave vectors shorter than the Brillouin zone (within the central dashed square) and scattering wavevectors which are larger than the Brillouin zone.

To extract the physical processes governing the QPI patterns measured, we calculate the joint density of state (JDOS) obtained by autoconvolving the band structure, $\rho_k(E)$, in momentum space at a given energy $JDOS(q, E) = \int dk \rho_k(E) \rho_{k+q}(E)$. The JDOS calculated based on the band structure shown in Fig. 10.3a is given in Fig. 10.3f. The blue, yellow, and green patterns correspond to scattering within the ellipse-like band, the bow-tie like band, and among them, respectively (as marked by corresponding colored arrows in Fig. 10.3a). Accordingly, these are all QPI patterns from scatterings among trivial bands. The red QPI patterns in Fig. 10.3f involve scattering with a Fermi-arc surface band. This includes intra-Fermi-arc scatterings (Q3 in Fig. 10.3a), inter-Fermi-arc scatterings (Q2) and scattering between a Fermi-arc, and a trivial band (Q1, Q4). Among all scattering processes in Fig. 10.3e, we identify only the leaf-like pattern that peaks beyond the ellipse-like pattern, given in greater detail and compared to calculated JDOS in Fig. 10.3g, as one involving Fermi-arcs (Q1).

The JDOS calculation captures only the scattering processes of wave vectors shorter than the size of the Brillouin zone. Clearly, to account for higher scattering wave vectors, higher Brillouin zones have to be considered. Those higher Brillouin zones are a direct consequence of the periodicity of the crystal as manifested by the Bloch wave function:

$$\psi_k(r) = \sum_G c_{k,G} e^{i(k+G) \cdot r} \tag{10.1}$$

where $c_{k,G}$ is a Bloch coefficient and G is a reciprocal wave vector. Accordingly, the translational invariant local density of states can be written as

$$\rho(r, E) = \sum_{G,G'} A_{G,G'} e^{i(G-G') \cdot r} \tag{10.2}$$

$$A_{G,G'} = \sum_k c_{k,G}^* c_{k,G'} \delta(E - E_k) \tag{10.3}$$

This means that even in the absence of a scatterer the local density of states will be modulated by $g = G - G'$ whenever more than a single dominant Bloch coefficient, c_G^k, exists. This is indeed observed in the modulated local density of states in the inset of Fig. 10.3d. On the same footing, the JDOS will now assume the form:

$$JDOS(q, E) = \sum_{G,G'} \int dk \rho_{k+G} \langle \sigma_{k+G} | \sigma k + q + G' \rangle \rho_{k+q+G'} \tag{10.4}$$

which will replicate QPI features contained within the first Brillouin zone to higher ones whenever several dominant Bloch amplitudes, c_G^k, appear [20].

Free particles can have pure plane wave-like behavior. However, electrons within periodically ordered material will be susceptible to some extent to the underlying potential imposed by the crystal. This will necessarily render higher Bloch coefficients to be non-vanishing. The relative strength of the different Bloch coefficients determines the structure of the electronic wave function within the unit cell. It can be predicted in ab initio calculation and measured by QPI. Representative calculated Bloch coefficients of the ellipse- and bow tie-like bands are given in Fig. 10.4a, b, respectively. We indeed find an anisotropic structure which corresponds to the replications of QPI patterns found in experiment (Fig. 10.3e). However, the topological Fermi-arc states, in contrast to dangling bond states, result from the bulk topology rather than the surface potential. It raises the question to what extent are the topological surface states susceptible to the underlying crystalline structure. Calculation of the Bloch structure of the Fermi-arc wave function, shown in Fig. 10.4c, indeed finds a single dominant coefficient. This intriguingly suggests that Fermi-arc states are plane wave-like. Accordingl, they will not be replicated to higher Bragg peaks

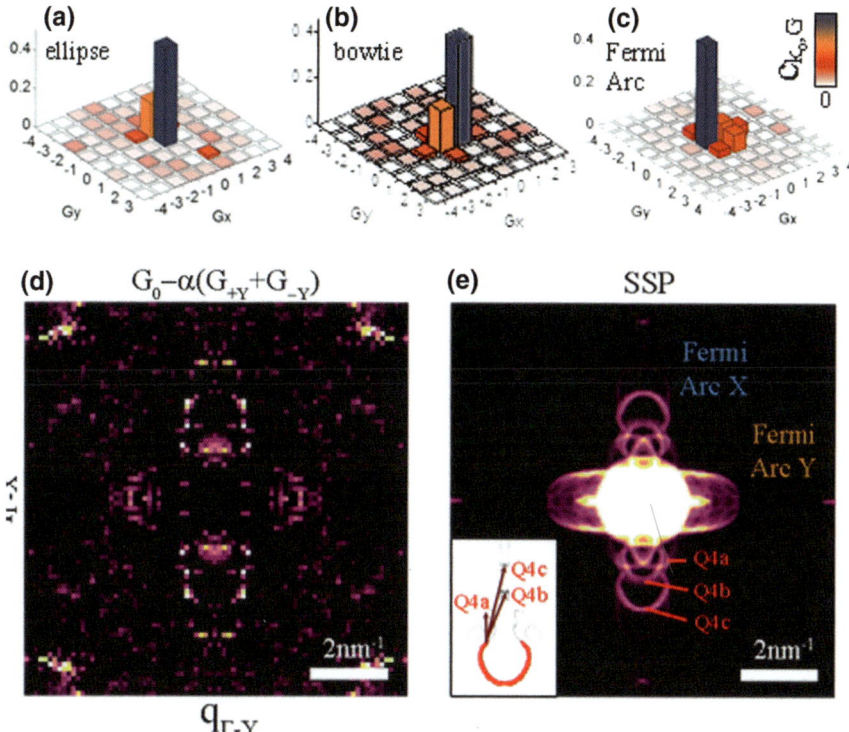

Fig. 10.4 Visualizing the Bloch wave function in QPI. **a–c** Calculated Bloch coefficients of three different bands. The ellipse and bow tie bands have a complex anisotropic Bloch structure while the Fermi-arc band is rather plane wave-like with a single dominant Bloch coefficient. **d** Subtracting the ellipse-like QPI of the Γ-Y Bragg peak from that at q = 0 reveals previously buried QPI pattern. **e** Calculated JDOS about q = 0 in the absence of ellipse and bowtie bands

in QPI. We indeed do not seem to detect any replication of the leaf-like structure, associated with scattering from a Fermi-arc state, to higher Bragg peak.

With this in mind, we perform a novel analysis on the QPI data. We subtract the QPI ellipse pattern that appears around $\Gamma_{\pm Y}$ from the QPI pattern that appears about q=0. By doing so, we are indeed able to eliminate the ellipse-like QPI pattern, as seen in Fig. 10.4d. In this procedure, the leaf-like QPI patterns in which Fermi-arcs are involved are indeed hardly changed, signifying that this QPI pattern is indeed not replicated. Remarkably, once the ellipse-like pattern is eliminated, we find a residual curly QPI structure that fits well the scattering pattern of the second Fermi-arc surface band along Γ-Y (see JDOS calculation in Fig. 10.4e). This demonstrates that generally QPI patterns involving trivial bands are replicated according to the structure of their Bloch wave function. The topological Fermi-arc states are found to be remarkably unsusceptible to the underlying crystal structure and be well approximated by a pure plan wave-like wave function.

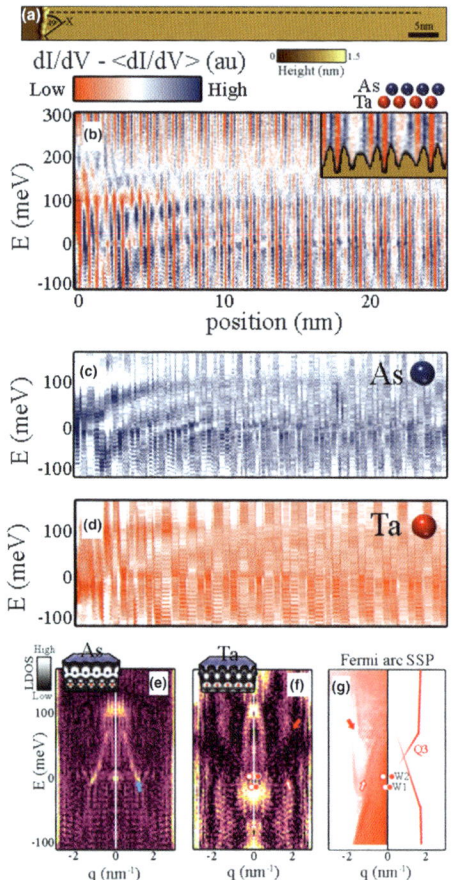

Fig. 10.5 Distinct atomic origin of trivial and topological surface bands. **a** Topographic image of a crystallographic step edge. **b** dI/dV linecut measured normal to the step-edge (along dashed line in **a**). **c, d** dI/dV linecut taken only on As **c** or Ta (**d**) atomic sites. **e** Fourier analysis of c finds the dispersion of the trivial ellipse band derived from the As dangling bonds. Inset shows that the calculated wave function distribution of the ellipse band is indeed highly localized on top-most As layer. **f** The QPI mode on the Ta sites differs from that seen on the As sites. Inset shows the calculated wave function distribution of the Fermi-arc band. **g** Calculated JDOS of intra-Fermi-arc scattering processes alone

10.5 Topological Bulk Origin of the Fermi-Arc States in TaAs

Additional information on the structure of the wave function of the Fermi-arc bands we obtain from their scattering properties off crystallographic step edges. A topographic image of such a step edge is shown in Fig. 10.5a. The step edge is oriented 49 degrees relative to the crystal axis and therefore scatters approximately along the Γ-M direction. The scattered electrons give rise to an intricate QPI pattern, shown in Fig. 10.5b. It comprises dispersing features as well as atomic modulation, highlighted by the inset. We therefore separate the dI/dV linecut into two subsets— the dI/dV measured on the As atoms and in the valleys between them where the top-most Ta atoms reside (Fig. 10.5c, d, respectively). Each of these subsets displays a distinct dispersing QPI pattern.

Indeed, Fourier transformation of each of them, given in Fig. 10.5e, f, displays two distinct sets of dispersing scattering modes. On the As surface layer, presented in Fig. 10.5e, we find the dispersion that corresponds to the ellipse-like QPI pattern of Fig. 10.5e. Calculation indeed verifies that the ellipse band results from the As dangling bonds and is accordingly highly localized on the top-most As monolayer, as shown by the calculated wave function distribution at the inset. In contrast, the QPI modes that originate from the local density of states in between the topmost As atoms, shown in Fig. 10.5f, find a completely distinct dispersing modes. Comparison with the calculated JDOS, shown in Fig. 10.5g, identifies them with scattering processes within the Fermi-arc located on the Γ-Y direction. They both disperse toward the energy and momentum at which the surface projection of the bulk Weyl node resides. The wave function distribution of that Fermi-arc band, presented at the inset of Fig. 10.5f, confirms that this topological band resides predominantly on the Ta sites and penetrates deeper into the bulk. Indeed, the bulk Weyl cones as well as the surface Fermi-arcs induced by them are derived mainly from the Ta orbitals.

10.6 Summary

We have shown that the chiral structure that screws dislocation lines and bulk electronic wave functions in a Weyl semimetal both have resulted in a rather analogous bulk-boundary correspondence. In both cases, unique surface modes are realized that terminate at a point in real or momentum space rather than forming a closed contour. We stress that this analogy is not exact as the two classifications differ in their dimensionality and topological index. Still, it lands an intuitive comprehension of the deep topological notion of bulk-boundary correspondence and the manner exotic modes are ensued. It would be interesting to identify topological defects and topological electronic classes that exhibit true equivalence. Beyond the pedagogical benefit, this may serve as a tool identify yet unknown electronic topological classes and perhaps even yet unresolved crystallographic defects.

We have used STM to visualize in real-space surface modes associated with screw dislocations in the form of step edges that uniquely terminate at a point on the surface. This point is the surface termination of the bulk screw dislocation line. From the surface curvature around the screw the Burgers vector can be extracted. We have further visualized in momentum space the surface modes associated with bulk Weyl nodes in the form of Fermi arcs that uniquely terminate at a point in the surface Brillouin zone. That point is the surface projection of the bulk Weyl nodes. We characterize the unique structure of the electronic wave function of Fermi-arc states compared to non-topological bands.

Acknowledgements HB acknowledges Rajib Batabyal, Noam Morali, Nurit Avraham, Yan Sun, Marcus Schmidt, Claudia Felser, Ady Stern, and Binghai Yan that participated in the study of TaAs, and Roni Ilan from many discussions as well as funding from the European Research Council (ERC) (Starter Grant no. 678702, TOPO-NW), the German-Israeli Foundation for Scientific Research and Development (GIF Grant no. I-1364-303.7/2016) and the Israeli Science Foundation.

References

1. A. Kelly, G.W. Groves, *Crystallography and Crystal Defects* (Addison-Wesley, Reading, 1970)
2. H. Kleinert, *Gauge Fields in Condensed Matter. Volume II: Stresses and Defects* (World Scientific Publishing company, Singapore, 1989)
3. N.W. Ashcroft, N.D. Mermin, *Solid State Physics* (Saunders, Philadelphia, 1976)
4. C.L. Kane, E.J. Mele, Phys. Rev. Lett. **95**, 226801 (2005)
5. B.A. Bernevig, T.L. Hughes, S.C. Zhang, Science **314**, 1757 (2006)
6. B.A. Bernevig, T.L. Hughes, *Topological Insulators and Topological Superconductors* (Princeton University Press, Princeton, 2013)
7. X.L. Qi, S.C. Zhang, Rev. Mod. Phys. **83**, 1057 (2011)
8. M.Z. Hasan, C.L. Kane, Rev. Mod. Phys. **82**, 3045 (2010)
9. J. Maciejko, T.L. Hughes, S.C. Zhang, Ann. Rev. Condens. Matter Phys. **2**, 31 (2011)
10. M.Z. Hasan, J.E. Moore, Ann. Rev. Condens. Matter Phys. **2**, 55 (2011)
11. Y. Ando, L. Fu, Ann. Rev. Condens. Matter Phys. **6**, 361 (2015)
12. S. Murakami, New J. Phys. **9**, 356 (2007)
13. C. Fang, Y. Chen, H.Y. Kee, L. Fu, Phys. Rev. B **92**, 081201(R) (2015)
14. S.M. Young et al., Phys. Rev. Lett. **108**, 140405 (2012)
15. X. Wan, A.M. Turner, A. Vishwanath, S.Y. Savrasov, Phys. Rev. B **83**, 205101 (2011)
16. H.B. Nielsen, M. Ninomiya, Phys. Lett. B **130**, 389 (1983)
17. V. Aji, Phys. Rev. B **85**, 241101 (2012)
18. M.F. Crommie, C.P. Lutz, D.M. Eigler, Nature **363**, 524 (1993)
19. L. Brgi, O. Jeandupeux, H. Brune, K. Kern, Phys. Rev. Lett. **82**, 4516 (1999)
20. B.G. Briner, Ph Hofmann, M. Doering, H.-P. Rust, E.W. Plummer, A.M. Bradshaw, Phys. Rev. B **58**, 13931 (1998)
21. J.E. Hoffman, K. McElroy, D.-H. Lee, K.M. Lang, H. Eisaki, S. Uchida, J.C. Davis, Science **297**, 1148 (2002)
22. H. Inoue, A. Gyenis, Z. Wang, J. Li, S.W. Oh, S. Jiang, N. Ni, B.A. Bernevig, A. Yazdani, Science **351**, 1184 (2016)
23. R. Batabyal, N. Morali, N. Avraham, Y. Sun, M. Schmidt, C. Felser, A. Stern, B. Yan, H. Beidenkopf, Sci. Adv. **2**, e1600709 (2016)
24. H. Zheng, S.-Y. Xu, G. Bian, C. Guo, G. Chang, D.S. Sanchez, I. Belopolski, C.-C. Lee, S.-M. Huang, X. Zhang, R. Sankar, N. Alidoust, T.-R. Chang, F. Wu, T. Neupert, F. Chou, H.-T. Jeng, N. Yao, A. Bansil, S. Jia, H. Lin, M.Z. Hasan, ACS Nano **10**, 137 (2016)
25. D. Ke, G. Wan, P. Deng, K. Zhang, S. Ding, E. Wang, M. Yan, H. Huang, H. Zhang, Z. Xu, J. Denlinger, A. Fedorov, H. Yang, W. Duan, H. Yao, Y. Wu, S. Fan, H. Zhang, X. Chen, S. Zhou, Nat. Phys. **12**, 1105 (2016)
26. J. Sangjun, B.B. Zhou, A. Gyenis, B.E. Feldman, I. Kimchi, A.C. Potter, Q.D. Gibson, R.J. Cava, A. Vishwanath, A. Yazdani, Nat. math. **13**, 851 (2014)
27. H. Weng, C. Fang, Z. Fang, B.A. Bernevig, X. Dai, Phys. Rev. X **5**, 011029 (2015)
28. Y. Sun, S.-C. Wu, B. Yan, Phys. Rev. B **92**, 115428 (2015)
29. S.-Y. Xu, I. Belopolski, N. Alidoust, M. Neupane, G. Bian, C. Zhang, R. Sankar, G. Chang, Z. Yuan, C.-C. Lee, S.-M. Huang, H. Zheng, J. Ma, D.S. Sanchez, B.K. Wang, A. Bansil, F. Chou, P.P. Shibayev, H. Lin, S. Jia, M.Z. Hasan, Science **349**, 613 (2015)
30. B.Q. Lv, H.M. Weng, B.B. Fu, X.P. Wang, H. Miao, J. Ma, P. Richard, X.C. Huang, L.X. Zhao, G.F. Chen, Z. Fang, X. Dai, T. Qian, H. Ding, Phys. Rev. X **5**, 031013 (2015)
31. L.X. Yang, Z.K. Liu, Y. Sun, H. Peng, H.F. Yang, T. Zhang, B. Zhou, Y. Zhang, Y.F. Guo, M. Rahn, D. Prabhakaran, Z. Hussain, S.-K. Mo, C. Felser, B. Yan, Y.L. Chen, Nat. Phys. **11**, 728 (2015)

Index

A
Action of an element, 15
AFM, 203
Aharonov–Bohm phase, 97
Ambiguities, 168
Andreev bound, 110
Angle-Resolved Photoemission Spectroscopy (ARPES), 226, 235
Anomalous conservation, 170
Antiskyrmions, 208
Atomic limit, 28
Average SOC, 200
Axial electromagnetic field, 155

B
$BaBiO_3$, 222
Ballistic, 122, 125, 142
Ballistic wire, 98
Band crossing, 18
Band representation, 1
Band theory, 64
Béri degeneracy, 95
Berry connection, 33, 37, 40, 46, 68, 183
Berry curvature, 168, 184
Berry phase, 33, 67–69, 94, 96
Berry potential, 68
Berry–Wilczek–Zee connection, 33
BiO_2, 222
Birefringence, 166
Bloch states, 64, 181
Bloch wave functions, 13
Boltzmann transport theory, 180
Bonding type, 215, 216
Bosons, 57
Boundary modes, 38
Bravais lattice, 2, 23

Bridgman method, 225
Brillouin Zone (BZ), 32, 103
Bulk–boundary, 93
Bulk-boundary correspondence, 41, 66
Bulk-edge correspondence, 66
2b Wyckoff positions, 6

C
Canonically conjugate variables, 183
Capacitances, 139
Carbon nanotubes, 100
Carroll–Field–Jackiw (CFJ), 166
Cd_3As_2, 225
CeSbTe, 240
Charge conservation, 105
Charge pumping, 72
Chemical potential, 99
Chemical strain, 238
Chern insulator, 39, 44, 49, 77, 160
Chern invariant, 68
Chern number, 40, 44, 49, 63, 67, 69, 81, 185
Chern–Simons action, 166
Chiral anomaly, 168, 202
Chiral charge, 187
Chiral chemical potential, 194
Chiral current, 170
Chiral edge states, 102
Chiral fermions, 169
Chiral gauge fields, 195
Chirality, 152
Chiral kinetic theory, 180
Chiral magnetic effect, 156, 194
Chiral mode, 106
Chiral symmetry, 36, 41, 46, 152, 159, 179
Circumference, 107
Classes, 9

Collision term, 181
Commutation relation, 17
Compatibility solutions, 22
Compensated ferrimagnets, 206
Compositeness, 28
Conductance, 99
Conduction or valence, 22
Configuration space, 181
Conformal anomaly, 195
Conformal invariance, 195
Connected set of bands, 21
Connectivity, 2
Conserved charge, 178
Conserved current, 178
Continuum, 140
Convergence options, 82
Corner modes, 50
Corner states, 58
Correlation length, 109
Coset decomposition, 24
Coset representatives, 4, 6
Coupling to the continuum, 122
Crystal, 23
Crystal growth methods, 224
Crystal orbitals, 7
Crystal symmetry, 2
Crystal wave-vector, 64
Current bias, 132, 136, 140
Current operator, 164

D
3D Dirac semimetal, 218, 220
Decompose, 14
Degeneracy, 14
Density, 123–126, 130, 142
Density matrix, 181
Density of states, 218
Detailed balance condition, 185
Dirac cone, 18, 49, 54, 56
Dirac equation, 109
Dirac fermion, 94, 95
Dirac Hamiltonian, 151
Dirac line node, 235
Dirac matrices, 150
Dirac point, 95
Disconnected set of bands, 22
Disorder, 100
Disordered wires, 98
Distribution function, 181
Distributive property, 28
Double group, 11
Double-valued, 9

Drude conductivity, 94
3DTI, 111

E
Edge states, 66
Effective action, 166
Effective field theory, 184
Electron acceptors, 212
Electron count, 213
Electron donors, 212
Electronegativity, 212
Electronegativity difference, 218
Electronic bands, 11
Electron rule, 212
Elementary band representations, 1
Emergence of topological superconductivity, 104
Emission linewidths, 141
End modes, 38
Energy bands, 64
Energy Distribution Curve (EDC), 228
Energy gap, 64
Even sequences of Shapiro response, 132

F
Fermi arc, 168, 205, 249–252, 255
Fermi golden rule, 188
Fermi level, 22, 184
Fermion doubling, 159
Fermi surface, 167
Fermi velocity, 96
Fiber bundle, 65, 67
First-principles calculation, 74
Floating zone method, 226
Flux growth method, 225
Fourier transformed Wannier, 12
Fractional Josephson effect, 121, 129
Fredholm alternative, 196
Free-electron-like final state approximation, 230
Fujikawa, 180
Fujikawa formalism, 167

G
Gapless Andreev bound states, 119
Gapless Majorana-Andreev bound states, 142
Gapless 4π-periodic Andreev bound states, 122
Gauge *covariant*, 34
Gauge invariance, 164, 178

Index 259

Gauge transformation, 97
Gauss law, 185
Generators, 3
Genus, 65
Geometric phase, 67
Ginsparg–Wilson (GW) fermions, 158
Graphene, 217
Gravitational contribution to the chiral anomaly, 195
Group of the crystal, 23

H
Haldane model, 77
Half-filled band, 221
Half-Heusler compounds, 199
Half-integral angular momentum, 10
Half-metallic, 202
Hall conductivity, 39, 101, 167
Hamiltonian, 109
Hamiltonian matrix, 73
Heisenberg uncertainly principle, 183
Helicity, 153
Higher-order topological phase, 60
Higher-order topological insulators, 50
High-Resolution Transmission Electron Microscopy (HRTEM), 235
High-symmetry lines, 20
Hilbert space, 12, 85
Hinge modes, 57
Hybrid Wannier Charge Center (HWCC), 70, 71
Hybrid Wannier orbitals, 71
Hybrid Wannier states, 35

I
Individual Chern numbers, 85, 86
Inelastic Mean Free Path (IMFP), 227
Inorganic Crystal Structure Database (ICSD), 216
Integral total angular momentum, 10
Integro-differential equation, 182
Inter-node scattering, 187
Inter-valley scattering, 190
Intra-valley scattering, 190
Inversion symmetry, 36, 44, 151
Inverted band structure, 200
Ising model, 60
Isolated bands, 66

J
Josephson emission, 121, 127, 129, 137

K
Kitaev, 103
k.p approximation, 73
k points, 13
Kramers pair, 57, 86, 87
Kubo formula, 181

L
Lagrangean, 182
Landau level, 102, 169
Landauer equation, The, 98
Landau–Zener Transitions, 122, 140
Lattice fermions, 158
Line element, 172
Line node material, 233
Little group, 19
Localized orbitals, 7
Local limit, 189
Local symmetry, 178
Longitudinal magnetoconductivity, 194
Lorentz breaking field theories, 156
Lorentz breaking QED, 150, 164
Lorentz breaking quantum electrodynamics, 156
Lorentz frame transformations, 155
Lorentz invariance, 157
Lorentz symmetry, 149, 155
Luttinger liquid, 178

M
Mach–Zehnder interferometer, 102
Magnetic length, 169
Magnetic phase diagram, 241
Magnetization, 157
Majorana, 95
Majorana bound states, 117
Majorana number, 103
Majorana zero mode, 54
Material development, 223
Maximal subgroup, 7
Maximal Wyckoff position, 7, 24
Mirror Chern number, 47, 55
Mirror plane, 5
Mirror symmetry, 45, 49, 54, 59, 86, 203
Mobility, 123, 124, 126
Monopole, 185
Mott insulators, 221
move_tol, 84
Multiplicity of a Wyckoff position, 24

N

Nanowires, 95
New fermions, 220, 241
Nielsen Ninomiya theorem, 41, 159, 179
Nodal lines, 157
Nodal line semimetals, 220
Nodal semimetals, 150
Noether's theorem, 177
Non-Abelian Wilson loops, 32
Non-collinear AFMs, 204
Non-interacting materials, 63
Non-local Dirac fermion, 160
Non-Minkowski metric, 171
Nonsymmorphic cluster compounds, 223
Nonsymmorphic symmetry, 220, 236
Non-trivial topological properties, 102
`num_lines`, 84

O

Orbital magnetic moment, 184
orbit of q, 3
Oscillatory pattern, 132, 134
Overlap matrices, 71, 74

P

Parallel transport, 67
Particle current, 182
Particle-Lorentz transformation, 155
Particle physics, 178
Path integral, 182
Pauli algebra, 58
Pauling rules, 215
Peierl's distortions, 221
Pfaffian, 32, 103
Phase winding, 104
Photon self-energy, 165
Physical Band Representations, 2
P6mm, 8
Polarization, 37, 44
Polarization function, 162
p_z orbitals, 7
Position operator, 34
`pos_tol`, 83
Precession Electron Diffraction (PED), 233
Projector, 29
Protected semimetal, 22
Proximity effect, 103
Proximity-induced superconductivity, 102
p-wave superconductor, 54, 103

Q

Quantum anomalies, 161, 178
Quantum electrodynamics, 184
Quantum Hall edge states, 101
Quantum Hall effect, 178
Quantum Hall states, 94
Quantum spin Hall effect, 94
Quantum well, 123, 124, 126, 129, 141
Quasiparticle density, 182
Quasi-Particle Interference (QPI), 250–255

R

Rarita–Schwinger Lagrangian, 172
Real-space topological states, 208
Reciprocal lattice vector, 168
Regularization, 161
Relaxation, 121
Representation, 8
Resistive shunt, 129, 136
Rotation group, 8

S

Scanning Tunnelling Microscopy (STM), 103, 235
Scattering matrix, 100, 108
Scattering states, 108
Schrödinger equation, 64
Schwinger model, 162
Screw dislocation, 245–250, 255
Semiclassical equations of motion, 182
Semimetal, 22
Set of relations, 25
Shapiro steps, 121, 127, 132, 134, 137
Shirley background, 228
Shunt resistor R_s, 131
Single-particle Hamiltonian, 64
Single-valued, 9
Small representation, 25
Solid-state chemistry, 211
Spinful, 11
Spin-gapless semiconductor, 203
Spinless, 11
Spin–Orbit Coupling (SOC), 102, 218
Spin–orbit interaction, 20
Spinor wavefuctions, 179
Spin up, 10
Split, 20
Spontaneous symmetry breaking, 60
Square-net compounds, 231
Stabilizer group, 3
Strained HgTe, 124
Structural motif, 216

Index 261

Subduction, 8
Subharmonic Shapiro steps, 137
Subharmonic steps, 132
Superconducting analogue to Klein tunneling, 120
Superfluid ^{3}He, 178
Surface state, 237
Su–Schrieffer–Heeger (SSH) model, 36
S-wave superconductor, 102
Symmetries, 94
Symmetry breaking, 139, 142
Symmetry-Protected Topological (SPT) phases, 57
Symmetry-protected topological classification, 86

T
Table of characters of the group C_{3v}, 10
Tight-binding model, 74
Tilt parameter, 172
Time-dependent variational approach, 182
Time-reversal, 31
Time-reversal symmetry, 85, 99, 151
Topological Andreev doublet, 120
Topological classification, 65
Topological crystalline insulators, 31, 44
Topological defect, 245, 246, 249, 255
Topological Insulators (TIs), 1, 93
Topological invariant, 65
Topological materials, 211, 216
Topological obstruction, 69
Topological phases of matter, 22
Topological phase transition, 22
Topological properties, 19
Topological Quantum Chemistry (TQC), 2
Topological quantum computation, 111
Topological semimetals, 81
Topological superconductors, 103
Total angular momentum, 10
Transfer matrix, 108
Transport time, 191
Triangle diagram, 168
Trivial spin degeneracy, 18
Type-II Weyl fermions, 171
Type-II Weyl semimetals, 157

V
Valence electron count, 214
Vapor transport method, 224, 233
Vector, 8
Voltage, 140
Voltage bias, 128, 129
Vortex, 104

W
Wannier orbitals, 71
Wannier representable, 22
Wannier states, 8
Wavepackets, 183
Weak antilocalization, 99
Weyl crossings, 241
Weyl Hamiltonian, 179
Weyl node separation, 154
Weyl point, 220
Weyl semimetal, 81, 150, 155, 220, 245, 246, 248–250, 255
Weyl semimetal phase, 167
Wilson fermions, 158
Wilson loop, 33, 37, 39, 41, 43, 53, 70
Winding number, 38, 70
WKB approximation, 182
Wyckoff position, 4, 6

Z
Zeeman coupling, 105
Zeeman splitting, 200
Zero bias peak, 111
$\mathbb{Z}_2$ index, 85
Zintl concept, 214
Zintl phases, 214
$\mathbb{Z}_2$ invariant, 87
ZrSiS, 230
Z2Pack, 63
ZPack Code, 72
z2pack.invariant.chern, 75
z2pack.invariant.z2, 88
z2pack.io.load, 76
z2pack.io.save, 76
z2pack.plot.chern, 75
z2pack.plot.wcc, 89
z2pack.surface.run, 75
ZrSiTe, 239

Printed in Great Britain
by Amazon